DENYING SCIENCE

JOHN GRANT

DENYING SCIENCE

CONSPIRACY THEORIES, MEDIA DISTORTIONS, AND THE WAR AGAINST REALITY

Prometheus Books
59 John Glenn Drive
Amherst, New York 14228–2119

Published 2011 by Prometheus Books

Cover design by Jacqueline Nasso Cooke

Inquiries should be addressed to
Prometheus Books
59 John Glenn Drive
Amherst, New York 14228–2119
VOICE: 716–691–0133
FAX: 716–691–0137
WWW.PROMETHEUSBOOKS.COM

15 14 13 12 11 5 4 3 2 1

Library of Congress Cataloging-in-Publication Data

Grant, John, 1949–
Denying science : conspiracy theories, media distortions, and the war against reality / by John Grant.
p. cm.
Includes bibliographical references and index.
ISBN 978–1–61614–399–2 (cloth : alk. paper)
ISBN 978–1–61614–400–5 (ebook)
1. Science—Political aspects. 2. Science—Social aspects. I. Title

Q175.5.G734 2011
500—dc23

2011014991

Printed in the United States of America on acid-free paper

For Thomas Alan Paul Crowther,
with apologies for the way we've left your world.

"If your beliefs fit on a sign . . . think harder."

—Sign at the Rally to Restore Sanity,

Washington DC, October 30, 2010

CONTENTS

ACKNOWLEDGMENTS

The staff of the West Milford Township Library have gone above and beyond the call of duty to obtain books for me from far parts and have done so with unflagging friendliness and helpfulness. My thanks to Elaine Bindler, Liz Frey, Beth Gamble, Bruce Gilliard, Joanne Grady, Kitty Heuer, Theresa McArthur, Debbie Maynard, Aimee Morrow, Elyse Schear, Sue Small, Maria Villecca, and Dale Warn.

Much information has been channeled my way by The Spammers, to whom my thanks yet again: Randy M. Dannenfelser, Bob Eggleton, Gregory Frost, Neil Greenberg, Wally Hayman, Jael, Stuart Jaffe, Karl Kofoed, Todd Lockwood, Aaron McLellan, Lynn Perkins, Ray Ridenour, Tim Sullivan, and Greg Uchrin. Others who have passed along useful stuff include Fragano Ledgister, and of course, Pam Scoville.

Forgive me if I've omitted anyone.

Introduction

UNLESS WE THINK, WE AREN'T

The trouble with most folks isn't their ignorance. It's knowing so many things that ain't so.

—Josh Billings, but usually (in keeping with the observation) attributed to Mark Twain

This world is a strange madhouse. Currently, every coachman and every waiter is debating whether relativity theory is correct. Belief in this matter depends on political party affiliation.

—Albert Einstein, letter to Marcel Grossmann, September 12, 1920

Denial of science is everywhere around us, not just in bars and offices and pulpits but on the lips of our broadcasters and legislators, our movers and shakers. There's denial that AIDS and HIV are connected and that traditional remedies can't cure AIDS. There's denial of the science showing vaccines don't cause autism. A century and a half after publication of Darwin's *Origin of Species* (1859) there's still denial that we came to be what we are through natural selection; and 180 years or so after publication of Lyell's *Principles of Geology* (1830–33), demonstrating beyond all possible doubt that the earth must be of great antiquity, there are plenty who'll tell you it's only a few thousand years old. And that our ancestors romped with dinosaurs. And that the sun and all the rest of the universe go round the earth.

There are those who argue that putting a zygote to use to offer potential cures to the currently uncurable is killing a child, whereas flushing that same zygote down the toilet isn't. There are those who deny the benefits of genetically modified foodstuffs in a world of soaring human population (and *there*'s another denial story in itself) and those who deny concerns that in some instances more research is needed before letting those modified organisms into the wild. The dangerously deleterious environmental impacts of intensive farming are so widely denied that in the average US supermarket there's likely to be no nonfactory meat, poultry, and fish on offer *at all*. And there are still plenty of people around who deny that smoking causes illness and death.

There are those who, because it offends their ideology or simply because they've been deluded by demagogues and industry shills, deny the reality of global warming even though the evidence is becoming drastically harder to ignore with every passing year.

And this is hardly to start the list of the science denials that are currently corrupting our culture and poisoning our public discourse—stifling our ability to make the right decisions for society's benefit and the well-being of our children and grandchildren.

The denial of science is a topic of enormous breadth and depth. When I first decided to write this book, I thought I'd be able to cover every aspect of it, if not exhaustively then at least in enough detail to give a fairly comprehensive idea of what was going on. It soon became clear to me this was impossible. If I didn't want to produce a book the size of *Encyclopedia Britannica* I was going to have to tone down my aspirations a trifle—which meant deciding to give some aspects of denialism scant treatment to allow space for others that seemed more important. Important? I'd say some of the areas I've chosen not to focus on—like stem cell research and genetically modified foods—are very important indeed, but they're still not *as* important as climate-change denialism. Because of our continued denial of global warming—or certainly our seeming political inability to address it—there's a good chance human civilization as we know it will have disappeared by the end of this century. Now *that*'s important.

How could the richest country on the planet, home to some of the world's finest universities and libraries, have reached the sorry state where denialism of science—of objective reality—is rampant? In the weeks after a deranged gunman, Jared Loughner, shot Congresswoman Gabrielle Giffords and eighteen others, Rightists in the US claimed Loughner was a Leftist while Leftists claimed he was a Rightist. In an essay on this situation, "Why We

Should Take Jared Loughner's Politics Seriously," educator Steve Striffler put his finger on one of the primary reasons there's so much antiscientific nonsense permeating US culture:

> Like Loughner, a significant portion of young people are, for very good reasons, profoundly anti-establishment, distrustful of anything they hear from the government or mainstream media. . . . [I]n a world where fragments of information come from so many sources, it often leads them to the odd place where *any explanation of the world is as good as any other, where there is no conceptual rudder for judging one theory or idea against another.*[1]

In other words, we have a bad case of GIGO—if you put enough garbage into people's minds, sooner or later too many people will believe too much of it, even if the bits they believe are mutually incompatible. Rather than being taught how to think analytically or constructively, people are being taught to *stop* thinking . . . all the while believing their disagreement with the views of "experts" to be really a case of "thinking for themselves."

This was the view that Don McLeroy, a Creationist, persuaded his colleagues on the Texas State Board of Education to adopt in 2010. That scientists said one thing while McLeroy's uneducated instinct said another meant the scientists were wrong. McLeroy used the phrase that gave its title to Orlando Wood's indie movie project *Standing Up to the Experts*. People who know what they're talking about are "lovely people, I'm sure," said McLeroy, seemingly close to tears of frustration and bafflement; so how could they persist in saying stuff his gut told him was false?

Is science denialism merely a product of ignorance or illiteracy? In many instances it certainly is. In their *Unscientific America* (2009), however, Chris Mooney and Sheril Kirshenbaum point out that there's no simple equation: Many antivaxers and AGW[2] "skeptics" have considerably more scientific knowledge than the average citizen, having researched extensively, usually on the Internet, to find everything they can that'll reinforce their convictions. The trouble is that, through ignoring all the rest of the information, they're essentially creating a false worldview—false because it's incomplete.

Obviously, the worldview of *anyone* is incomplete; the difference is that many of us recognize our areas of ignorance and rely on experts to fill in the gaps—just as most of us call a plumber rather than attack that burst pipe ourselves. There's no shame in being ignorant about something—in fact, about lots of things—because it's essential for our survival that we don't try to be "jacks of all trades, masters of none." The danger lies in being unaware of our areas of ignorance or in thinking ignorance deserves an equal voice in debate as expertise.

Of relevance here is the Dunning–Kruger effect, first formally described in 1999 in "Unskilled and Unaware of It: How Difficulties in Recognizing One's Own Incompetence Lead to Inflated Self-Assessments"[3] by Justin Kruger and David Dunning: "We argue that when people are incompetent in the strategies they adopt to achieve success and satisfaction, they suffer a dual burden: Not only do they reach erroneous conclusions and make unfortunate choices, but their incompetence robs them of the ability to realize it." In various tests Dunning and Kruger discovered this hypothesis amply supported: The "lack of skills" of "unskilled" people not only makes them incapable of realizing how inept they are but in fact makes them *overestimate* their abilities—they're like the guy in the famous photo with his "Get a Brain, Morans!" sign. Able people, by contrast, either accurately estimate their own abilities or, quite often, underestimate them.

A corollary is that "unskilled" people are far more certain about things than "skilled" ones; further, "unskilled" people regard it as a weakness when others use statements like "all the evidence suggests" rather than the—always false—"we know for a fact." In the hothouse of US politics, any lack of willingness to state false certainties is an electoral liability—indeed, politicians often gain popular and media approval for stating with certainty things that are palpably false.

A fine place to see the Dunning–Kruger effect in full pomp is in the comments section of any online article about a scientific topic that's a focus of public "skepticism." Here people who are by any definition ignorant and/or foolish let rip at those who are too STUPID!!!! to realize the earth is less than ten thousand years old, or whatever the particular bugbear might be. Often their beliefs can be very funny; at the same time, it's sobering to realize that, because of the Dunning–Kruger effect, our public decisions are influenced by people who are not just ill informed but aggressively determined to remain that way.

The sciences are particularly affected, since it's in the sciences that people are most likely to be poorly educated. It's hardly surprising that legislative bodies at both the state and the federal levels have long track records not just of coming to imbecilic science-related judgments—which would be bad enough—but of attempting arrogantly to impose these upon the scientists concerned; for instance, there are cases of legislative assemblies in effect forcing physicians and law-enforcers to treat quack medicine with respect.[4]

So, what are these things we do not know?

According to the National Science Board's *Science and Engineering Indicators* (2000), only about half of US adults are aware the earth travels around the sun annually;[5] about one in five think the sun orbits the earth daily. A 2002 *Time*/CNN poll found that nearly 25 percent of the US population believes the terrorist attacks of September 11, 2001, were predicted in the Bible. A 2009 poll in New Jersey found that 18 percent of self-identified conservatives thought President Obama was the Antichrist; another 17 percent—the postmodernists?—were "not sure."

Among 2,455 US adult respondents in a 2007 online Harris Interactive poll,[6] 79 percent believed in miracles, 75 percent in Heaven (though only 62 percent in Hell), 74 percent in angels, 69 percent in the survival of the soul after death, 62 percent in the Devil, 41 percent in ghosts, 35 percent in UFOs, 31 percent in witches, 29 percent in astrology, and 21 percent in reincarnation. Acceptance of evolution (42 percent) just beats out Creationism (39 percent). Other results indicate some measure of cognitive dissonance; for example, only 12 percent of respondents accepted the Torah as entirely the word of God though 35 percent believed that the Old Testament is entirely the word of God. The results of online polls must always be treated with extra caution, but other polls—including earlier Harris polls with the same or similar questions—have given broadly similar results.

A 2008 survey carried out by the National Institutes of Health (NIH) and the Centers for Disease Control and Prevention (CDC) indicated that in 2007 about four in ten US adults resorted to "unconventional medical approaches" at least some of the time, with one in nine children being treated likewise.[7]

A 2010 Gallup poll showed that Americans ranked global warming the least concerning among eight environmental issues.[8]

"When compared to children in the rest of the US, a Texas child is 93 percent more likely not to have access to healthcare, 33 percent more likely not to receive mental healthcare services, 35 percent more likely to grow up poor, and 16 percent more likely to drop out of school."[9] A University of Texas/*Texas Tribune* poll done in early February 2010 found that only 35 percent of adult Texans know humans have evolved from earlier species; 51 percent of respondents believed that humans have always existed in their current form, while 14 percent were uncertain.[10] Confronted by the statement, "God created human beings pretty much in their present form about 10,000 years ago," 38 percent of the respondents agreed. And 30 percent of adult Texans believed that humans and dinosaurs lived alongside each other. The liberal arts professor who'd devised the questions, David Prindle, commented: "[Comedian Lewis Black] did a standup routine a few years back in which he said that a significant proportion of the American people think that *The Flintstones* is a documentary. Turns out he was right."

In response to a 1999 poll's result that 45 percent of adult Americans believe the Bible's account of Creation should be taken literally, it occurred to distinguished physicist Robert L. Park—as recounted in his book *Superstition* (2008)—that "it would be interesting to ask people if they agree with scientists who say dinosaurs lived millions of years ago, omitting any reference to Genesis" (p. 49). He approached a pollster at the *Kansas City Star* and suggested such an experiment. Astonishingly, when the experiment was conducted, 81 percent of the respondents agreed with the scientists. In other words, more than half the people who thought the world was just a few thousand years old *also* believed that dinosaurs existed millions of years ago.

The fact that rates of belief in Creationism are far higher in the US than in other developed nations would suggest that something is seriously amiss with US education—both in schools and colleges and later. So do reports such as one that showed that 37 percent of the citizens in Washington DC have a third-grade reading level or lower.[11] Writing in *Salon*, Alfred McCoy offered gloomy predictions for America's future, not least because

> the US education system, that source of future scientists and innovators, has been falling behind its competitors. After leading the world for decades in 25- to 34-year-olds with university degrees, the country sank to 12th place in 2010. The World Economic Forum ranked the United States at a mediocre 52nd among 139 nations in the quality of its university math and science instruction in 2010. Nearly half of all graduate students in the sciences in the US are now foreigners, most of whom will be heading home, not staying here as once would have happened. By 2025, in other words, the United States is likely to face a critical shortage of talented scientists.[12]

How deep does the educational problem go? Here's a website discussion reported by James Robert Brown in *Who Rules in Science?* (2001, p. 83):

> *A:* A pen always falls when you drop it on Earth, but it would just float away if you let go of it on the Moon.
> *B:* What? A pen would *fall* if you dropped it on the Moon, just more slowly.
> *A:* No it wouldn't, because you're too far away from Earth's gravity.
> *B:* You saw the APOLLO astronauts walking around on the Moon, didn't you? Why didn't they float away?
> *A:* Because they were wearing heavy boots.

What are the effects of a scientific education on the ability to reason critically? Elaine Howard Ecklund's book *Science vs. Religion* (2010) contains interesting figures on the religious (or nonreligious) attitudes of scientists working at major US research universities; she derived these between 2005 and 2009 while conducting the Religion among Academic Scientists (RAAS) study under a grant from the John Templeton Foundation. "The study involved 1,646 survey respondents, achieving a response rate of 75 percent, and 275 in-depth interviews with scientists."

Perhaps unsurprisingly, religious believers are in a minority among them; more unexpected is the extent to which this is true. Almost exactly one-third of those surveyed (34 percent) agreed with the statement "I do not believe in God," while almost the same number (30 percent) chose "I do not know if there is a God, and there is no way to find out" (p. 16). Ecklund found that just 8 percent of her scientists chose the God option. The percentage of Americans who're either atheist or agnostic is uncertain, but the summary report of the American Religious Identification Survey (ARIS), carried out between February and November 2008 by Barry A. Kosmin and Ariela Keysar, concluded that "roughly 12% of Americans are atheist (no God) or agnostic (unknowable or unsure)" (Ecklund, p. 2). Other surveys put the level of atheism/agnosticism among the general population even lower, at about 6 percent. We can conclude that scientists are anywhere between four and ten times as likely as nonscientists to discount or be dubious about the existence of a god.

Of course, scientists themselves are not immune to science denialism. We've already noted Don McLeroy, a dentist who denies evolution. Likewise, there are all the dentists, engineers, surgeons, economists, meteorologists, and so forth, who dispute the conclusions of climate scientists concerning imminent climate change—rather as if climate scientists should advise surgeons and dentists about surgery or dentistry.

This is not a new situation. There was huge resistance among scientists when Louis Pasteur announced his germ theory of illness. Resistance to Galileo came largely from his fellow scientists, who couldn't reproduce his results. Earth scientists for decades denied all the evidence favoring Alfred Wegener's theory of continental drift. Ignatz Semmelweiss's discovery of the importance of hygiene in hospitals led to his being ostracized from the medical profession. Sir Arthur Eddington, father of astrophysics, suppressed Subrahmanyan Chandrasekhar's prediction of black holes. Leibniz denied gravitation because he was squabbling with Newton. As late as 1636, the scholars brought together for the formation of New College—later Harvard—rejected the Copernican hypothesis, opting to stick with Ptolemy.

Until the 1980s it was accepted that peptic ulcers were a consequence of excess gastric acid, accompanied by or perhaps caused by stress. When it was proposed by Barry Marshall and Robin Warren that there could be a bacterial cause—specifically, that *Helicobacter pylori* could be responsible—the medical community was largely skeptical. Nowadays, although in the case of stomach ulcers another frequent diagnosis is overuse of anti-inflammatory drugs like aspirin and ibuprofen, the default treatment for an ulcer is a course of antibiotics.

What's not often realized is that Marshall and Warren weren't the first to make the discovery about *H. pylori*. That honor goes to Greek physician John Lykoudis, who began treating patients thus in 1958. Here the medical establishment went into lockdown denialism. *The Journal of the American Medical Association* refused the paper he submitted, a disciplinary committee handed him a heavy fine, and so on. What's startling is that no one seems to have thought it worth replicating his work to find out if he might be right.

Perhaps the most prominent contemporary scientist to reject Darwin's proposed theory of evolution by natural selection was Lord Kelvin. His insistence that the earth couldn't possibly be more than 100 million years old—not enough for Darwinian evolution to have produced modern life-forms—

while justified within the prenuclear physics available to him, was inspired and enhanced by Kelvin's personal religious convictions.

Even Nobel laureates can fall into the trap of denying science in favor of unorthodox theories in areas outside their expertise. The later career of Linus Pauling was marked by profound scientific crankdom. He became convinced that massive doses of vitamin C kept cancer at bay. The shame is that Pauling is now remembered more for this than for his earlier, genuine scientific achievements. Nikolaas Tinbergen set some kind of a record for the descent into crankery by espousing in his Nobel acceptance speech the notion that autism is a stress disease.[13] Louis J. Ignarro, whose 1998 Nobel was for work on nitric oxide (NO), now claims the gas is a cure for conditions as diverse as heart disease, impotence, and cancer. And Luc Montagnier, winner of the 2008 Nobel Prize in Physiology or Medicine for his discovery of the human immunodeficiency virus (HIV), has taken to claiming there's a scientific justification for homeopathy.

Denying the abilities of technology is common even among science fiction writers. To take just a single example, in his famous story "The Dead Past" (1956) Isaac Asimov makes much of the fact that in his hi-tech twenty-first century there's no need to retain a printed sheet of paper any more; you can store it digitally on computer. Fine so far, except that the computer concerned, MULTIVAC, fills half a block—the notion of personal computers hadn't occurred to the Good Doctor. And, should you want a fresh copy of that page, MULTIVAC can spew it out for you in "mere minutes." Asimov was not alone. An unsigned article, "Brains That Click," in the March 1949 issue of *Popular Mechanics* maintained, "Where a calculator like the ENIAC today is equipped with 18,000 vacuum tubes and weighs 30 tons, computers in the future may have only 1,000 vacuum tubes and perhaps weigh only 1½ tons" (p. 258).

The list could be extended indefinitely.[14] It's arguable that most of these misguided predictions have arisen because the society in which the predictors were embedded failed to appreciate that, always, the future will be quite extraordinarily different from the past. This was not at all apparent even a few decades ago (despite augurs like Alfin Toffler, with his 1970 book *Future Shock*). Yes, it was assumed there'd be an onward and upward march of technology, but no one appreciated the steepness of the gradient.

Science comes to conclusions, but they're always provisional, open to revision as fresh evidence comes in. Unfortunately, as noted, this rational diffidence is often interpreted by audiences and the news media as a sign of uncertainty, weakness, or vacillation. Obviously scientists would do well to make themselves a bit cannier about how what they say will be heard by lay people; at the same time, it might make sense for lay people to educate themselves just enough to realize that emphatic assertion and shouting-down are usually signs someone doesn't know what they're talking about.

In the field of astronomy/cosmology, there are large areas still under active discussion. How does one differentiate denialism from worthwhile debate? A certain number of professional astronomers—notably including Halton Arp, Eric Lerner, and Hermann Bondi—have disputed the Big Bang theory. These scientists have still been operating within science. Compare and contrast any of their writings on the subject with the "Big Bang" entry in Conservapedia®, which spends more wordage promoting Creationist dissent than it does on the theory itself. Here is the entirety of that article's "Scientific Criticism" section:

> It should be noted that the Big Bang theory has received criticism because it ignores the theory of an oscillating universe. Also, no first cause from the Big Bang has ever been successfully identified. Furthermore, critics of the Big Bang point out that not everything in the universe is actually moving apart from everything else as some galaxies have collided with other galaxies in the past, although this could be explained through understanding of classical mechanics.

Small wonder Conservapedia feels justified in billing itself as "The Trustworthy Encyclopedia."

The Science Wars, as they've been called, focus on a rivalry between certain sociologists/philosophers, on the one hand, and scientists, on the other, as to what science is and the validity of its conclusions—a dispute between amateurs and professionals, in other words.[15] This sounds innocuous enough. Where the bloodletting becomes more frenzied, however, is in the significant skirmishes that center on the sociologists' appropriation and often abuse of scientific concepts and terms of whose meaning, all too frequently, it's obvious they're ignorant. Most of the offenders come from the sociological school called social constructivism; for historical reasons they're generally called, within the context of the Science Wars at least, postmodernists.

The book that served as the scientists' declaration of war was *Higher Superstition: The Academic Left and its Quarrels with Science* (1994) by Paul R. Gross and Norman Levitt. This unfortunately commits a piece of partisan political framing from which subsequent discussions have yet fully to escape. As Mario Bunge observes in his "In Praise of Intolerance to Charlatanism in Academia" (1996), "[I]rrationalism, in particular the distrust of science, has no political color; it is found left, center, and right."

To be irritated by Gross and Levitt's politicization is not to say the bloc they identify doesn't exist, and isn't a menace. Worse still, some of its general ideas have diffused into the wider society, promoting, for example, such New Ageist notions as that all truth is merely relative, scientific truth included. It was not a New Ager but an unnamed Bush administration official who came out with that most "postmodern" of all recent political pronouncements:

> That's not the way the world really works any more. We're an empire now, and when we act, we create our own reality. And while you're studying that reality—judiciously, as you will—we'll act again, creating other new realities, which you can study too, and that's how things will sort out. We're history's actors . . . and you, all of you, will be left to just study what we do.[16]

If Gross and Levitt launched the opening salvoes, the start of hostilities in earnest came in 1996. Acknowledging that professional scientists were hitting back at the invasion of their territory by unqualified critics, the editors of the journal *Social Text* announced a special issue—titled "Science Wars"—to be devoted to the matter. One submission they accepted was from the physicist Alan Sokal, entitled "Transgressing the Boundaries: Toward a Transformative Hermeneutics of Quantum Gravity." Bizarrely, this paper wasn't sent out for peer review. Citing many of the social constructivist heroes as if reverentially, Sokal gaily lampooned, for example, the prevailing view among postmodernists that all truth is relative (that what we perceive as scientifically revealed reality is in fact merely a social construct, with different cultures and communities discovering different "realities" that are all equally valid) with such statements as "The pi of Euclid and the *G* of Newton, formerly thought to be constant and universal, are now perceived in their ineluctable historicity" and "the discourse of the scientific community, for all its undeniable value, cannot assert a privileged epistemological status with respect to counter-hegemonic narratives emanating from dissident or marginalized communities."

A matter of days after the appearance of his essay in *Social Text*, Sokal 'fessed up to the hoax with an essay in the magazine *Lingua Franca*, "A Physicist Experiments with Cultural Studies." The resulting cacophony extended far beyond the academic world, with articles in major newspapers around the world.

Sokal thereafter expanded his attack in the form of the book *Fashionable Nonsense* (1998; with Jean Bricmont). In one example (p. 27) they cite the French psychoanalyst Jacques Lacan equating the male "erectile organ" with the irrational number *i*, the square root of –1.[17] A further delight is French sociologist Bruno Latour's analysis of Einstein's Special Theory of Relativity, "A Relativistic Account of Einstein's Relativity" (1988), in which he—Latour, not Einstein—reveals that it's not about what Latour calls "the electrodynamics of moving bodies," as Einstein thought, but about long-distance travelers. Inexplicably, some have questioned Latour's description of himself as a "Darwin of Science."

Another part of the mix is the so-called Strong Programme, advanced in the mid-1970s by Edinburgh University sociologist David Bloor and others. The reason for this approach's name is to contrast it with the weak, or ordinary, sociological approach to studying the way the sciences operate (that the behavior of scientists and their institutions are fair game for consideration as social structures, but that scientific theories and results are independent of social factors). Bloor and his disciples maintained—and maintain—that, au contraire, even science itself should be regarded as a social construct and open to deconstruction on that basis. It's not hard to see how one can get from this viewpoint to the notion that $E = mc^2$ is a sexed equation, as Belgian sociologist Luce Irigaray claimed in "Sujet de la Science, Sujet Sexué?" (1987): Had it been derived by a female scientist within the context of a predominantly female discipline, who knows what the relationship between the three quantities might have been.

A further product of this sort of approach, although in fact preceding Bloor's enunciation of the Strong Programme, is Paul Forman's 1971 essay on the genesis of the quantum theory titled "Weimar Culture, Causality, and Quantum Theory, 1918–1927: Adaptation by German Physicists and Mathematicians to a Hostile Intellectual Environment." Forman's thesis is that Heisenberg and the rest devised quantum theory—all indeterminate and "mystical" as it is—in order to regain their social prestige in a Germany which at the time was avidly embracing Romanticism and rejecting science. Dismissed as immaterial by Forman are the notions that the German scientists had good scientific reason to develop quantum theory—it solved a number of outstanding scientific problems and rapidly showed itself capable of solving many more—and that this has proven to be one of the most successful and important theories in science's history. It's difficult to understand how this could be so if quantum theory were just a faddish social construct.

Mario Bunge is less diplomatic even than Sokal and Bricmont in his assessment of the postmodernist fad; the title of his bracing essay "In Praise of Intolerance to Charlatanism in Academia" (1996) makes this clear:

> These are not unorthodox original thinkers; they ignore or even scorn rigorous thinking and experimenting altogether. Nor are they misunderstood Galileos punished by the powers that be for proposing daring new truths or methods. On the contrary, nowadays many intellectual slobs and frauds have been given tenured jobs, are allowed to teach garbage in the name of academic freedom, and see their obnoxious writings published by scholarly journals and university presses. Moreover, many of them have acquired enough power to censor genuine scholarship. They have mounted a Trojan horse inside the academic citadel with the intention of destroying higher culture from within.[18]

Another of Bunge's targets is the US political scientist Samuel P. Huntington, who had a habit of using pretend mathematics. The mathematician Neal Koblitz exposed this phoniness in the essay "Mathematics as Propaganda: A Tale of Three Equations; or, The Emperors Have No Clothes" (1988);[19] as a result, the US Academy of Sciences declined Huntington's induction. The "mathematical equations" that created all the fuss came in Huntington's book *Political Order in Changing Societies* (1968) and read thus:

- social mobilization/economic development = social frustration
- social frustration/mobility opportunities = political participation
- political participation/political institutionalization = political instability

Try putting a numerical value to any of those terms. And how *do* you divide social mobilization by economic development?

Huntington—whose views in their day were taken very seriously by governments—is merely one of the most prominent examples of political scientists and economists using the pretense of mathematics where plain English might reveal the paucity of reasoning.

Overall, the general pattern of the postmodernist assault on science seems to be to create a sort of random spaghetti of scientific terminology that makes no sense precisely *because* random, and then to attack it as nonsensical for one "sociological" reason or another. This would be merely bad if the attacks on the straw men themselves were coherent; alas, they aren't, presumably because the "postmodernists" concerned have not the remotest understanding of the scientific concepts whose names they so merrily jumble; it's rather like the way so many pseudoscientists and pseudomysticists—and their followers—believe it gives their particular crankery bonus street cred to add the adjective "quantum": "Quantum Healing," "Quantum Theology," even "Quantum Buddhism." The end result of the postmodernists' efforts is a gross corruption of reasoning—not to mention a gross deception—that's both a denial of science and a tool devised to promote a more universal denial of science. Also, as philosopher of science Meera Nanda points out, with its notion that science is merely a culture-specific social construct, postmodernism has been "a blessing for *all* religious zealots, in all major faiths, as they no longer feel compelled to revise their metaphysics in the light of progress in our understanding of nature in relevant fields."[20]

Science isn't alone in suffering the onslaughts of "relativism." In *Not Out of Africa* (1996), her stirring defense of history against the corruptions of it embraced by so-called Afrocentrist writers—such as that Aristotle stole his philosophy from the books of black writers he found in the Library at Alexandria (a "fact" that continues to be promulgated by Afrocentrists even though the library wasn't built until some decades after Aristotle's death)—Mary Lefkowitz (p. xiv) states the problem:

> There is a current tendency, at least among academics, to regard history as a form of fiction that can and should be written differently by each nation or ethnic group. The assumption seems to be that somehow all versions will simultaneously be true, even if they conflict in particular details. According to this line of argument, Afrocentric ancient history can be treated not as

> pseudohistory but as an alternative way of looking at the past. It can be considered as valid as the traditional version and perhaps even more valid because of its moral agenda.

This does of course cut both ways. Until relatively recently, US textbooks omitted much about the attainments of the Native American civilizations that were here when the white man arrived and suffered near genocide at his hands. Just think of the myth of the Mound People—the mighty long-lost civilization that left all the architecture and artifacts that surely, surely, surely the Red Man was too stupid and wastrel and savage to accomplish. That myth endured in textbooks well into the twentieth century.

Lefkowitz's thesis, that the history our youngsters are taught should be as accurate as the best endeavors of research can make it, would seem so monumentally obvious as to be barely worth stating. And yet, during the writing of her book and the article that was its precursor,[21] Lefkowitz, a humanities professor at Wellesley College, Massachusetts, was subject to considerable abuse—including accusations of racism—in both private and, more worryingly, professional life.

This pattern of threatening the bearers of unwelcome truths is common throughout all fields of science denialism.

Some people get paid a very great deal of money to deny basic science. On January 4, 2011, David Silverman, president of the American Atheist Group, appeared on *The O'Reilly Factor* to talk about his organization's ad campaign telling the public that religion's a scam:

> *O'Reilly:* I'll tell you why it's not a scam, in my opinion: tide goes in, tide goes out. Never a miscommunication. You can't explain that.
>
> *Silverman:* Tide goes in, tide goes out?
>
> *O'Reilly:* See, the water, the tide comes in and it goes out, Mr. Silverman. It always comes in, and always goes out. You can't explain that.[22]

God moves, I suppose, in mysterious waves.

As a general rule, if the mass of mainstream journalists says one thing and the mass of scientists working in the relevant field says another, believe the scientists. Unfortunately, swaths of the public see it the other way round.

In a 2002 paper called "Bringing Female Scientists into the Elementary Classroom: Confronting the Strength of Elementary Students' Stereotypical Images of Scientists,"[23] Gayle A. Buck et al. recorded that many young children refused to believe working scientists could be young, friendly, and female. The kids *knew* from their TV watching that scientists are middle aged (or older), nasty, and male!

One factor in the modern fad for science denialism may be the collapse, at least in the US, of the mass middlebrow culture, a phenomenon that came to an end by, arguably, the early 1970s. Middlebrows read books copiously (even though many of the books they read were looked down upon by self-styled highbrows), followed news and current affairs with a keen interest, listened avidly to "improving" radio broadcasts, displayed an enthusiasm for rationalism, and, most relevantly, had an abiding and prodigious near reverence for the discoveries of science and the accomplishments of technology. Added to this was the zeal for self-education: If the *Reader's Digest* fast-food meal of astronomy or medicine wasn't enough to sate the appetite, there would likely be books on the subject in the public library, not to mention a portfolio of middlebrow scientific digests, from *Scientific American* to *Popular Mechanics* and beyond, available at the newsstand.[24]

It was because of the prevalence of middlebrow ideals in the country that John F. Kennedy's successful presidential candidacy in 1960 stressed not just his youthfulness and charm but his literacy, his culture, and his intellectual aspirations. The middlebrows of America clearly thought it was a good idea to have a president who was intelligent. Nearly half a century later, by contrast, a similarly youthful and charming and probably *more* intelligent presidential candidate, Barack Obama, had to do a fair amount of spinning to make sure his image was suitably dumbed down for fear of his seeming "elitist." After all, eight years earlier, the likewise intelligent Al Gore had lost what should have been a safe bid for the presidency through being perceived by the electorate as insufficiently stupid.

Another way of looking at today's widespread phenomenon of scientific denialism is that it is, in essence, a sort of large-scale bull session gone horribly wrong. In *On Bullshit* (2005, pp. 36–37), Harry G. Frankfurt points out that the bull session (and its female equivalent, the hen session) has characteristics unlike those of normal conversations:

> What tends to go on in a bull session is that the participants try out various

> thoughts and attitudes in order to see how it feels to hear themselves saying such things and in order to discover how others respond. . . . Each of the contributors to a bull session relies, in other words, on a general recognition that what he expresses or says is not to be understood as being what he means wholeheartedly or believes unequivocally to be true.

Yet this is surely only half the story. While most bull sessions, most of the time, are as Frankfurt describes, we've surely all witnessed instances where, for some reason or no reason, a line is crossed and views that were until then just genially supported suddenly become entrenched; tempers can spill over and violence break out over matters that are entirely trivial—and that were, minutes before, *regarded by all concerned* as entirely trivial. Worse, it's quite possible, as Frankfurt implies, that even those who're now prepared to fight in defense of a viewpoint didn't believe in it until that very moment.

It seems plausible that at least some science deniers go through a similar arc. At first they're just having fun being bloody-minded by playing devil's advocate to perceived wisdom but, as their ideas are repeatedly rebuffed—and especially since the rebuffers likely do so rather tiredly, having encountered this "wildly original idea" myriad times before—what started off as just an enjoyable bit of bullshitting becomes a set of rigid beliefs. At this stage, like one of the participants in that deteriorating bull session, the more the falsity of the viewpoint becomes evident, the greater the denialist's vehemence—perhaps even to the extent of violence: There have been countless threats made against, for example, prominent climate scientists.

Historian Deborah E. Lipstadt was talking about Holocaust deniers when she made this remark, but the more general application is obvious:

> My refusal to appear on . . . shows with deniers is inevitably met by [TV] producers with some variation on the following challenge: Shouldn't we hear their *ideas*, *opinions*, or *point of view*? Their willingness to ascribe to the deniers and their myths the legitimacy of a point of view is of as great, if not greater, concern than are the activities of the deniers themselves. What is wrong, I am repeatedly asked, with people hearing a "different perspective"?[25]

A common grumble of denialists, whatever the subject of their denial, is that their "dissident view" is being censored by the scientific authorities—through refusal of publication in peer-reviewed journals, perhaps, or through lack of inclusion in international reports like the IPCC's on climate change. Yet, ironically, a very frequent technique used by denialists is a form of censorship. This is the cherry-picking of quotes, a practice whose use was exemplified par excellence in the whipped-up furor over Climategate (see pages 246ff.). To cherry-pick those pieces of an opponent's work that can be made to seem, out of context, as undermining that work is also to censor it: You are gagging the opponent's expression of her or his actual views.

A similar argument could be extended to those bizarre shouting matches that pass for "balanced" debate on so much television these days. Almost inevitably, the louder shouter drowns the argument of the more reasoned voice, making sure its argument cannot be heard and thus in effect censoring it. Some of the shouters have radio or TV shows of their own, where they can either shout down views they dispute or censor them out entirely.

Other common elements of denialist rhetoric include the ad hominem attack and the trick of focusing on details and pretending they represent the whole story. A disagreement among climate scientists as to the anticipated rate of retreat of Himalayan glaciers is trumpeted as a fundamental error that must surely bring established climate science to its knees. A particular case in point is that of Paul Ehrlich, who in denialist circles is regularly held up for special ridicule as the exemplar par excellence of the "scientific alarmist" who "so spectacularly got it wrong"—this because, in books like *The Population Bomb* (1968), he made some rather gloomy assessments of how much longer the planet's ecosystem can tolerate an exponentially increasing human population. So far, each of Ehrlich's assessments has been proven overly pessimistic, as human ingenuity has warded off the evil moment; but he's perfectly correct to state that the crunch moment will eventually come unless we curtail population growth. His critics are like the guy who jumps off the Empire State Building and, after falling the first thirty stories, says, "See? I told you I'd be okay."

A frequent semantic tool used by denialist campaigners is confusion over the word "theory": "It's only a theory," we're told of Darwin's theory of evolution by natural selection, though obviously in this context the word "theory" implies something as close to fact as science will ever claim.[26]

In their paper "Chiropractors and Vaccination: A Historical Perspective"

(2000)[27] James B. Campbell, Jason W. Busse, and H. Stephen Injeyan identify a set of bogus arguments used by those chiropractors (about one-third) who reject vaccination:

- immunizations are not effective
- vaccines can be harmful
- there is disagreement about, and even opposition to, immunizations among medical experts
- immunization policy is governed by the medical-pharmaceutical complex and motivated by greed
- any compulsory medical treatment is unacceptable
- vaccinations are unnecessary
- acceptance of vaccination is to repudiate chiropractic philosophy

For his discussion of the denial of evolution in *The Making of the Fittest* (2006), Sean B. Carroll adapts this list to produce one that's applicable to the tactics of science denialists in general. Carroll's version (pp. 231–33) offers a very useful analytic tool:

- doubt the science
- question the motives and integrity of scientists
- magnify disagreements among scientists, and cite gadflies as authorities
- exaggerate potential harm
- appeal to personal freedom
- acceptance repudiates key philosophy

It's not entirely clear to me why Carroll should have omitted an equivalent to Campbell et al.'s sixth item, "Vaccinations are unnecessary." AGW denialists are forever telling us we don't need to take action to mitigate climate change, Creationists tell us evolution is an unnecessary hypothesis because we were made just the way we are, and so on. That aside, it's depressingly easy to see Carroll's template being cynically applied by the leaders of all the major denialist movements today.

One of the differences between science and pseudoscience is that science accepts the results it doesn't like—the results that are flatly counterintuitive. Intuition is a very useful tool in many different circumstances, but the fact that something's counterintuitional doesn't automatically mean it's wrong.

For millennia intuition told us the sun went round the earth; there are still those who prefer that intuitively derived model. In this instance, their delusions are probably not harmful (except to their children). When, however, people's intuitions tell them the MMR vaccine causes autism or that antiretroviral drugs cause rather than cure AIDS, the results can be catastrophic.

In their *Devenez Sorciers, Devenez Savants* (2002), Georges Charpak and Henri Broch point to a related cause of science denialism: the fact that the laws of probability very often seem to fly in the face of commonsense. It's surely illogical that a party need have only 57 guests before it's a 99 percent certainty that two will have the same birthday, even though there are 365 days in the year? Or, even if the probability involved is strictly commonsensical, we tend to see not the *real* probability involved but only a superficial effect. One example they cite is the television trick whereby a "psychic" asks the audience at home to turn on the lights around them; he will use his powers to blow those bulbs. Sure enough, by the end of the program the studio switchboard is jammed—because, if a million people have their lights on, quite a few bulbs will inevitably blow before the show's over.

In *Denying AIDS* (2009), Seth Kalichman (p. 14) draws some pointed parallels between denialists and the "suspicious thinker," as defined by David Shapiro in the latter's *Neurotic Styles* (1965), who

> "does not pay attention to the apparent facts but, instead, he or she pays sharp attention to any aspect of them or their presentation that lends confirmation to his original suspicious idea." The suspicious person constructs a subjective world based on "significant" clues with a complete loss of appreciation for the context. Shapiro also discusses the suspicious person as having encapsulated delusions, limited in content and type. Encapsulated delusions fit what we see in denialism, where a person can be grounded in reality in nearly every facet of his or her life and yet have a circumscribed entrenched belief system that is not reflective of reality and not refutable by facts.
>
> The insights offered by Shapiro are that denialists are not "lying" in the way that most anti-denialists portray them. . . . Rather, denialists are trapped in their denialism.

Further, Kalichman argues that just as "suspicious thinkers" are especially prone to conspiracy theorizing, so are denialists. The idea of a global conspiracy of climate scientists to help establish a New World Order is the kind of preposterous scenario you might expect to find as, say, the plot of a Michael Crichton novel, but surely no one could possibly take it seriously? Well, no one except a "suspicious thinker" . . . and of course a denialist. Yet something very like this is believed by literally millions of Americans today.

There are also misconceptions as to what science actually is and how it is done. George E. Webb, in *The Evolution Controversy in America* (1994), discussed the arguments offered by Creationists. One common theme was the image of the scientific pursuit "as the collection and organization of data. From this perspective, any theory that attempted to extrapolate from the data to answer larger questions represented nothing more than a guess" (p. xi).

What sticks in the craw of many Creationists and others is the fundamental scientific principle called methodological naturalism. No hypothesis that attempts to explain physical phenomena is worth investigating if it requires the action of a supernatural agent. While the tools of science may be used, however fruitlessly, in attempts to find the supernatural, the supernatural itself offers no useful explanation of real-world observations. It is this principle that most upsets the members of the Discovery Institute, and impels its drive to market the concept of Intelligent Design (ID) to the gullible. Although the superficial purpose of the institute is to counter Darwinism, in fact its war is with methodological naturalism, and beyond that with materialism as a rational means of explaining the universe. This is made explicit in the infamous Wedge Document (see page 180).

In his entertaining *Spin This!* (2001), Bill Press jeers at the Bush administration for its use of the "not yet settled" sophistry to deny the evidence of science. He contrasts the administration's justifications for its policy concerning climate change and its attempt to resuscitate a version of Ronald Reagan's long-ago discredited Strategic Defense Initiative, "Star Wars." As Press summarizes (p. 145): "The Bush rule is: *More science on global warming, where none is needed; and no more science on missile defense, where much more is needed.* That's the spin. And, as spin, it makes perfect sense."

The Bush administration took the corruption and denial of science in US government to heights probably never seen before; for just some accounts see Seth Shulman's *Undermining Science* (2006), Chris Mooney's *The*

Republican War on Science (2005), or even the relevant chapter in my own *Corrupted Science* (2007). Yet the seeds had been sown far earlier. In 1973, when Richard Nixon was told by his scientific advisers that various of his pet technological schemes, such as the Supersonic Transport Program, were scientifically unsound, he responded by firing the advisers. While Jimmy Carter earned world headlines by claiming to have seen a UFO, at least he didn't try to insert ufology into the nation's policy. Ronald Reagan, both as California governor and US president, promoted Creationism as at least a viable alternative hypothesis to evolution; he also ignored or derided those scientists—the vast bulk of *relevantly qualified* scientists—who told him the purported technological underpinnings of "Star Wars" were outright fanciful. Meanwhile, the White House had consultant astrologers.

Even during the presidency of Bill Clinton, with Congress under the control of Newt Gingrich's Republican Party from 1994, the denial of science continued. Although climate scientists were already warning about AGW, the nonscientists of Congress believed they knew better and rejected the information—as they also tried to do with the smaller but then far more immediate problem of the depletion of the ozone layer. Attempts were made, under the guise of good fiscal housekeeping, to shut down such federal agencies as the US Geological Survey. Crazily, in 1995 Congress *did* abolish the Office of Technology Assessment (OTA), thereby at a stroke removing the most important overseer of the quality of the scientific advice coming into Congress. Clinton did not institute these measures, but he must have realized the importance of stopping them, something it was in his power to do. Yet, given the choice, he opted consistently for political expediency.[28]

A decade and a half later, Louisiana Governor Bobby Jindal would provide an ironic echo of such doings in his response to President Obama's February 25, 2009, State of the Union address. As an example of wasted expenditure in the Economic Stimulus Bill, Jindal picked out the $140 million allocated to the Geological Survey for "something called volcano monitoring. . . . Instead of monitoring volcanoes, what Congress should be monitoring is the eruption of spending in Washington." Less than a month later, the Alaskan volcano Mt. Redoubt erupted, covering Anchorage and its surroundings in ash. Fortunately, the people living in the Mt. Redoubt area were able to take measures to ameliorate the effects, having been warned as early as February 6 of an imminent eruption, thanks to . . . volcano monitoring.

A measure of the level of importance that science denial has reached among our political rulers came with the 2008 presidential election. The organization ScienceDebate2008, set up to demand that the various candidates debate what is after all a crucial topic, received support from the National Academy of Sciences, the American Association for the Advancement of Science, a plethora of distinguished scientists including many Nobel laureates, and even politicians from both sides of the political spectrum . . . yet was spurned entirely by *all* of the major campaigns, including that of soon-to-be-president Obama,[29] not to mention the mainstream media: Of roughly three thousand questions posed to the main candidates by frontline political pundits, exactly *six* addressed policy on—or even mentioned—the most urgent scientific topic of the age, climate change.

Through much of 2010, as I was laboring on this book, the world was being kept entertained by a tragicomedy that could be regarded as an example *sans pareil* of the incapacity for rational thought that both characterizes and inspires science denialism—and does immeasurable harm to this nation. This was the Dunning–Kruger effect writ large.

In February 2010, the Texas State Board of Education began discussing the social studies curriculum guidelines. Led in theory by newspaper editor Gail Lowe but in fact by dentist Don McLeroy, the board took a document that teams of professional experts had spent a year devising and proceeded to mangle it; the experts themselves had been excluded from the process since the preceding November. Among the amendments imposed by McLeroy in what the *New York Times*'s Russell Shorto described as "a single-handed display of archconservative political strong-arming"[30] was that students be required to "describe the causes and key organizations and individuals of the conservative resurgence of the 1980s and 1990s, including Phyllis Schlafly, the Contract With America, the Heritage Foundation, the Moral Majority and the National Rifle Association." The children's author Bill Martin Jr. was expunged from the standards because a completely different Bill Martin had written a book called *Ethical Marxism*. Watchdog Kathy Miller, cited by Shorto, commented of McLeroy and another board member: "It is the most crazy-making thing to sit there and watch a dentist and an insurance salesman rewrite curriculum standards in science and history. Last year, Don McLeroy believed he was smarter than the National

Academy of Sciences, and he now believes he's smarter than professors of American history." In a previous assault on Texas's educational standards, McLeroy had managed to remove from the textbooks any reference to the age of the universe—a datum that offends his own young-earth Creationism.

Weeks and months passed. Howlers abounded. Thomas Jefferson, disliked because he penned the Constitution's provision concerning the separation of church and state, was removed in favor of Thomas Aquinas, John Calvin, and William Blackstone. The Enlightenment was eradicated. A standard concerning the First Amendment was altered to include discussion of the right to bear arms, which is of course the subject of the *Second* Amendment. *Texas Freedom Network*, live-blogging the discussions on March 11, 2010, recorded: "The Texas State Board of Education today refused to require that students learn that the Constitution prevents the US government from promoting one religion over all others. They voted to lie to students by omission."[31] Study of Isaac Newton was dropped in favor of the requirement to examine the scientific advances military technology has brought about. The standard concerning the Civil Rights movement was altered to stress the role of the Republican Party in passing the relevant legislation; in fact, no southern Republican voted for either the 1964 Civil Rights Act or the 1965 Voting Rights Act. Board member Barbara Cargill demanded there be a standard requiring students to "explain three pro-free market factors contributing to European technological progress during the rise and decline of the medieval system."[32] The word "capitalism" was censored throughout because some of the board members thought it sounded a bit pejorative; "free enterprise" was inserted in its stead. The phrase "democratic societies" was replaced by "societies with representative government." Most shocking of all in this category was the replacement of "slave trade" with "Atlantic triangular trade." There was an amendment seeking to restore the reputation of Senator Joseph McCarthy, whom McLeroy has declared "vindicated."

Board member Terri Leo, part of the conservative bloc pushing through all this, went on the offensive against the media:

> Voters "get their information from the papers, which is mostly inaccurate," says Leo. "Most reporters are lazy, and they don't do their homework."[33]

In September, the conservative bloc on the board was back at it, this time crusading to remove the "pro-Islam bias" from the state's history books. As many cried, *What* pro-Islam bias? It was hard to see this as other than an attempt to institutionalize the anti-Muslim bigotry that was then sweeping the country.

The 2010 elections in Texas—and also the Republican primaries—considerably shifted the political balance of the Texas State Board of Education, and already efforts are being made to limit the damage done by the previous regime. But, because of the size of Texas, the actions of this rogue group are likely to affect US education nationally—affect it for the worse. The blogger at *Gin and Tacos* saw the fiasco as merely the tip of the iceberg:

> We have systematically devalued and dismantled education in this country to the point that the Japanese, Europeans, and so on aren't just beating us at math and science. They can beat us at essentially anything, because most of us can't comprehend things we read, retain simple facts, or construct an argument that adheres to the basic rules of logic. We are ignorant of the past, the present, and even our own professed belief systems. We often bemoan apathy, our national lack of desire to understand the government, law, economy, or politics. But the problem is not simply that we don't want to know; if our slipshod grasp of the few things in which we do profess an interest are any indication, we wouldn't get it even if we tried.[34]

1

GOD TOLD ME TO DENY

Wandering in a vast forest at night, I have only a faint light to guide me. A stranger appears and says to me: "My friend, you should blow out your candle in order to find your way more clearly."

—Denis Diderot, *Addition aux Pensees Philosophiques*

In a sense, religious belief is intrinsically a denial of science. Religion, like science, was born out of the universal human urge to produce explanations for phenomena. When the major religions began, a lot of phenomena were absolute mysteries. Why does wind roar and fire burn? How does the land produce the water that keeps the river flowing? What moves the sun in its daily path across the sky? Religion offered answers to these and plenty of other questions, and can thus be thought of as humankind's earliest stumbling attempt at science. There were forces in the world that were obviously far more powerful than anything mere humans could conjure—earthquakes that shook the land, volcanoes that filled the world with ash and fire and noise, storms that made the heavens tremble—so it must have seemed obvious these events were engendered by superhuman beings: gods.

The function of religions, then, was to explain the inexplicable. But along came people with inquiring minds who weren't satisfied with the "gods did it" catchall explanation. The shadow of the inexplicable began to be pushed back on every side, and—barring various dark ages—shrank ever more swiftly, until today there are only shreds of it left. And yet people go on believing in religious claims and denying scientific ones, even in instances where the latter have been proven beyond all rational doubt. Why?

In their book *The Unreality Industry* (1989, p. 127) Ian Mitroff and Warren Bennis offer a succinct answer:

> Our minds are set up to simplify, to bring order to a world that often is as chaotic as it appears. To do this we not only throw out a great deal of information that is presented to us but we lock in, long after they have seemed to retain [any] usefulness or validity, older patterns and pieces of information.

We're all guilty of this offense. Plenty of perfectly intelligent older people still think of the atom as a cluster of big pool balls around which smaller ones orbit, because that's the model they were taught in school and there has never seemed any compelling reason to update their thinking. But in other instances it *matters* if we cling to old, wrong explanations.

A difficulty facing anyone who seeks to discuss scientific denialism with the devout concerns the latter's deceptive tendency to label as science all sorts of things that patently aren't; it's difficult for someone who's merely read the label to accept that "Creation Science" isn't science at all. Whatever the motives of the person who wrote the label, the con trick is very often effective. Taner Edis sums this up well in the Islamic context in his *An Illusion of Harmony* (2007): "Fundamentalists sincerely think of themselves as supporting science while endorsing views that would cripple scientific practice" (p. 75). How many Americans who support the inclusion of Intelligent Design (ID) in the school science curriculum on the grounds that this would be "teaching the debate" realize how such a measure would cripple school science education? Many such advocates are not fundamentalists; some aren't even religious. They've been misled into believing that pseudoscience is science.

It seems adherents of traditional belief systems, confronted by the undoubted achievements of science and technology, have three available ways of reacting:

- They can cling yet more fervently to the traditional belief system, thereby denying science and reasserting their irrationalism.
- They can adapt their traditional beliefs so that these accord—or at least seem to—with scientific reality. Thus, when geology required the history of our planet to be hugely longer than allowed for in *Genesis*, many Christians readily adapted their beliefs.
- They can reject the traditional belief system—at least as a description of physical reality, even though it may still have value for them at what we could call a spiritual level. For example, many who reject the

supernatural aspects of Christianity nevertheless abide by an approximation of its moral code, such as tolerance and "good works."

In *The Great Derangement* (2008), Matt Taibbi describes, among much else, his experiences as a mole amid the fundamentalist Cornerstone Church in Texas, part of the John Hagee Ministries. One episode (pp. 177–80) serves well to suggest an explanation as to why the reactions of so many US Christian fundamentalists seem so frequently divorced from reality. The occasion Taibbi describes is a sermon by Matthew Hagee, "both dumber and more vicious than his dad," in which the preacher seriously misrepresents various environmental concerns before lambasting them as anti-American, the relevant activists as puppets of Satan, and so forth. So far, so customarily loony. But then Hagee (p. 178) went a step further:

> Encouraged, the portly pastor now looked down at his pulpit and read from a bunch of paper sheets.
>
> "*Time* magazine says that the Sierra Club and others met with environmental leaders in Brazil in 1992 to discuss how to use the environment to reduce the American population from 175 million to 75 million, to control us. This was 1992. How many environmental laws have they passed since then?"

Later, Taibbi combed through the entirety of *Time*'s coverage of the 1992 Earth Summit and discovered there was no such reference.

Next Hagee cited a law that Congress had attempted to pass a few days earlier, a law intended to encourage abortions through promoting fetal screening for genetic defects. Again, Taibbi checked. The act being lied about was the Genetic Information Nondiscrimination Act, whose purpose was to prevent discrimination by employers and insurers on the basis of individuals' genetic information as revealed through medical tests.

Large swaths of US Christians have as their main source of authoritative information on scientific issues—authoritative because beamed down directly from God—the accounts given to them by the preachers they have come to trust. It's hardly a wonder that all these individual fundamentalists "know" a version of the sciences that bears little relation to reality.

But it's not just at the level of preacher/congregation that false science is being disseminated. There are fundamentalist institutions, from faux universities to supposedly educational publishing houses, dedicated to the propagation of righteous falsehood. The second edition of *Science 4 Student Text* was published by the Bob Jones University Press in 2004. Here is its "explanation" of electricity:

> Electricity is a mystery. No one has ever observed it or heard it or felt it. We can see and hear and feel only what electricity *does*. We know that it makes light bulbs shine and irons heat up and telephones ring. But we cannot say what electricity itself is like.
>
> We cannot even say where electricity comes from. Some scientists think that the sun may be the source of most electricity. Others think that the movement of the earth produces some of it. All anyone knows is that electricity seems to be everywhere and that there are many ways to bring it forth.

The question obviously arises: Surely many of the parents whose children are expected to learn from this textbook, however devoutly Christian they might be, must *know* that stuff like this is twaddle . . . so why haven't they battered down the doors of the Bob Jones University Press to tell the august editors to do a little better?

As Bertrand Russell wrote, in "Is There a God?" (1952):

> If I were to suggest that between the Earth and Mars there is a china teapot revolving about the sun in an elliptical orbit, nobody would be able to disprove my assertion. . . . But if I were to go on to say that, since my assertion cannot be disproved, it is an intolerable presumption on the part of human reason to doubt it, I should rightly be thought to be talking nonsense. If, however, the existence of such a teapot were affirmed in ancient books, taught as the sacred truth every Sunday, and instilled into the minds of children at school, hesitation to believe in its existence would become a mark of eccentricity. . . .[1]

Major Christian religious figures in the US like Jerry Falwell and Pat Robertson have blamed disasters such as the 9/11 attacks and the 2005 flooding of New Orleans on the nation's permissiveness, specifically the widespread tolerance of homosexuality. In April 2010, a few weeks after the US Senate passed the first healthcare reform bill in decades, the Icelandic volcano Eyjafjallajökull erupted, sending a plume of ash into the high atmosphere, where it streamed in the direction of Europe and caused perhaps the biggest disruption of plane traffic ever. It takes a nimble mind to make a connection between healthcare reform and a volcanic eruption, but in his show on April 16, 2010, Rush Limbaugh was up to the challenge:

> You know, a couple of days after the healthcare bill had been signed into law Obama ran around all over the country saying, "Hey, you know, I'm looking around. The earth hasn't opened up. There's no Armageddon out there. The birds are still chirping." I think the earth has opened up. God may have replied.[2]

Just a few weeks earlier, Pat Robertson had claimed the Haitians had brought the 2010 Port-au-Prince earthquake upon themselves because long ago they'd made a pact with the Devil. And a few days after Limbaugh's broadcast the US online political magazine *WorldNetDaily* was promoting the ideas of Mark Biltz, pastor of El Shaddai Ministries in Bonney Lake, Washington, concerning biblical astronomy. The Second Coming is likely to be in 2014 or 2015, Biltz believes, because a recently observed gamma ray burst from the direction of the constellation Boötes, which contains the bright star Arcturus, is a message from God to this effect: "The word 'assemble' is the same word that is translated as 'Arcturus' in Job," he said. "So it means the same thing, to assemble, to come. And if you'll notice the word 'come' is 'bo,' which is the name of this constellation: 'Bo-otes.'"[3]

At the time, *WorldNetDaily* had just initiated a lawsuit against the White House Correspondents' Association for failing to recognize it as a serious news organization.

It was Hindu mathematicians who discovered the concept of zero, possibly in the sixth century; they transmitted it in due course to the Arabs, and thence, a very long time later, it reached Europe. The seventh-century Brahmagupta knew not only of the concept of infinity but of its properties. Sripati, in the eleventh century, worked out the means of solving what came in the West to be called Pell's equation: $y^2 - nx^2 = 1$. The European name for the equation honors the English mathematician John Pell, whose work trailed Sripati's by a small matter of six centuries. By the ninth century, Indian mathematicians could figure pi to ten decimal places and had worked out the rudiments of coordinate geometry. Attainments in other areas of science were far from negligible. Yet, just as happened in Islamic science (see below), Hindu science thereafter went into a long period of eclipse.

The Indian cosmologist J. V. Narlikar was fond of observing that when the Steady State theory of the universe was widely accepted, Hindu scholars found plenty of passages in the Vedas that supported this cosmology. When the Steady State theory was ousted by the Big Bang theory, those same Hindu scholars discovered, *often in the very same passages*, that . . .

Narlikar is just one of the great scientists India has produced. Among many others have been the physicist and polymath Jagadish Chandra Bose, the physicist S. N. Bose, the astrophysicist (and Nobel laureate) Subrahmanyan Chandrasekhar, the molecular biologist G. N. Ramachandran, the physicist (and Nobel laureate) C. V. Raman, the mathematician Srinivasa Ramanujan, the chemist P. C. Roy, and the astrophysicist Meghnad Saha. Yet, despite so many Indian contributions of distinction to modern science, all is not well within Indian science, according to Meera Nanda in a series of articles and books like *Prophets Facing Backward* (2003) and *The God Market* (2009). She points to an alliance of rightwing Hindu nationalists/traditionalists (the Hindutva, organized as the Sangh Parivar), left-wing postmodernists in both the West and India, and the presence within Indian society of a "fashionable denigration of science."[4]

Because of the efforts of the Sangh Parivar, there's an industry in reinterpreting—or pretending to reinterpret—the Vedas such as to maintain that they contain, albeit sometimes in coded or capsule form, information that just happens to match up with the latest scientific discoveries. As Nanda is keen to emphasize, "[A] scientific understanding of nature completely and radically negates the 'eternal laws' of Hindu dharma which teach an identity between spirit and matter."[5]

Of course, "Vedic science" isn't limited to India; later (see page 169) we'll look at the work of US writers Michael Cremo and Richard L. Thompson on Vedic Creationism. We can note the irony of Western woomongers seeking universal scientific truths in the Vedas even as the inheritors of the Vedic tradition themselves—outside the ranks of the Sangh Parivar and their converts—are looking to modern science and technology as bases for their booming economy, while Indian scientists have, as we've seen, shown themselves to be at home among modern science's top echelons.

A second prong of the Hindu fundamentalist assault on science is to claim that traditional "sciences" like astrology and palmistry are of equal merit with modern Western science. Here the fundamentalists are abetted by the postmodernist school, with its idea that all knowledge is purely relative (see pages 20ff.).

At the same time, the Sangh Parivar are mounting an assault on modern science with the intent of devaluing it to the point where it can be safely ignored. Again, there are two prongs to this assault. One is to denounce modern science as a Western construct, forced on conquered peoples by their colonial overlords. Those who speak out in favor of modern science and against superstition are guilty of "colonized" thinking. The other meme relies on, again, ideas borrowed from the postmodernists. Science is by its nature "deconstructivist": Faced by a major problem, its general approach is to break down that problem into its various bits in the belief that, by investigating the bits separately and then in groups, eventually a solution will be found to the original major problem. The Sangh Parivar dispute that the deconstructivist approach pays dividends. Their Vedic science is far superior in that its approach is not piecemeal but *holistic*. A wise Hindu will look not just at the bits but at the *whole picture*. The trouble is that this method was tried in the West and discovered to fail: Plato's philosophy may be fascinating but, so far as its ventures into science are concerned, it's brilliant tripe.

In *Science and the Indian Tradition* (2007) David L. Gosling describes how such "holistic" thinking essentially led Hindu traditionalists down a cul-de-sac. The Hindu concept of universal unity, while philosophically satisfying, bears—beyond the idea of unification—no real relation to unifying theories of science.

In modern India, as Gosling points out, one finds cutting-edge science and technology—like the widespread use of solar and other environment-friendly energy technologies—alongside prevalent belief in superstitions like reincarnation and even the eclipse demon! And our opinion of the work of the pioneering Indian physicist Jagadish Chandra Bose (see above) must be colored by the knowledge that he did considerable research into how plants experience pain, convincing himself this was a real phenomenon. As Meera Nanda observes, although the hypothesis was "falsified and rejected by mainstream biology in his own life-time [it] is still touted as India's contribution to world science in Hindutva literature."[6]

> *[I]t must be admitted that there remains an outstanding puzzle: Muslim science lasted for nearly six centuries and this . . . is longer than Greek, medieval Christian, or even modern science, has lasted. How individuals could have sustained science for this immense period is indeed something that no one understands.*
>
> —Pervez Hoodbhoy, *Islam and Science* (1991), p. 94

Modern Islam is not entirely without distinguished scientists. In 1979 the Pakistani–UK physicist Abdus Salam was the first Muslim to become a Nobel sciences laureate. He was followed in 1999 by the Egyptian–US physicist Ahmed Zewail, now a member of President Obama's Presidential Council of Advisors on Science and Technology (PCAST). Of course, there are other Muslim scientists who're as distinguished but less recognized; yet a total of two Nobel science prizes for adherents of one of the world's major religions seems very slim pickings—and there is no mileage in the pretense that the Nobel Committee is discriminating against Muslims.

Arab science was the light of the world for roughly the period 750–1350 CE.[7] Although there was much overlap of influences, in simplistic terms the Arabs had inherited science from the Greeks, Egyptians, and Babylonians, mathematics and medicine from India, literature from Persia, and philosophy and logic from Greece. The caliphs of the age, swayed by the enlightened Mu'tazila movement, called not just for the nurturing of this existing knowledge but also the investigation of new areas; for example, alchemy was really an Arab invention, even if it did have precursors in the ancient world. (The name comes from the Arabic word *al-kimia*.) The end of this golden age came about at least in part because the Mu'tazila movement, having become just as corrupt and despotic as any before it, was swept aside by a rejuvenated spirit of orthodoxy, which cracked down on such concepts as free will and reason. In reality, this seems to have occurred on a local scale all through the period in question; on glancing at the biographies of the leading Arab scientists of the era, it's notable that many suffered at least occasional persecution.

At the height of this golden age, at a time when the best libraries in Christian Europe contained at most a few hundred books apiece, those in Baghdad and Cordoba each held, it is claimed, over forty thousand. When Baghdad was overthrown by the Mongols under Hulegu Khan in 1258, the libraries were sacked and their contents destroyed; according to one rather fanciful account, so many books were hurled into the Tigris that a horseman could cross the river on top of them. The world's oldest degree-granting university, founded in 859, is the still-extant University of Al Karaouine in Fez, Morocco.

A relic of the ascendancy of Muslim astronomy is the number of stars that bear Arabic names—for example, Achernar, Aldebaran, Algol, Altair, Betelgeuse, Deneb, Fomalhaut, Nekkar, Rigel, Thuban, and Vega, plus a whole host of lesser-known ones.

The extent to which Nicolaus Copernicus relied upon Islamic precursors has become more and more obvious in recent years. It seems no Islamic scholar made the conceptual jump from the inherited Ptolemaic cosmology to a heliocentric arrangement, but at least one, Ali Qushji, put forward a proof that our astronomical observations would be the same regardless of whether the earth were rotating or the celestial lights were circling it.

Even so, many still dismiss the early days of Islamic science as merely a matter of translating Greek, Persian, and Indian texts into Arabic, and thereafter annotating them. Yet there must have been Arab science in existence before then, as otherwise the language would not have been capable of the translation. While much new technical terminology had to be devised, much was already there, which could only have been the case had there been a use for it. There's plenty of other evidence of a rich tradition of Islamic science before the translation and absorption "program" got underway, but this linguistic clue seems persuasive. It should also be noted that, when we talk of the work of the Arab scientists in this "program" as mere translation, we're seriously underestimating what they did. As an example of the extent to which they also adapted, expanded, and annotated the original works, Dioscorides' *De Materia Medica* contained discussion of some six hundred plants; its first known Arabic versions covered about fifteen hundred. Moreover, as Edward Grant points out in *Science and Religion, 400 B.C.–A.D. 1550* (2004), the Byzantines enjoyed the same inheritance yet essentially did nothing with it.

Typical of the West's amnesia about the Arab achievements in science, and the important role played by the medieval Islamic empire in science's history, is Karen Armstrong's *Islam: A Short History* (2000), which seems quite willfully to avoid any mention of the subject. When the important physician Ibn Sina (Avicenna) appears, for example (p. 83), we're given a mere dozen lines covering his attitudes toward religion, prophets, Sufism, mysticism . . . and nothing whatsoever of his science. And Ibn Sina is lucky: most other significant Arab scientists aren't mentioned at all.

One hindrance to the later progress of Islamic science was that for some long while after the Renaissance, and even up to the Industrial Revolution, the Arab nations continued to believe the Europeans were barbarians whose near-nonexistent culture could most sensibly be ignored . . . except, obviously, when their armies loomed with slaughter in mind. Had the Arabs paid attention to the tectonic changes occurring in the European cultures as modern knowledge erupted, they could well have shared the benefits; even more important to the broader picture, there is much they could have contributed to the process.

The Arab world instead fell prey to the "religious sciences" (i.e., the various techniques worked out to index and interpret the Qur'an and the Sunnah). These began to develop in Islam even before the Prophet was dead, and thus were well entrenched before the emergence of other fields of intellectual endeavor, such as the natural sciences. At the same time, the two are regarded in Islam as intimately interlinked. Just as the verses of the Qur'an are *ayat* (signs), so are the features of the natural world around us—they are *ayat* sent by God as signs of his intervention in the world. A consequence is that the laws of nature must be immutable, since they're the laws of God. As God has not only put those laws in place but designed them such that they can be discovered, it follows that the study of nature is really a means of better understanding the will and intentions of God—in other words, a holy duty. In Islam, all realms of knowledge are, as Muzaffar Iqbal has put it, "branches of the same tree."[8]

It's thus hardly surprising that many imams make a habit of scouring the Qur'an and the Sunnah for passages that can be "interpreted" in such a way as to seem to prefigure modern scientific discoveries; Hindus do it with the Vedas and Christians with the Bible, but the Muslim scholars seem even more eager than their counterparts. Oddly enough, one of the most assiduous proselytizers in this vein has been not an Arab but a Frenchman, Dr. Maurice Bucaille, who died in 1998. In books like *La Bible, le Coran et la Science* (1976; trans. as *The Bible, the Koran, and Science*) and *Réflexions sur le Coran* (1989) he demonstrated, at least to his own satisfaction, that the Qur'an contains countless statements of scientific fact and certainly nothing that contradicts modern science, while the Bible is by contrast a barrel of unscientific hogwash.

Qur'anic cosmology states in essence—strictly in concurrence with the discoveries of modern science, Bucaille would have you understand—that the universe was created by Allah and is still sustained by him from moment to moment (occasionalism); this situation will not subsist indefinitely because he has already determined that at some particular moment in the future he will destroy the universe as we know it, although the timing of that event is as yet concealed from us. After the destruction of all that we know, there will be a universal resurrection to create what will be in effect a new *kind* of universe, with new laws.

An implication of occasionalism was, according to the eleventh-century Abu Hamid Muhammad ibn Muhammad al-Ghazali, that even the notion of cause-and-effect was heretical—and impossible. Since Allah re-created the world afresh in each new instant, it was clearly ludicrous to think the events of one instant might affect the events of the next. If your house burned down it was legitimate to observe that the flames and the burning occurred simultaneously, but not to deduce that the flames *caused* the burning. The cause of the burning was Allah. Fire, as a dead entity, could be the cause of nothing.

In his 1988 article "They Call it Islamic Science,"[9] the Pakistani physicist Pervez Hoodbhoy described a few recent papers presented by self-styled Islamic scientists, several of them addressing the October 1987 International Conference on Scientific Miracles of Qur'an and Sunnah. Hoodbhoy attended much of this conference, but had to miss the program item called "Panel Discussion on Things Known Only to Allah"—what *could* they have discussed?

Among these supposedly scientific papers was one from Mohammed Muttalib of the Department of Earth Sciences at Egypt's Al-Azhar University. This geologist advanced his theory that the deep roots of mountains act like pegs holding in place the earth's outer layers, which would otherwise be thrown off into space by the planet's spin.

At a similar conference the previous year in Karachi, the International Seminar on Qur'an and Science, Arshad Ali Beg of the Pakistan Council for Scientific and Industrial Research announced he'd derived a mathematical formula suitable for calculating the amount of hypocrisy present in any given society. At the same conference, Salim Mehmud, the chair of SUPARCO (Pakistan's NASA), used Relativity to explain how the Prophet had ascended to heaven for a meeting with God and been able to return within mere moments. Hoodbhoy observed wryly that the seventy or so papers scheduled for presentation at the Islamabad conference were all vetted beforehand by a panel of scholars to ensure their theological acceptability; not one was checked by a scientist.

When expanding his article as an appendix for *Science and Islam*, Hoodbhoy attached a brief exchange of letters that had appeared in the *Herald*, a Karachi monthly, subsequent to the article's publication. S. Bashiruddin Mahmood, at the time the chair of the Holy Qur'an Research Foundation and once a senior director of the Pakistan Atomic Energy Commission,[10] complained bitterly that Hoodbhoy had misrepresented him: In his book *Mechanics of the Doomsday and Life after Death* (1980)[11] he had been comparing not the evolution of the universe but the transformation of the soul to the passage of an electric current. As Hoodbhoy observed in his response, while he had indeed got this wrong, Mahmood was merely replacing one piece of foolishness with another; and he noted that this was the same Mahmood who, some years earlier, had been keen to investigate the possibilities of communicating with djinns. His interest wasn't philosophical or spiritual: Since God had made djinns from fire, they must be packed with energy. Mahmood's notion was to construct djinn power plants and thereby solve Pakistan's energy problems at a stroke.[12]

Mahmood isn't the only modern Islamic scientist with a penchant for djinn studies. Hoodbhoy directs us to the paper "Dichotomy of *Insan* and Jinn & Their Destiny" by Safdar Jang Rajput. This appeared in the journal *Science and Technology in the Islamic World* in March 1985. According to Hoodbhoy, Rajput contends that djinns, as fiery entities, must be formed from methane or similar hydrocarbons because these burn without smoke (after all, there are no reports of anyone ever having seen a djinn giving off smoke); that djinns and men must be of the same genotype because the houris of heaven were created for their joint use; and that "the jinns are the white races."[13]

Taner Edis, in his *An Illusion of Harmony* (2007), discusses the view of djinns' role in science that's been put forward by the Turkish religious leader Fethullah Gülen. It's Gülen's belief that djinns are likely responsible for such mental illnesses as schizophrenia (here we have something close to the Christian view of psychological illness as demonic possession), and he states also that there are instances of not just mental illnesses but cancer being driven out by prayer (or, as Christians might say, exorcism). He sees a possibility for djinns to be put to useful work.

The writer and broadcaster Ziauddin Sardar, Honorary Visiting Professor in the School of Arts at the City University, London, initiated during the 1980s in *Nature* and *New Scientist* a discussion of the problems facing Islamic science. He described the results in "Islam and Science," a lecture he gave to the Royal Society on December 12, 2006:

> The debate itself was reduced to two components.
>
> The first derives from the fundamentalist idea that all knowledge, including scientific knowledge, can be found in the Qur'an. . . . Backed by a lavishly funded Saudi project—"Scientific Miracles in the Qur'an"—this tendency has sprouted a whole genre of apologetic literature. . . . From relativity, quantum mechanics, big bang theory to the entire field of embryology and much of modern geology has been "discovered" in the Qur'an.
>
> Meanwhile, "scientific" experiments have been devised to discover what is mentioned in the Qur'an but not known to science—for example, the programme to harness the energy of the jinns. . . . This reductive fundamentalism now embraces Creationism and is generating a growing movement for "Intelligent Design" in the Muslim world. . . .
>
> The second component is best described as mystical fundamentalism. . . . The material universe is studied as an integral and subordinate part of higher levels of existence, consciousness and modes of knowing. Thus, science is not a problem-solving enterprise and socially objective inquiry; it is a mystical quest for understanding of the Absolute. . . .
>
> These two trends, the fundamentalist and the mystical, suggest that real science has almost evaporated from Muslim consciousness. In a recent survey, *Nature* noted, today's Muslim states "barely register on indices of research spending, patents and publications." And it concludes the situation is not just bad; it is set to get worse.[14]

Bearing in mind that for some centuries Arab science was the lamp that stayed alight while Europe suffered the Dark Age, this descent into irrationality is especially depressing.

> *After we came out of church we stood talking for some time together of Bishop Berkeley's ingenious sophistry to prove the nonexistence of matter and that everything in the universe is merely ideal. I observed, that though we are satisfied*

> *his doctrine is not true, it is impossible to refute it. I shall never forget the alacrity with which Johnson answered, striking his foot with mighty force against a large stone, till he rebounded from it, "I refute it thus."*
>
> —James Boswell, *Life of Johnson* (1791)

A frequent meme among steadfast deniers of science is that science—or a particular branch of science, usually evolution—is itself a religion: that there is no difference between a scientist's belief in natural selection and a religionist's belief in divine intervention. In part this contention is simply trivial and bogus, a play on two different meanings of the word "belief." The other root of the contention is probably to be found in the postmodernist notion that scientific conclusions are not so much a depiction of reality as part of a culturally determined narrative. In this context it would certainly be possible to portray the scientific community as having many characteristics of a religion. The trouble, of course, is that the premise is dubious.

One could, however, make a very good case that science denialism has many of the hallmarks of a religion, what with its counter-evidential beliefs, its refusal to shift its stance in light of new information, its reluctance to pry too closely into the integrity and motivations of its "prophets" or the truth of what those "prophets" are saying, and so on. Why did it have to be scientists who pointed out the astonishing levels of misinformation in the "teachings" of a Christopher Monckton? Why did no AGW deniers raise the alarm? If a university science lecturer committed as many egregious errors, sooner or later some of the students would be raising flags. Could it be that Monckton's audiences aren't so much students as . . . "disciples"?

2

"THE LAW IS A ASS"

"If the law supposes that," said Mr. Bumble, "the law is a ass—a idiot."

—Charles Dickens, *Oliver Twist* (1837–39)

Of course, it's not so much the law itself that's a ass—although there are plenty of stupid laws on the statute books—as the people who apply it, from judges and lawyers and cops and juries down to you and me. The courts have a bad record of listening to pseudoscience and ignoring the real thing. The main way you and I tend to exercise our denial of science is through the ballot box, where we support ludicrously high incarceration rates and draconian punishments in the cause of "being tough on crime" even though there's scientific evidence galore to show these measures aren't solutions but part of the problem. The other major contribution members of the public make to denial in the nexus of science and law takes the form of the panics that periodically paralyze the country even though their scientific basis is thin to nonexistent. All too many of these baseless panics end up in the courtrooms, sometimes with massive class actions being successful—at immense cost not just to manufacturers but eventually to the community, which is deprived for no good cause of a valuable, perhaps life-saving piece of technology.

For example, a few high-profile judgments based on dodgy scientific evidence were enough to drive the morning-sickness drug Bendectin (Debendox) off the US market in 1983; the manufacturer, Merrell Dow Pharmaceuticals, was unwilling to continue producing a drug that brought it nothing but frivolous lawsuits and hefty penalties. Bendectin may have been a casualty of an often-overlooked consequence of the Thalidomide disaster of the 1950s and 1960s (as if that catastrophe needed to be any more heartrending): that people were thereafter far too ready to attribute any misfortune to whatever prescription pharmaceuticals they might be taking. At the same time, because of the obfuscations of the UK manufacturers of Thalidomide, Distillers Company Limited,[1] juries became overly ready to believe people claiming to be victims of side-effects of a prescription drug.

A major flaw in the US legal system is that the supposedly expert scientific witnesses called are not there to produce scientific evidence. A presentation of the scientific evidence would involve something like a survey of the literature on the subject, with conclusions drawn therefrom—an account of the scientific consensus, in other words. By contrast, the testimony the "expert" witnesses are asked for is personal speculation on the subject. If the witness is responsible, the evidence given will be in accordance with the consensus. If not, it's quite likely to be crackpot.

The result is that there are scientists, often with minimal and sometimes zero qualifications, who make a good living testifying in trial after trial on subjects in which their understandings are so paltry or their ideas so loopy that they can't get actual *jobs* in the fields concerned. Often the only authority they can cite in support of their views is themselves, describing research and research results that the jury cannot directly examine; sometimes, later, it becomes evident the research never existed, but by then it's too late—and the legal profession, presumably on the grounds of protecting its self-regard, is astonishingly remiss about prosecuting these "expert" perjurers afterward.

Expected to make judgments based on evidence they're not allowed to scrutinize and which they're generally unqualified to evaluate, juries are, of course, being put in an impossible position. It's a principle ripe to be corrupted by the unscrupulous, which is exactly what happens. Sharkish lawyers present phony science and phony scientists without the slightest possibility of ever being held accountable for the dishonesty.

One way out of this mess might be to make it the responsibility of judges, acting in cooperation with bodies like the American Association for the Advancement of Science, to appoint neutral expert scientific witnesses in any trial. In fact, US judges already have a right to do this should they so choose; it's just that they very rarely do. It's time they woke up.

The "science" of clinical ecology was born in the 1960s, largely as a result of the ideas of the US physician Theron Randolph, who pioneered the idea of multiple chemical sensitivity (MCS), which has it that we're far more susceptible than we think to trace chemicals in our environments, and that some people can develop allergy-like symptoms that may be extremely damaging. He founded the Society for Clinical Ecology in 1965; in 1984, by which time the term "clinical ecology" had earned itself something of a bad reputation, the name changed to the American Academy of Environmental Medicine. Today it's a "science" that exists largely only within courtrooms.

While the premise seems not absurd—think of how we can be affected without realizing it by passive smoking, or lead in gasoline—in practice it becomes far too easy to attribute any group of unexplained symptoms to environmental influences that might as well have been chosen at random, from printers' ink to high-voltage cables and junction boxes. We've had scares about levels of radon in the atmosphere so low as to be near-homeopathic. One early panic was that microwave ovens might emit dangerous levels of radiation. The astronomer James Van Allen, of Van Allen radiation belts fame, calmed many fears in his native state of Iowa—the Iowa city of Amana was a center of microwave manufacture—when he offered publicly in 1973 to "sit on top of my Amana Radarange for a solid year while it is in full operation, with no apprehension as to my safety. . . . [I]n my judgment its hazard is about the same as the likelihood of getting a skin tan from moonlight."[2]

The pinnacle of clinical ecology perhaps came with its establishment—not in peer-reviewed journals but in the courtroom—of a whole new medical syndrome, "chemical AIDS." The trouble is that no one has ever been able scientifically to define what MCS and chemical AIDS *are*. And, since all manner of conditions and symptoms have been ascribed to these two mysterious syndromes in the courtrooms, studying legal records doesn't make investigators any wiser. There's no doubt people who're diagnosed by clinical ecologists as having MCS are (fakers aside) suffering. Orthodox physicians are, however, likely to assume the vague, inexplicable symptoms have some kind of psychosomatic origin, such as stress. It's a diagnosis that doesn't go down too well with many patients, who assume they're being fobbed off with a variant of the "it's your own fault" diagnosis. The clinical ecologist's suggestion that the sufferer's immune system has been sparked into hyperactivity by contact with some or other chemical in the environment is a lot more comforting, and might even be expected to have some placebo effect.[3]

If clinical ecologists could establish a cure rate for MCS that was any better than that achieved by orthodox treatment, the discipline might be taken more seriously. In his booklet *A Close Look at "Multiple Chemical Sensitivity"* (1998)[4] Stephen Barrett of Quackwatch cites the Board of the International Society of Regulatory Toxicology and Pharmacology (ISRTP):

> Current scientific information reports no clinical, laboratory, or other objective support for the proposition that MCS represents a clinically definable disease entity. The theories claiming to unify this condition as a toxicologically mediated disorder transgress basic principles of toxicology and clinical sciences.

Breast Implants: America's Silent Epidemic

—Mercola.com headline, June 6, 2001

A striking example of the law acting in complete denial of the available science occurred in the 1990s in the US, in the years following the placing of a moratorium by the Food and Drug Administration (FDA) in 1992 on the cosmetic use of silicone gel breast implants. A very full and perceptive account of the subsequent legal travesty is given by Marcia Angell in *Science on Trial* (1996).

For years the FDA regarded the implants, invented in 1961 by two Texas surgeons, Thomas Cronin and Frank Gerow, and marketed initially through the Dow Corning Corporation, as undeserving of scrutiny: The science indicated they were safe so long as the cosmetic surgeons doing the implanting were competent. Even so, there had been scattered reports of discomfort and complications. In 1982 the FDA decided that, if the prosthetics were to continue to be licensed, the manufacturers—by this time there were others involved in the lucrative trade beside Dow Corning—had to show they were safe.

It took six years, until June 1988, for the FDA to implement its own proposal; the deadline it then gave to the manufacturers for the production of safety evidence was July 1991. Dow Corning presented some documentation in time for the deadline, but the FDA's new chief, David Kessler, concluded this was mere flimflam: No serious studies had been done by Dow Corning and the other manufacturers. The companies' claim was that they'd unearthed no evidence to show the prosthetics were dangerous; the FDA retorted that neither had they produced any evidence to show they were safe. The FDA permitted the implants to remain on the market while giving the manufacturers a little extra time to produce something a bit more convincing.

The moratorium Kessler introduced in 1992 affected only implantation for cosmetic purposes; surgery using silicone-gel implants for reconstruction after a mastectomy, for example, or to correct deformity—or indeed to replace earlier implants that had been damaged—was still permissible, so clearly the FDA wasn't hugely troubled by the safety issue. On the other hand, even if the risks were small, the numbers of women opting for breast augmentation were rising annually; it was legitimate to call a temporary halt. Dow Corning and the others, after a brief period of flailing around, at last began the process of complying with the FDA's requirement.

There was no reason at all for a panic.

So there *was* a nationwide panic, as many of the estimated one to two million women who'd received silicone implants, their fears fanned by irresponsible reporting, promptly assumed the only reason the FDA would have issued a "ban" must be that the prosthetics were dangerous. And, naturally, such being the way the human mind works, many women who'd hitherto been perfectly happy with their implants suddenly began to notice pain and discomfort . . .

Was it possible leaky implants could cause cancer? That was the first cry. Although it had first been sounded as early as the 1980s by Ralph Nader's Public Citizen Health Research Group, it soon fell from favor. Much more persistent was the rumor that errant deposits of silicone could cause connective tissue diseases. Connective tissue diseases are genuine enough—examples are rheumatoid arthritis, dermatomyositis, polymyositis, and scleroderma—and they can be spectacularly unpleasant. They're believed often to be a consequence of an autoimmune reaction; something persuades the immune system that tissues belonging to the body itself present a threat, and so it tries to eliminate them, causing painful inflammation, among other possible symptoms. The assumption became widespread that faulty silicone implants might cause the condition of *mixed* connective tissue disease, in which more than one of these autoimmune malfunctions could be operational at once. The advantage of this condition, so far as the hypochondriac or the tort lawyer is concerned, is that it's impossible to be prescriptive as to the precise symptoms that must be present to satisfy a diagnosis; one can pick and mix.

The FDA's moratorium was not the sole cause of the flap. Years earlier, in 1984, a woman called Maria Stern, who'd been diagnosed with systemic autoimmune disease, suspected there might be a connection with her breast implants and accordingly sued Dow Corning. Her attorney, Nancy Hersh, filed for disclosure of Dow Corning's internal documents, and Hersh's assistant, Dan Bolton, came across plentiful evidence of the company's employees displaying a somewhat cavalier attitude toward consumer welfare. There was not the slightest evidence that Dow Corning or any other qualified authority thought their implants could cause medical problems of any kind, let alone systemic autoimmune disease, but within the US legal system that problem could be easily fixed: Stern's attorney introduced into court various "experts" who theorized aloud to the jury as to how wayward silicone *might* cause an abreaction from the immune system. Taken in conjunction with the Dow Corning employees' indisputably objectionable behavior, this was enough for the jury, who awarded Stern $211,000 in compensatory damages and a further $1.5 million in punitive damages.

As part of the deal, the scientific evidence, including the offending memos, was sealed. In 1990, a congressional hearing into the safety of silicone implants made play of the fact that much of Dow Corning's scientific information on the matter could not be disclosed to the public because of this; a widespread conclusion was that Dow Corning had muzzled the Stern case participants as part of a damage-limitation exercise. There could hardly be more conclusive proof that silicone gel caused connective tissue disease, could there?

The congressional hearing roughly coincided with the airing of one of those investigative television programs that fails to investigate the actual science of a situation, instead encouraging the dissemination of anecdotes and innuendo to draw a predetermined conclusion. This was an episode of *Face to Face with Connie Chung* in which the personable broadcaster conducted tear-jerking interviews with women who'd both received breast augmentation and suffered illnesses.

By the end of 1991, Dow Corning was facing 137 individual lawsuits in connection with the company's implants. A new record was set for the size of the damages extracted from the company in consequence of the illness their product had supposedly produced: $7.3 million, paid to Mariann Hopkins. The Hopkins judgment was appealed all the way up to the Supreme Court, which declined to take any action.

Following the declaration of the moratorium in 1992, the pace of the lawsuits began increasing frenetically, and the damage awards rose likewise. In December 1992 Pamela Jean Johnson was awarded $5 million in damages plus $20 million in punitive damages against Bristol–Myers Squibb—this although it was generally agreed the sum total of her symptoms amounted to no more than a "bad flu." In March 1994 the company 3M was ordered to pay three implant recipients $12 million in compensatory damages and a further $15 million in punitive damages.

In that same month the final touches were put to a class-action settlement that was to date the largest in US legal history: $4.25 billion. The major contributors were Dow Corning, 3M, Bristol–Myers Squibb, and Baxter. Women claiming recompense from the fund were under no obligation to prove their illnesses had anything to do with their implants—or, really, to prove they had any illnesses at all—because a cottage industry soon sprang up, in liaison with some of the lawyers concerned, of physicians prepared to give appropriate diagnoses on request.

Around this time, properly conducted scientific investigations began to be published, and they indicated there was no detectable medical risk involved in the use of silicone-gel breast prosthetics. The first results to appear, in the *New England Journal of Medicine* for June 16, 1994, were those by Sherine Gabriel et al. of the Mayo Clinic.[5]

The effect on proceedings in the courts holding the manufacturers accountable for health risks of which Gabriel and her team had found no trace was initially zero. The effect on the legal profession as a whole was not quite zero, however. Gabriel received legal subpoenas from some of the tort lawyers who were making a killing in the implant cases (notably the Houston lawyer Charles E. Houssiere, who was representing two to three thousand women in their suits against implant manufacturers) demanding vast amounts of documents from her, including all kinds of personal details of the test subjects—details that were strictly confidential. Of course, the lawyers didn't want the documents; they merely wanted to harass, and to intimidate other scientific researchers from publishing their own work revealing the implants to be harmless. According to Angell (p. 30), "Authors of two other epidemiologic studies were served with similar subpoenas."[6] The intimidation arises not just from the massive logistical effort to supply all the documentation demanded—years of potentially life-saving research time can go down the tube as scientists labor at this completely pointless task—but also from the breach of trust involved between subjects and researchers: The latter may never be able to conduct an epidemiological study again.

Other published reports came to the same or similar conclusions as Gabriel's: There was no evidence to suggest the implants were other than harmless. In October 1995 the American College of Rheumatology was moved to issue a statement that included:

> Two large studies now have been completed. The first was conducted on all women in a single county in Minnesota who received implants between 1964 and 1991.[7] At a mean follow-up of 7.8 years, there was no association between breast implants and connective tissue or rheumatic disease. . . . The second study was a follow-up of the Nurses Health Study. Among this very large cohort, after 14 years of follow-up, no evidence existed for an association between silicone breast implants and connective tissue diseases.[8]
>
> The American College of Rheumatology believes that these studies provide compelling evidence that silicone implants expose patients to no demonstrable additional risk for connective tissue or rheumatic disease.[9]

In that same month Charlotte Mahlum was awarded $3.9 million in compensatory damages and a further $10 million in punitive damages against Dow Chemical, Dow Corning's parent company.

By the end of the year there had been twenty studies/abstracts published that showed a lack of any causal relationship between the implants and a spectrum of autoimmune-related diseases. The American Medical Association had joined the American College of Rheumatology in rejecting any connection between silicone implants and increased risk of autoimmune disease. And at last, like a brachiosaur bitten in its rear, the legal profession began to turn. In April 1996 two New York federal judges appointed an expert panel to examine the scientific issues, effectively stalling for the while any further settlements based on pseudoscientific evidence. In September of that year the California Court of Appeal upheld the dismissal of eighteen hundred breast-implant suits against Dow Corning, and in December, in Oregon, a judge who'd taken scientific advice forbade lawyers from presenting pseudoscientific claims in his court and dismissed seventy current and pending suits. More and more studies poured in to show the implants were harmless; more and more judges threw out implant cases. The bonanza was over for the tort lawyers and their clients, many of whom were actually ill, even if not because of their breast implants. In this light, it seems bizarre that in July 1998 Dow Corning paid out $3.2 billion in settlement for tens of thousands of claims as part of the company's bankruptcy reorganization.

And then, completely ignoring all that had gone before, a jury in a Washington federal court awarded an attorney who said her breast implants had given her scleroderma a whopping $10 million in compensatory damages against Bristol–Myers Squibb. Now *there*'s denial of science for you!

Like everything else, the implantation of silicone breast-size enhancements, while overwhelmingly safe if moderation and responsibility are involved, can be dangerous if the ingredient of human stupidity is added. In January 2011, fans were mourning the death of twenty-three-year-old German porn and reality TV star Carolin "Sexy Cora" Berger, who died after suffering cardiac arrest during an operation to enlarge her breasts for the sixth time in rapid succession, this time to increase each implant from 500g to 800g; since she weighed just 48kg (about 106 pounds), the implants would have accounted for more than 3 percent of her body weight.

The FDA's ban on silicone breast implants was finally lifted on November 17, 2006.

For a time it was believed by many involved in social work and a few psychiatrists that victims of sexual abuse often repressed their memories of the assaults for years or even decades, but that skilled questioning could elicit those memories, to the psychological benefit of the victim. (Sometimes the memories are not of abuse but of abductions by aliens. The psychological phenomenon involved is the same.) Recovered Memory Syndrome was hijacked into the service of the criminal justice system, which continued to rely on such evidence long after science had rejected the concept. At the same time, there was a widespread fad, fueled by sensationalist media and often enough fomented by self-serving Christian leaders, for believing that organized Satanic cults were everywhere, and that the ritual abuse of children—including not just sexual abuse but torture and murder—was a mainstay of their practices.

The combination of these two items of false belief spelled disaster, with a distressingly high number of gross miscarriages of justice being perpetrated through ignorance, superstition, and fear.

Among the most celebrated instances was that concerning Margaret Kelly Michaels of the Wee Care Nursery in New Jersey, who in 1985, shortly after leaving to take up a job at a different daycare center, was accused of hundreds of instances of child abuse, Satanic molestation, and so forth. She was jailed in 1988 for 47 years for 115 instances of child abuse, the details so fantastic—and so *childishly* fantastic—that no one outside a court of law would have believed for one minute they were genuine; one boy claimed that "we chopped our penises off" and other children claimed Michaels had put cars on top of them.

On appeal, it was revealed that the prosecutors had used suggestion to a quite extraordinary degree in their priming of the child witnesses, and the appeal went uncontested. A sample:

> *Investigator 1:* Just tell me—show me what happened with the wooden spoon. Let's go.
> *Child:* I forgot.
> *Investigator 1:* No, you didn't. I'll tell you what, let's just go to the doll, we won't waste any time.
> *Investigator 2:* Now listen, you have to behave.
> *Investigator 1:* Do you want me to tell him to behave?
> *Investigator 2:* Are you going to be a good boy? Huh? You have to be good. Yes or no?
> *Child:* Yes.[10]

Another:

> *Investigator:* If you don't help me, I'm going to tell your friends that you not only don't want to help me, but you won't help them.[11]

The prime movers in the lead-up to the prosecution of Michaels were Sara McArdle, an assistant district attorney in Essex County, Peg Foster of the Children's Hospital of New Jersey, and Lou Fonolleras of New Jersey's Department of Youth and Family Services. They brought to bear aggressive interrogation that in due course sent most of the children into counseling, from which many have reportedly never emerged, and they regaled the parents—not all of whom were gullible enough to believe them—with mythology in the guise of proven science:

- Children who said they'd been sexually abused were invariably telling the truth.
- Children who said they *hadn't* been sexually abused were lying because terrified of the abuser. The more they refused to say they'd been abused, the more terrified they must be—and the greater must

> have been their suffering. Empathetic interrogators could, however, finally get the truth out of them. . . .

This is, of course, like something out of a medieval witch trial—if the accused drowned she was innocent and if she floated she was obviously a witch and thus promptly sentenced to death.

Michaels was finally exonerated. Some might regard this as a happy ending; others might point out that she lost five years of her life and suffered the destruction of both reputation and livelihood simply because career lawyers recognized her jury would almost certainly swallow pseudoscience whole. Needless to say, no one has ever been held to account. Michaels has never in fact been formally acquitted. No apology has ever been extended to her. When, in May 1999, she tried to claim some sort of restitution, her suit was dismissed.

As celebrated as the Wee Care case was the 1983 McMartin Pre-School case in Manhattan Beach, California. Virginia McMartin had founded the daycare center some while earlier; by the time of the scandal the daycare was mostly run by her daughter, Peggy McMartin Buckey, and the latter's son, Ray Buckey. At one stage all three were being accused of, in total, 208 counts of abuse involving 41 children; eventually this was whittled down to "just" 65 charges involving 11 children. Among the accusations were not only sexual abuse but also animal sacrifice and Satanic rituals. The case stretched over seven years, went through six judges, and cost California's taxpayers in excess of $13 million. It cost McMartin and the Buckeys more than that: "I've gone through hell and now we've lost everything," said Peggy McMartin Buckey in the wake of her and her son's acquittal.[12] Even then, a vengeful prosecution was not done; the jury had deadlocked on eight of the counts concerning Ray Buckey, so he faced a retrial—which was eventually abandoned. In 1991 the three defendants plus Peggy McMartin Buckey's daughter Peggy Ann were successful in a slander suit against one of the affected children's parents; so apologetic were the courts to these people who'd seen their lives destroyed that the damages awarded came to just $1.

It took until 2004 for John Stoll, convicted on seventeen counts of child molestation in 1985 in Bakersfield, Kern County, California, as one of a supposed forty-six-strong ring of child abusers, to be released from prison, his conviction overturned. No physical evidence had been presented at the trial, even though, with sodomitic rape included among the charges, a medical examination might have seemed imperative.

Paul R. Ingram of Olympia, Washington, actually pled guilty in 1988 to six counts of third-degree child rape involving his daughters Ericka and Julie, who by the time of his conviction were twenty-one and eighteen, respectively. The case, the subject of the book *Remembering Satan* (1994) by Lawrence Wright, appears to have begun when Ericka attended a Pentecostalist youth retreat and was told by a self-styled prophetess that she'd suffered sexual abuse by her father. Ericka couldn't remember any such thing, but accepted counseling from a therapist who subjected her to Recovered Memory Therapy; thereafter, oh my, she could recall clear as a bell how her father had abused her regularly from toddlerhood until just a year ago; he had even given her an STD, for which she'd had to be treated. Soon afterward Julie, too, began to claim she'd been sexually abused by her father and his poker buddies.

The daughters weren't the only ones to be persuaded they'd been repressing memories of Satanic abuse by their father; Paul Ingram came to believe the same himself. Once again, the facilitator was Christian belief, this time in the form of the church he attended, the Church of Living Water in Olympia. Among the tenets preached by this Christian splinter were that Satan can make you do terrible things and then wipe your memory of them, and that recovered memories are always accurate because God wouldn't allow them to be otherwise. Subjected to sleep deprivation and suggestion techniques, Ingram eventually "remembered" his guilt. He "remembered" lots of other things too, including that he'd murdered a prostitute and been involved in the Green River Killings.

In the UK the Satanic abuse craze happened a little later and mercifully didn't last very long. The case that gained the most publicity—and which in due course stopped the whole nonsense in its tracks—centered on South Ronaldsay in the Orkney Isles, far to the northeast of northern Scotland. It began with, seemingly, a reasonable suspicion of child sexual abuse, for which the father of the "W" family was imprisoned in 1986. Then, in late 1990, the "W" family's eight children were abruptly removed from their mother's care, the older six being placed in various foster homes or institutions on the mainland while the youngest two, bizarrely, were told their mother had died and were placed with adoptive parents. The real headlines began, though, a few months later when one of the "W" children supposedly "remembered" how children on the island were being abused by Satanists led by the local Presbyterian minister, the Rev. Morris MacKenzie—who just happened to be in the vanguard of those campaigning to have the "W" children returned to their mother.

Predawn raids were mounted on the minister's home and those of four other families who'd been supporting his campaign. Nine more children were removed to the mainland. When, within a couple of months, the case concerning these nine children was brought to court, Sheriff (Judge) David Kelbie listened to the first day's evidence and threw the case out as "fundamentally flawed"—i.e., as patent nonsense. Among the more remarkable claims by the social services was that the toy plane one child had made out of two sticks was "a wooden cross"—quite obviously Satanic!

These children were promptly returned to their parents, but progress was far slower for the "W" children. In October 1992 a panel chaired by Lord Clyde presented its report into the Satanic abuse case, which could hardly have been more damning about the poor judgment and behavior of the Orkney Social Services, and even about the very concept of Satanic ritual abuse. Even so, it took six years before the last of the "W" children was permitted to rejoin the rest of the family.

Among the victims in all these cases of "recovered memory" of sexual abuse—and many more besides, not just in the US and UK but in New Zealand, Canada, Australia, Ireland, Italy, the Netherlands, and on and on—were indeed the children, but, as has repeatedly been observed, the abusers were not those accused but the interrogators who subjected the children to something little short of the third degree, who implanted false memories in their minds, and who tore families apart—sometimes irrevocably. Minnesota's attorney general, Hubert H. Humphrey III, reporting in 1985 on a 1983 case, stated with dismay: "The mother of [one] child indicated that her daughter had been interviewed at least thirty and possibly as many as fifty times by law enforcement or Scott County authorities. A large number of other children also were repeatedly interviewed."[13]

In 1997 an Illinois woman, Patricia Burgus, sued the psychiatrist who had persuaded her, with the aid of drugs and hypnosis, that she remembered belonging to a Satanic cult, being tortured, taking part in a ritual murder, and more.[14] In 2006 one of the "W" children sued the Orkney Social Services.[15] In 2006 a British Columbia woman, Donna Marie Krahn, sued those who had, during the course of Christian-based pastoral counseling, persuaded her that her husband was a member of a Satanic group that had sexually abused her and her children.[16]

In 2005 a Scottish woman, Katrina Fairlie, sued the hospital trust NHS Tayside because one of their psychiatrists, Alex Yellowlees, had in 1994 used Recovered Memory Therapy with her to "help her remember" having been raped by her father, Jim Fairlie, a former deputy leader of the Scottish National Party, and having seen him beat a six-year-old child to death with an iron bar. She'd also been deceived into accusing him and seventeen other men of conducting a pedophile ring. Although Katrina Fairlie had withdrawn the allegations the following year and the police had abandoned their investigation soon after, her father's political career was destroyed and, clearly, the damage done to the family was immeasurable. Jim Fairlie's own attempt to gain legal redress against the hospital trust failed on a technicality, but Katrina's succeeded. She accepted a large out-of-court settlement just before the case came to trial.[17]

Victims, too, are those falsely prosecuted and convicted on the grounds of others' implanted false memories. For far too long these people seemed to have no recourse. Eventually, however, some of the worms began turning. In 2003 an Ontario retiree was awarded $150,000 by the police force that had charged him in 1999 with having repeatedly raped his son and the son's playmate when they'd been children in the 1970s. The retiree spent eighteen months and his life savings proving the charges were impossible. During those eighteen months he suffered humiliation, restrictions on his travel, public loathing, and much else. Compensation totaling $150,000 seems inadequate.

And there are still some people in prison whose convictions rest solely on "recovered memories." Among these is the North Carolina man Patrick Figured, who was convicted in 1992 of having molested three toddlers some years earlier at the daycare center run by Polly Byrd, the mother of his girlfriend, even though Byrd was adamant that on his rare visits to the center he'd never at any time been alone with the children. It was admitted at the trial that the children hadn't mentioned any abuse by Figured until it was suggested to them by the investigating social worker, Nancy Berson, who was convinced Satanic ritual abuse was a real and widely prevalent phenomenon.

The supposedly scientific basis for believing the North Carolina children had been sodomized was the phenomenon known as anal winking, whereby some people's anuses, on being touched, exhibit a reflex action; for a while it was thought anal winking was symptomatic of past sodomy, and it was on this phenomenon that, in the UK, the notorious Cleveland "child abuse" case of 1987 was based. By the time of Figured's arrest in 1989, however, it had already been recognized that anal winking was a perfectly natural occurrence. It was also claimed the children's anuses showed areas of discoloration consonant with healing from injury such as might have been caused by sodomy with a screwdriver. Ignored was the fact that such injuries would have been agonizingly painful and would likely have caused extensive bleeding . . . yet no one had noticed anything at the time.

Likewise still in prison is Robert Halsey of Lanesboro, Massachusetts, a school bus driver convicted in 1993 of repeatedly sodomizing two of the young passengers, twin brothers, on his bus on their way to or from school. The two boys also told how, before raping them, he would sometimes catch a few fish, which he might or might not stick up the children's rear ends. Yet neither parents nor teachers ever complained of the two boys being late for school or getting home. Halsey has seen appeals against his blatantly unsafe conviction turned down in 1996, 2005, and 2006.

These are only two instances of the legal system's refusal to recognize its error in accepting the now-discredited Recovered Memory Syndrome as hard evidence.[18] Some of the people falsely accused have died in prison, and at least one, although his case had been thrown out by the courts in 1990—Kaare Sortland of Tacoma, Washington—was murdered two years later by a vigilante.[19]

The case of Cameron Todd Willingham has been reported all over the world. One night in 1991 the family home in Corsicana, near Dallas, burned down with Willingham's three children trapped inside. Arson was suspected, so Willingham was brought to trial. The medical/psychiatric "expert" evidence presented to the court was of roughly the status of voodoo—for example, he was probably a sociopath because he had an Iron Maiden poster—but it was the arson evidence that truly scuppered him. He was convicted on the basis of testimony from Deputy State Fire Marshall Manuel Vasquez and Corsicana Assistant Fire Chief Douglas Fogg that the fire had been deliberate. One piece of evidence offered to the court was that some of the glass in the home had been crazed, uniquely symptomatic of high temperatures, which indicated fire accelerants must have been involved. In the eyes of the jury, this and other scientific indicators of arson were enough to find Willingham guilty in the teeth of considerable evidence to the contrary, and he was sentenced to death—even though no one could offer anything resembling a convincing motive.

Some while later, the Innocence Project (www.innocenceproject.org), an organization founded in 1992 that reassesses potentially unsound cases and has rescued a number of innocent people from Death Row, commissioned five arson experts to check the evidence. So far as they could establish, there was no reason at all to suspect the fire had been anything other than accidental. The evidence Vasquez and Fogg had offered, while doubtless they themselves assumed it scientific, consisted of lore passed down from one investigator to another without anyone ever taking the time to test it. The "telltale" of crazed glass, for example, while widely recognized among arson investigators, proved on experimentation to be nonsense. What produces crazing in glass is not heating but rapid *cooling*, as might happen when a fireman directs a blast of cold water against it. Further, fires without accelerants burn just as hot as fires with.

As the date for Willingham's execution grew closer, the Innocence Project passed the information so far gathered to Texas Governor Rick Perry, pointing out that at the very least the case against Willingham was rocky. Perry, seemingly worried about the political repercussions of appearing "soft on crime," ignored the evidence—deriding the "supposed experts" as purveyors of "anti–death penalty propaganda"—and on February 16, 2004, Willingham was executed.

The case came firmly to public notice in 2009 when an article in the *New Yorker* by investigative journalist David Grann[20] demonstrated not just Willingham's near-certain innocence but that the Texas authorities, aware of this, had gone ahead and killed him anyway. A few weeks before Grann's article, a report[21] requisitioned by the Texas Forensic Science Commission as part of Texas's first-ever state review of an execution concluded there was no sustainable evidence of arson. The *Chicago Tribune*, reporting this in August 2009, observed: "Over the past five years, the Willingham case has been reviewed by nine of the nation's top fire scientists. . . . All concluded that the original investigators relied on outdated theories and folklore to justify the determination of arson."[22] Immediately before the official presentation of the report to the commission by its author, Maryland fire investigation expert Craig Beyler, Governor Perry shuffled the commission's membership, seemingly to shift its balance toward hawkishness. In September 2010, Willingham's family filed in Travis County for a court examination of whether the state had neglected properly to consider contrary evidence before killing Willingham, and State District Judge Charles Baird agreed to give them a hearing. Lowell Thompson, who had prosecuted Willingham, promptly demanded that Baird recuse himself, the grounds offered being hazy.[23] In December 2010 Texas's Third Court of Appeals found in Thompson's favor.[24] And so the tortuous saga of the Texas coverup drags on, as good science is publicly derided by the authorities there while the bad science that killed Cameron Todd Willingham is upheld as sacrosanct.

One could say that it was in the year 1991 that the study of arson really became a science. In March of that year, investigators into the case of Gerald Wayne Lewis, indicted after a 1990 house fire in Jacksonville, Florida, that killed his wife and four of their children, were permitted to re-create the incident in the similar, condemned house next door. They found that signs up until then considered by arson investigators to be telltales of accelerant use in fact occurred naturally. Lewis was exonerated. Later in that same year, a brush fire in California that destroyed some three thousand homes provided a bonanza for the new breed of fire investigators, who were able to find out far more than had ever before been known about the way non-arson house fires behaved. In the process, they demolished many of the most cherished beliefs of traditional fire investigators. That year also saw the publication of *NFPA 921*, a guide published by the National Fire Protection Association that rigorously explained why many of the traditional, nonscientific theories of fire investigation were not just unproven but demonstrably false. Many old-time "professionals" in the field resisted the information in *NFPA 921*, and continue to do so to this day.[25] As do politicians: It will not have escaped notice that *NFPA 921* was published a full fourteen years before Governor Rick Perry, denying there was anything wrong or superseded in the "science" presented at Cameron Todd Willingham's trial, permitted the man to go to the death chamber.

Phony forensics in the courtroom are not limited to arson investigations, and nor of course are they limited to the US. In a widely reported Dutch case a nurse, Lucy de Berk, was acquitted in 2010, having been convicted in 2003

of murdering seven people in her care and attempting to murder three others, all the alleged victims being disabled babies or elderly and frail. That the deaths were murders was claimed at her trial based on the testimony of a statistician produced by the prosecutors. He said the odds against the string of deaths being a coincidence were 342 million to one. In fact, as emerged during a review instigated in 2008 by the Dutch Supreme Court, the deaths *were* all from natural causes—if you care for disabled infants and sick oldsters, you expect a fairly high death rate—and the statistician had simply pulled his "calculation" out of thin air.

In 2009, the National Research Council's Committee on Identifying the Needs of the Forensic Sciences Community released a devastating report titled *Strengthening Forensic Science in the United States: A Path Forward*, which concluded that the country's forensic services were essentially in a mess.[26] Discussing the report on Reason.com, Radley Balko summarized the essential problem:

> Most forensic disciplines were invented by police investigators, not scientists. Courts have allowed these disciplines to be admitted into evidence before they've been subjected to any serious scrutiny from the scientific community. . . . Yet when a forensic specialist testifies in the courtroom, his testimony usually carries the weight and veneer of actual science.[27]

To take a single example, in their section on forensic odontology (pp. 173ff.) —i.e. bite mark analysis—the committee reported something near chaos:

> The guidelines of the [American Board of Forensic Odontology] for the analysis of bite marks list a large number of methods for analysis. . . . The guidelines, however, do not indicate the criteria necessary for using each method to determine whether the bite mark can be related to a person's dentition and with what degree of probability. There is no science on the reproducibility of the different methods of analysis that lead to conclusions about the probability of a match. This includes reproducibility between experts and with the same expert over time. Even when using the guidelines, different experts provide widely differing results and a high percentage of false positive matches of bite marks using controlled comparison studies. (p. 174)

A further problem is that, in many states, the forensics laboratory is part of the Department of Justice, and thus regarded as being "on the same team" as the district attorney's office. Analysts are thus likely to regard their job as being to assist the prosecution in the task of getting a conviction rather than pursuing the truth as revealed by the science. Even if they don't see their role quite so blatantly, they're still vulnerable to pressure from their "colleagues" in the DA's office.

On occasion this situation is even explicit. After an investigation into South Carolina's forensics laboratory it was discovered that the prescribed standard practice was for the lab simply to issue no report when a result was, so far as the prosecution case was concerned, negative. In an instance cited by Balko in the article quoted above, Greg Taylor was released from prison after serving sixteen years for murder when it was revealed that what the police had discovered in his car and thought was blood had been found by the lab to be not blood at all; the lab and the prosecution had simply clammed up about the negative result, allowing Taylor to be convicted on the basis of what they knew was false evidence.

Taylor was lucky. In Texas, Claude Jones was executed in December 2000 for a 1989 liquor-store killing. The crucial evidence against him was a fragment of hair found on the counter near the victim's body. A state scientist, Stephen Robertson, was asked to examine this sample, but said it was too small for analysis, an evaluation later confirmed by an independent analyst. However, Robertson testified at the trial that the hair was Jones's. When asked at a separate hearing why he'd changed his mind, he said he didn't know. In 2007 the sample was subjected to DNA testing and discovered not to have come from Jones. Clearly this doesn't prove Jones innocent; it does, however, demonstrate that the forensic evidence that brought his conviction was a sham.[28] A rummage through the Innocence Project's online files reveals far, far too many similar cases.

Sometimes, as in the Willingham case, it's not that the science hasn't delivered or is being suppressed but that politicians and/or prosecutors choose to deny it. Jerry Hobbs, charged with the double murder of his daughter and another infant, spent five years in jail in Illinois before being released in August 2010, just a few months before his planned October trial. A full four years earlier, DNA testing had revealed that the semen discovered on his daughter's body was not Hobbs's. The DA kept him in jail anyway, and was willing to bring him to trial on the basis that, soon after the crime, Hobbs had been browbeaten into making a swiftly recanted false confession. The only reason for Hobbs's eleventh-hour release was that a DNA match was made with a man picked up on suspicion of rape in Virginia.[29] There was no apology from the DA's office, whose attitude was that Hobbs should never have been so silly as to make that false confession.[30] That stance in itself represents a denial of science, since there's abundant evidence that false confessions are extremely common—leading to about 25 percent of wrongful convictions and nearly 60 percent of wrongful murder convictions, to judge by the statistics of DNA exonerations.[31]

One of the basic rules of the modern enterprise of science is that, should an individual or entity attempt to settle a scientific matter through the legal system rather than through the normal processes of peer review and open debate, then, whatever the verdict of the courts, the plaintiff is deemed to have lost the scientific argument.[32] Of course, there are exceptions—there can be legitimate lawsuits over patents and precedence, for example—but they're few and far between. The reason's obvious. Lacking peer review and open debate, with all participants knowing they can give their honest opinions without fear of repercussion, science cannot sensibly progress.

A disturbing recent trend has been the use of the antiquated UK libel laws by moneyed individuals or organizations to silence criticism. These laws effectively put the onus on the critic to prove his or her innocence, something that can cost huge amounts of time and money that may never be recompensed; often merely the threat of litigation can be enough to silence further dissent. The most alarming aspect of this corruption via legal system is that the cases we hear about are the ones where the critic decides not to yield under the impressive intimidation but to fight for the right to criticize—and for what they perceive as the truth. It is hard to think of a single example in which that perception has been false.

Vitamin seller Matthias Rath, who sells vast quantities of his products in South Africa based on his claim that they can cure AIDS (see page 137), in 2007 sued the UK physician and journalist Ben Goldacre for describing those claims as spurious and deadly. In this instance, the plaintiff backed off, but only after fifteen months during which Goldacre had time to do little other than consult with lawyers to prepare his defense; the legal bill incurred would have long before obviated any hope of his defending himself had not the *Guardian* newspaper dug deep into its pockets on his behalf. (The story is told in the expanded edition of Goldacre's book *Bad Science* and can also be read free online; use search term "The Doctor Will Sue You Now.") The UK plastic surgeon Dalia Nield was cited in a 2010 newspaper article as saying she thought it "highly unlikely" the breast enlargement cream Boob Job, manufactured by Rodial Ltd., actually worked. Rodial threatened her with a libel action.[33]

It's not only Brits who have to fight wary of the UK libel laws. In 2007 the Swedish researchers Francisco Lacerda and Anders Eriksson published in *International Journal of Speech, Language and the Law* a paper called "Charlatanry in Forensic Speech Science," which surveyed those lie detectors that are based on voice risk analysis (VRA) and questioned whether any could be relied upon. The authors and the journal were threatened with legal action by the Israeli company Nemesysco, which manufactures VRA devices, unless the paper was pulled; the journal complied. A complication here was that the UK government's Department for Work and Pensions (DWP) was investing a considerable amount of money on a pilot scheme with the Nemesysco gadget as a means of trying to catch fraudulent benefits claimants. In November 2010 the DWP announced it was abandoning plans to use VRA technology. Of twenty-three pilot studies, the VRA device had been judged successful in just five.[34]

The UK consultant cardiologist Peter Wilmshurst might be regarded as something of a professional whistleblower. His acceptance address for the 2003 HealthWatch Award[35] details his travails when his team's research into the drug amrinone indicated that, far from helping patients recuperating from heart failure, as advertised, it could well be harmful to them. This led him into a running battle with the drug's manufacturers, Sterling–Winthrop. In the end, in 1986, Sterling–Winthrop had to back down and withdraw amrinone from the market.

His current legal battle is with NMT Medical of Boston, manufacturers of a device called STARFlex®. People who've suffered conditions involving a right–left shunt of blood between the two atria of their hearts, and who have had that shunt closed, often report a decrease in incidence of migraine headaches. The STARFlex device is designed to block off this shunt. Clinical trials of STARFlex were mounted in the UK in early 2005, with Wilmshurst and the headache specialist Andrew Dowson as team leaders. To their surprise, the results of the trials were unfavorable. In a subsequent interview in 2007 with the specialist website *Heartwire*, Wilmshurst speculated as to why this might be, including that some of the subjects treated might not have been suffering from shunts in the first place, and that possibly STARFlex was not as effective at cutting off shunts as the manufacturer believed. NMT Medical sued Wilmshurst for libel and slander.[36] Since then, the case has rumbled on . . . and on.

The most celebrated instance in recent years of an organization using the UK libel laws in an attempt to stifle criticism concerns the British Chiropractic Association (BCA) and the science writer Simon Singh. The BCA is supposed to be the body regulating the practice of chiropractic in the UK, and its members are expected to conform to defined professional standards. In actuality, as Singh and Exeter University's Edzard Ernst, his coauthor on the book *Trick or Treatment* (2008), discovered, the BCA was failing to restrain those of its members who advertised that chiropractic was efficacious in treating children who suffered colic, sleeping difficulties, feeding difficulties, chronic ear infections, asthma, prolonged crying, and more. Since Ernst had studied over seventy trials and found a general conclusion that chiropractic might be useful in relieving back pain but little else, Singh felt perfectly confident when, in a 2008 article in the *Guardian* newspaper, he described these exaggerated claims as bogus and criticized the BCA for failing to clamp down on its members.

The BCA sued him for libel and, at first bite, found an ally in Mr. Justice Eady, who ruled that Singh's use of the word "bogus" had ruled out any defense that his remarks were fair comment; "bogus," Eady decided, implied a conscious intent to deceive, and thus Singh's comments were libelous.

Singh had the option of cutting his losses—he was already out of pocket to the tune of £100,000 in legal fees—or risking a lifetime of debt by appealing Eady's ruling. Heroically, he chose the latter, and on April 1, 2010, the UK High Court ruled that he had the right to appeal and to use, in that appeal, the defense of fair comment. The scientific community breathed a sigh of relief that could be heard around the world; as BBC science correspondent Pallab Ghosh commented, "Had Justice Eady's ruling stood, it would have made it difficult for any scientist or science journalist to question claims made by companies or organisations without opening themselves up to a libel action that would be hard to win."[37] Almost immediately the BCA dropped the case. It's to be hoped that Singh will eventually recover his legal expenses in full.

But the case had far greater repercussions. Irate that an organization should be using financial clout in order to impose the acceptance of pseudoscience upon the scientific community, individuals and groups all over the world responded by contributing to Singh's cause. The organizations Index on Censorship, English PEN, and Sense About Science set up the Libel Reform Campaign (www.libelreform.org) to press for permanent reforms in the UK libel laws so that this sort of travesty of the right to fair comment will not be permitted to occur again. Senior members of the Coalition government that took power in the UK during 2010 were strongly supportive, and the relevant legislation is promised soon.

Although the situation has not been so bad in the US, a couple of recent cases suggest the decline has started. In June 2010 Doctor's Data Inc. filed suit against Stephen Barrett (who runs the excellent site Quackwatch, scourge of bogus medicine), Quackwatch Inc., and the National Council Against Health Fraud Inc. to the tune of $10 million. Doctor's Data is in the business of providing testing for "prevention and treatment of heavy metal burden"—which is to say, in diagnosing whether patients have elevated levels of mercury and other heavy metals, using what's called provoked urine testing, a technique whose merit has been widely questioned. The situation is of particular interest because many people who believe there's a connection between vaccination and autism think heavy metals could be involved.[38]

Barrett, through his various websites, has been among the more prominent critics of the use of provoked urine testing. It's of interest that, during the weeks leading up to the issue of the suit, he had asked the lawyers for Doctor's Data Inc., Augustine, Kern & Levens Ltd., to itemize those comments of his that their client deemed inaccurate or objectionable. Answer came there none.[39] The case is still in progross as this book goes to press.

Legal action can be used to intimidate or punish whistleblowers, too. In April 2009 two nurses in Kermit, Texas, Vicki Galle and Anne Mitchell, reported to the Texas Medical Board (TMB) concerns they had about one of the physicians at Winkler County Memorial Hospital, Dr. Rolando Arafiles, who they said was abusing his position to encourage patients in the hospital's emergency department and the rural health clinic to buy his own herbal medicines. The two were promptly fired and then indicted by the county attorney, Scott Tidwell, for "misuse of official information" (their confidential report to the TMB had necessarily included some patient information), a third-degree felony that can carry a jail term of two to ten years. On June 12, just after the closing of the sixty-day window they could have used in their defense as proof of retaliation, they were actually arrested.

Galle was dropped from the case just before trial. On February 11, 2010, Mitchell was acquitted by the jury. In August 2010 the two nurses received $750,000 as settlement of a civil suit against Winkler County. In December 2010, the Texas attorney general's office arrested Arafiles for the crimes of retaliation and abuse of confidential information; in July 2011, perjury was added to the charges against him. A little earlier, in June 2011, Sheriff Robert Roberts, who had played his part in the persecution of the nurses and who had been Arafiles's business partner in selling "complementary" medicines, received four years' jail time for his crimes. And, at the time this book goes to press, Scott Tidwell is scheduled to appear in court in September 2011.

The denial of climate change has brought with it a whole new raft of reasons why people should wish to game the legal system in order to suppress scientific criticism.

Christopher Monckton (see pages 293ff.) is modern AGW denialism's master of the rhetorical technique known as the Gish Gallop (named after Creationist debater Duane Gish), whereby the debater throws out a plethora of "facts" that his opponent has no possible chance of checking on the spot; the effect is to make the "galloper" seem omniscient while his opponent is stumbling ineptly and ignorantly in his wake. In the modern age, while this still works well with untutored live audiences, it's now possible for the performance to be recorded and analyzed afterward. His testimony to the Select Committee on Energy Independence and Global Warming of the US House of Representatives on May 6, 2010, prompted a team of climate scientists to issue in September a detailed report titled *Climate Scientists Respond*, which listed the errors and misinterpretations he had promoted on that occasion.[40] Arthur Smith, manager of the Database Group at the American Physical Society, went through a 2008 article by Monckton called "Climate Sensitivity Reconsidered"[41] and found in its scant few pages over 120 errors, irrelevancies, and illogicalities; his full report is available online.[42] Others have performed similar exercises.

But the most extravagant was that performed by John Abraham, a professor in thermal sciences at the University of St. Thomas, Minnesota.

Appalled by a presentation Monckton had given at the nearby Bethel University in October 2009, Abraham determined to go through every single one of Monckton's references and check with the original; where he wasn't certain himself of the precise implications of an original, he contacted the author(s) of the paper concerned for clarification. The result was an eighty-four-minute presentation, "A Scientist Replies to Christopher Monckton,"[43] posted online at the end of May 2010. It is a demolition.

Monckton's response was "Monckton: At Last, the Climate Extremists Try to Debate Us!":

> So unusual is this attempt actually to meet us in argument, and so venomously *ad hominem* are Abraham's artful puerilities, delivered in a nasal and irritatingly matey tone (at least we are spared his face—he looks like an overcooked prawn), that climate-extremist bloggers everywhere have circulated them and praised them to the warming skies.[44]

No ad hominem nastiness there!

Instead of addressing the scientific points made in Abraham's rebuttal, Monckton issued a list of 466 "pertinent questions" attacking Abraham's credentials, and so forth. He then used media appearances and a posting on the massively trafficked denialist site *Watts Up With That* to encourage his legion fans to inundate the president of St. Thomas University with demands for retribution.[45]

The university was not amused, or perhaps privately it was. Its response was . . . forthright. And so Monckton threatened Abraham with legal action—an action that has so far failed to materialize.

And the name of the defense fund set up on Abraham's behalf?

Prawngate.

Far more concerning is the use by those in power of the legal system in attempts to silence scientific information that conflicts with their ideology.

In 2010 the recently appointed attorney general for Virginia, Ken Cuccinelli, decided to take up the cudgels on behalf of climate-change denialism and persecute climate scientist Michael Mann (see page 249). Cuccinelli had already, during his brief tenure as AG, raised rationalist hackles and eyebrows by attacking the EPA, by telling public universities they did not have the legal right to forbid discrimination on their campuses based on sexual orientation, by filing a lawsuit to challenge federal healthcare reform, and even by adopting a version of the Virginia state symbol in which the left breast of the victorious Virtue, ordinarily chastely exposed, was saucily covered up. (Cuccinelli, faced by public ridicule on this last, claimed it was only a joke, honest.)

While at the University of Virginia during 1999–2005 Mann had received about a half million dollars in state grants. On the pretext that there were grounds for suspecting Mann of professional fraud, on April 23, 2010, Cuccinelli demanded from the university that by May 27—just five weeks later!—he be furnished all documentation associated with Mann's tenure there with the possible exception of any crosswords he might have done. Note that all considerations of the confidentiality of the other parties involved in any correspondence were blithely ignored.

It takes no great leap of deduction to recognize that Cuccinelli knew full well the demands he was making were impossible for the university to fulfill; even a partial fulfillment would be possible only at quite extraordinary expense, which expense would naturally be met in due course—like the costs of Cuccinelli's pursuit of this ideological vendetta—by the Virginia taxpayers. His sole conceivable purposes were to harass Mann and to warn all of Mann's future employers plus the climatology community in general that, unless their conclusions were in accord with Cuccinelli's own denialist ideology, they could expect similar treatment.

The assault was eventually rebuffed by the university, but Cuccinelli has promised to persist, so no one really knows if a further attack may suddenly materialize.

History will not treat Cuccinelli and his ilk kindly, but by then it will be far too late to make a difference.

3

THOROUGHLY UNCOMPLEMENTARY

[O]ne of the greatest challenges we face is the widespread public belief in the scientific method. . . . We're too reliant on the scientific method, and it stands in our way of forging ahead.

—naturopath Daniel Rubin[1]

All professions have assholes, including medicine, and their existence does not invalidate science.

—comment by "madder" on *Respectful Insolence*[2]

In 2000 it was estimated that in the UK the annual expenditure on alternative medicine was at least £1.6 billion (then about $2.25 billion). It's believed a comparable figure for the US would be about $27 billion. According to a slightly earlier estimate, done in 1997, the UK has about fifty thousand practitioners offering alternative medicine, whether registered with a professional body or not, plus perhaps ten thousand orthodox physicians who offer some or other alternative therapy in addition to the norm. These are frightening figures when you bear in mind that there's no reason to believe these therapies work *at all* beyond any placebo benefit. After all, many of the remedies now described as complementary or alternative medicine (CAM)—prayer, therapeutic touch, herbal medicine—are those that didn't do so well against the Black Death. Techniques we today call CAM were unsuccessful for millennia against smallpox; it took modern scientific medicine mere decades to wipe smallpox out entirely.

Yet still people insist powdered pond scum is a cure for cancer and essence of goose dropping the sure way to restore flagging libido. It seems a single anecdotal report, no matter how vague, no matter how far distanced from the direct experience of the individual repeating it, outweighs any number of deeply researched double-blind trials and peer-reviewed scientific conclusions.

Does this mean people who believe the bogus claims for alternative therapies are stupid? No. Well, obviously *some* are; but the vast majority are victims of a media environment in which—in matters relating to other sciences as well as medicine—it's deemed more desirable to promote baloney than to offer a reality check. Someone's remarkable remission of seemingly terminal cancer, if attributed to an infusion of bladderwort or a mail-order TM course, can make front-page news in the tabloid press. But if someone who lives up the road from you responds well to chemotherapy, that's not news. And even less is it news if someone who's been trying to fight cancer with magnetotherapy or reflexology succumbs to the disease—or to the lack of effective treatment for the disease.

In other words, when we make our medical choices we're very often not just underinformed but actively *mis*informed by our news media . . . and quite a number of us die every year because of it.

The people who rely on CAM may not always be the ones you expect. A 1998 study[3] revealed that in 1997 as many as 42 percent of the US population used at least one alternative therapy and that more women (49 percent) than men (38 percent) made use of CAM. Mildly unexpected was that the age group most likely to turn to CAM was the 36–49 bracket. More startling was that use of CAM was highest among people with a college education, fully 51 percent of whom turned to CAM during the survey year 1997, and among people with incomes over $50,000 (48 percent). Thus, the very people best able to afford healthcare and educationally best fitted to recognize its value were the ones making most use of CAM. Possibly we're looking at a consequence of the '70s generation's later-life New Agery; more likely the problem lies in the media's adulatory coverage of quackery.

A particular hotbed of this is to be found among the far-right websites/ezines like *Newsmax*, *WorldNetDaily*, and *Personal Liberty Digest*, all of which bombard their hapless subscribers not just with spun news and dodgy looking investment schemes but with promotions of fringe therapies, cancer cures being a particular favorite. Here's the selection of articles offered by *Newsmax* on a random date (November 24, 2010):

- Dr. Wascher: Proven Ways to Live Cancer-Free
- Special: Interviews with Doctors Who Are Quietly Curing Cancer
- Mix of Cardio, Strength Training Best for Diabetics
- Vera Tweed's Nutrition Tip: Holiday Eating Strategies
- Dr. Blaylock: Body Scanners Riskier Than Feds Admit
- ALERT: Thyroid Deficiency Caused by One Common Food Additive
- Orange, Dark Green Veggies Tied to Longer Life
- Doctor Uses a Deck of Playing Cards to Boost Brain
- Can Prayer Heal You? Dr. Crandall Discovered the Truth

The answer to that last question is, obviously, YES INDEEDY . . . and Dr. Crandall is hawking a book to prove it, available at a !!SPECIAL PRICE!! to *Newsmax* readers.

CAM gains support from some surprising quarters. In the UK there has been much controversy over governmental funding of CAM research. Some US insurance companies will underwrite their clients' treatment with homeopathic or other alternative therapies. Is this a vindication of CAM claims or, more sinisterly, a matter of fiscal calculation? Recipients of CAM die sooner—and that's where the companies' profits lie.

Even so, the companies' approach can be seen as part of a growing acceptance of CAM—as assisting the fulfillment of the CAMsters' quest for respectability. Many schools of CAM, in hopes of persuading more of the public of their responsible approach to medicine, have set up their own regulatory authorities—for example, the British Chiropractic Association (see page 72), the Israeli Association of Classical Homeopathy, the American Association of Acupuncture and Oriental Medicine—to supervise the activities and claims of practitioners. Well, maybe. The trouble is that, if a therapy's bogus, it really doesn't much matter if it's regulated or not: It's still bogus. Another approach is to adapt the format of the peer-reviewed journal to promote a denialist worldview. An example is *Journal of American Physicians and Surgeons* (see page 117), which publishes fringe-science papers and favorably reviews denialist books, while also advancing a politically conservative agenda.

It's not just therapies that can be alternative; diseases and conditions can be as well.

A relative newcomer to the scene of dubious medical conditions is "chronic Lyme disease." But what exactly *is* it? Lyme disease is spread by ticks carrying the *Borrelia* bacillus; in North America the culprit is *B. burgdorferi*, while in Europe *B. afzelii* and *B. garinii* are also involved. Although there are many indications of a lingering, little understood post–Lyme disease syndrome, or Stage 4 Lyme disease, there's no known mechanism that would allow for a chronic form. There seems, in short, considerable doubt that any such ailment as chronic Lyme disease exists.

Yet some people believe themselves to be suffering from it, and some physicians either believe likewise or at least are willing to go along with their patients' self-diagnoses. The supposed remedy involves long-term treatment with antibiotics, even though there's no evidence *Borrelia* is present or active. Four controlled trials (to date) have found no indication that long-term antibiotics are of use in such cases. The Infectious Diseases Society of America, the American Academy of Neurology, and the National Institutes of Health are among the academic institutions to have decided the practice is useless and potentially harmful; they are countered by the International Lyme and Associated Diseases Society (ILADS), an apparently grassroots organization supposedly concerned with patients' well-being. In "Inaccurate Information about Lyme Disease on the Internet"[4] James D. Cooper and Henry M. Feder Jr. reported that the ILADS site was inaccurate about the diagnosis, serology, and treatment of Lyme disease, plus much more; of the eight categories of information Cooper and Feder tested, the ILADS site discussed five, and in all five instances its information was inaccurate.

If the antibiotics simply had a placebo effect, this would be just a matter of people being ripped off, but long-term antibiotic use can have quite nasty health consequences. What makes matters far, far worse is that medically ignorant politicians have placed this likely nonexistent medical condition on the statute books. Connecticut, California, Massachusetts, and Rhode Island have laws that protect physicians, no matter how negligent or incompetent, who diagnose this mystery disease and prescribe long-term antibiotics for it from being sued by patients or their families. These are quacks' charters.

Mystery syndromes are nothing new. One unusual medical offering is *All About Radiation* (1957) by L. Ron Hubbard and others.[5] The text of this book, largely cobbled together from a quartet of lectures Hubbard gave on the subject, has been subject to some editing through its various editions. In their *Devenez Sorciers, Devenez Savants* (2002) Georges Charpak and Henri Broch reproduce a salient paragraph (p. xiii of US edition):

> At what point is radiation harmful to the human body? No one knows, but we can state the following: a wall fifteen feet thick can't stop a gamma ray. On the other hand, a body can. Which leads us to pose this medical question, of the greatest significance: why can gamma rays go through walls but not through the body? Clearly, a body is less dense than a wall. As we do not find an answer in the material domain, we must therefore enter the mental domain.

This is, obviously, nonsense. Earlier versions of Hubbard's text continue: "I can fortunately tell you what is happening when a body gets hurt by atomic radiation. It RESISTS the rays! The wall doesn't resist the rays and the body does." Hubbard's most astonishing claim is that radiation sickness (and indeed some cancers) can be cured by dosage with a tonic called Dianazene, devised by one L. Ron Hubbard.

Hubbard is described on the cover of later editions of *All About Radiation* as a nuclear physicist—indeed, on the dust flap of some editions we're told he was "one of America's first nuclear physicists"—curious, in that Hubbard wasn't any kind of physicist at all. In 1931 he briefly attended a course in nuclear physics at George Washington University; he received an F and dropped out. His medical qualifications consisted of a "PhD" from "Sequoia University"—a notorious diploma mill that in 1984 would finally be closed down by court order. In 2009 a Freedom of Information request in the UK revealed that Sequoia University had in fact been owned by . . . Hubbard himself.

It's a sad but generally accepted fact of medical science that in only about 15 percent of cases is there evidence that the "scientific" part of the treatment has actually done any good.[6] The rest is down to the placebo effect, or to the body's ability to cure itself of most ailments, or to the patient's trust in the doctor, and so forth. Thus, a purveyor of fake medicine may be able to record a modest success rate in treating certain ailments.

In blind trials of new drugs, it emerges pretty consistently that about 30 percent of the people who've been receiving the placebo feel better as a result. Thus, the average CAMster can expect a roughly 30 percent success rate even before other factors kick in. However, despite the power of the placebo effect, we really don't understand it. For example, it's been found that the effectiveness of a placebo is influenced by the form in which it's delivered: pill, capsule, liquid or, generally best of all, injection (there are cases of military surgeons in war zones who, bereft of anesthetics, have injected saline water to great anesthetic effect). With pills and capsules, such seemingly ephemeral matters as color and shape are important. Taking two placebo pills a day works better than taking just one. And as a general rule expensive placebos work better than cheap ones.

One of the most startling results to emerge in recent years is that Valium® (diazepam), for many years the mainstay of outpatient anxiety treatment—the "Mother's Little Helper" of the Stones song—works only if the patient *knows* it's being administered;[7] its beneficial results in relieving anxiety would thus appear entirely the product of placebo, even though untold millions will attest that Valium worked for *them*.

There was some media excitement over the 2010 paper "Placebos without Deception: A Randomized Controlled Trial in Irritable Bowel Syndrome,"[8] for which a team gave thirty-seven of eighty sufferers from IBS a placebo while the other forty-three sufferers received no treatment. What made this study unusual was that the patients receiving the placebo were *told* this was what they were being given; the pills were actually labeled "placebo." According to the team's public pronouncements (simplifying the more cautious claim made in their paper), nearly 60 percent of the patients receiving the placebo felt better, while just 35 percent of the others reported an improvement.

On face, this seems pretty convincing evidence that the placebo effect works even with patients who know what's going on. There were, however, some problems with the study.[9] First, the researchers recruited subjects using advertisements that talked of "a novel mind–body management study of IBS"—so the subjects were likely to be predisposed to a favorable result. Second, the efficacy of the treatment was judged merely by asking the patients if they felt better. Third, by its very nature, the study couldn't be blinded—far less double-blinded. And then there was the matter of how the placebo group was introduced to the concept of placebos:

> [P]atients were told that "placebo pills, something like sugar pills, have been shown in rigorous clinical testing to produce significant mind–body self-healing processes."

The basis for the placebo effect in ordinary circumstances is that patients are being given pills they're told will help them. This is exactly what the researchers did here. What they studied, then, was really little more than a standard case of the placebo effect at work.

A 1985 estimate[10] suggested that 35 to 45 percent of medical prescriptions offer placebos, whether the prescribing doctor knows this or not. Some are for drugs that are valid, but not for the patient's particular condition; others are for harmless compounds. Many GPs (general practitioners) openly admit that a large part of their practice is "placebo medicine" . . . although it's probably a good idea not to ask your own if this is the case. In addition, we all prescribe medicines for ourselves for minor ailments—aspirin from the bathroom cabinet, perhaps—that may often have only a placebo function. In so doing, we perhaps relieve just slightly the pressure on public healthcare systems whose restricted resources might be better spent elsewhere. And it's in this very limited sense, too, that CAM can be useful: Thanks to the placebo effect, a herbal infusion is likely to do as much for your cold as a prescription drug, and you'll be saving your GP some time. Although, of course, if you're wrong about your ailment being trivial . . .

Saying that CAM often works just as well as scientific medicine because of the placebo effect is one thing, portraying scientific medicine as a dangerous killer is another, yet this is exactly what Gary Null, Carolyn Dean, Martin Feldman, Debora Rasio, and Dorothy Smith did in a report called "Death by Medicine"[11] that was published by the online magazine *Life Extension* in March 2004 and later expanded as the book *Death by Medicine* (2010) by Null and Feldman. They produce some impressive-looking figures for the casualties caused by medical intervention, from diseases caught by hospital patients to bedsores. *Bedsores?* Yes, sometimes people in hospitals and nursing homes suffer from bedsores. Of course, bedridden patients tended at home by untrained family members are even more likely to develop bedsores. And merely pointing out that drugs can have side effects while ignoring all the *beneficial* effects they can have is like wringing your hands about how many people drown annually without mentioning that water is in many other respects jolly useful stuff.

Although the problem of iatrogenic infections is not a trivial one, it tends to be blown up out of all proportion by sensationalist media coverage. In *Bad Science* (2009) Ben Goldacre describes (pp. 279–89) the 2005 scare fomented in the UK by the tabloid media over MRSA (multidrug-resistant *Staphylo-*

coccus aureus), a particularly nasty bug that was said to be rife in the country's hospitals. Investigative reporters, in search of scoops, would infiltrate hospitals and take swabs of likely looking surfaces. Traces of MRSA were found aplenty . . . but only when the swabs were sent for analysis to a particular lab, Chemsol Consulting, which proved little more than a one-man operation run by a good-intentioned amateur out of his garden shed. Swabs sent elsewhere almost without exception came back with a clean bill of health. The response of the tabloid journalists was not to query Chemsol's competence but to stop sending their swabs to the other labs!

Still rumbling is the panic over fluoridation of water supplies, a measure proven to reduce caries in both adults and children. Part of the concern is over the fact that fluorine is a highly reactive element; do we really want to be drinking its compounds, the fluorides? Since fluorides occur naturally in many water supplies, this would seem a bit of a nonconcern; so far as fluorine's reactivity goes, sodium and chlorine are likewise highly reactive elements, yet we happily season our foods with sodium chloride. There can be, especially if the fluoridation in a local water supply has been overenthusiastic, a side effect called fluorosis, whereby the teeth become mottled, but this isn't especially common and seems a small price to pay. Other theorists claim fluoride is *really* a mind-control drug designed to make us all more vulnerable to the brainwashing efforts of those intent on imposing a One World Government . . . This was approximately the attitude of the John Birch Society, one of whose founder members was none other than the father of David and Charles Koch, who today are in the forefront of efforts to deny climate change (see pages 274ff.).

In April 1999 the Centers for Disease Control's *Morbidity and Mortality Weekly Report* listed the fluoridation of drinking water as one of its "Ten Great Public Health Achievements—United States, 1900–1999"[12] alongside such other items as vaccination and recognition of smoking as a health hazard.

The original snake oil, *shéyóu*, was probably fairly efficacious. It was oil from the fat sac of a particular type of sea snake, extracted by Chinese physicians for use in the treatment of arthritis. Sea snake oil, like fish oil, is rich in omega-3 fats, and ingesting these has a beneficial anti-inflammatory effect.

However, this is not the snake oil whose use has given the language a synonym for quack medicine. The entrepreneur concerned, one Clark Stanley—the so-called Rattlesnake King—presented his Stanley's Snake Oil at the 1893 Chicago World Fair, demonstrating its means of preparation: Take your rattler, slit its belly open, throw it into boiling water, cream off the fat that rises to the surface, and, bingo, you have a liniment suitable for treating a range of joint ailments, toothache, sore throat, frostbite . . . The trouble is that this isn't how Stanley's Snake Oil was prepared; he was just being a showman. When the feds finally got around to testing the stuff in 1917 they found it contained no snake oil; it was a mixture of ingredients like turpentine, camphor, red pepper, mineral oil, and a small amount of animal fat (probably lard). Curiously, this might have had some useful effect as a liniment—unlike actual rattlesnake oil. As rattlesnake oil is deficient in omega-3s, there's no point in swallowing it, either.

Of course, nowadays none of us are fool enough to fall for things like snake oil; we opt for CAM therapies instead. Even if we know enough to steer clear of those, many of us, distrusting the activities of Big Pharma, will gobble vitamin pills and supplements. Rather than listen just to our GP, we'll quite happily accept the advice of nutritionists who recommend—and often then proceed to sell us (not cheaply)—their own lines of vitamins and supplements. What we don't realize is that, whether or not they're really the baddies beloved of screen and printed thrillers, the folks in Big Pharma aren't necessarily mugs. Chances are the company that makes those very vitamins and supplements you're buying is owned by Big Pharma.

Perhaps the most dangerous US law ever passed has been the Dietary Supplement Health and Education Act of 1994. In consequence of this act, the FDA has no control over the contents of that brightly colored bottle of tablets or capsules you find in your supermarket. It is solely the responsibility of the manufacturer as to whether their pills contain what the label says, in the quantities indicated on the label, and prepared with any acceptable level of hygiene. Quite literally, vitamin tablets could contain cyanide, and could be *known* to contain cyanide, and yet the FDA would be powerless to stop the marketing exercise until people started dying.

There is one area of very slight control. When the labels make claims about the capabilities and efficacy of their products—"Effective against All Known Forms of Cancer," for example—they must include also a caveat, somewhere in the blur of small print, to the effect that "[t]hese statements have not been evaluated by the Food and Drug Administration. This product is not intended to diagnose, treat, cure, or prevent any disease." But who reads the small print? Besides, the manufacturers try to make sure you don't. To get an accurate version of the wording, I grabbed a bottle produced by those friendly folk at Sundown. I had to use a magnifying glass just to *find* the caveat.

The net result is that, while prescription drugs are (rightly) highly regulated and can be pulled off the market as soon as it's discovered they're toxic, and "traditional" medicines and supplements can contain, to pick a single example, heavy metals known to accumulate in the body and eventually cripple and kill (a 2004 report published in the *Journal of the American Medical Association*[13] discovered that about 20 percent of the Ayurvedic herbal medicines sold in Boston-area South Asian grocery stores contain lead, mercury, and/or arsenic at potentially harmful levels), large sections of the US public believe the unregulated stuff safe and the prescription drugs dangerous. A further unrecognized consequence is the death toll of consumers bamboozled into believing they're effectively treating their ailments through supplements and therefore refusing medical treatments that might have saved their lives.

The Dietary Supplement Health and Education Act, which was heavily lobbied by the supplements industry, was presented to the US Senate by Orrin Hatch (R-UT), Harry Reid (D-NV), and Frank Murkowski (R-AK). No one can guess how many lives have been lost as a result of it. As Dan Agin remarks in *Junk Science* (2006, p. 120), "We have struggled through thousands of years of medical history, acquiring scraps of medical knowledge piece by precious piece, and to have this struggle and its fruits tossed aside for the sake of a so-called free market economy is an intellectual travesty that ultimately causes hardship, pain, and death."

> *So the "double helix" of information in the heart of all reproductive cells is made up from 64 hexagrams, as in the I-Ching. Could this really be just coincidence?*
>
> —Colin Wilson, *From Atlantis to the Sphinx*, 1996

In *Not in Kansas Anymore* (2005) Christine Wicker tells of a magician called Daniel who first became seriously interested in magic during his schooldays because of a local woman who could heal cuts: "It was a very eerie thing to see, more so to actually feel if she was working on you. If you had a deep cut, she could hold her hands over it and chant a sort of prayer or bunch of syllables and the bleeding would stop, and it would sort of pull itself back together over the course of ten minutes or so" (p. 196)—all of which sounds rather impressive until you realize that minor cuts usually stop bleeding fairly soon if merely left to their own devices. But clearly Daniel *preferred*, consciously or unconsciously, the magical explanation to the mundane one. The woo-ness of the "sort of prayer or bunch of syllables" was actually a factor in convincing him there was something genuine at work. Over and over again when reading about medical woo we see this; the stark implausibility is one of the attractions. "You'll barely believe this, but old Mr. Benson could hardly walk because of his sciatica until he ate a pack of virgin tarot cards under the full moon, and he's not felt a twinge since."

In his book *Superstition* (2008) Robert Park discusses "energy-healer" Adam McLeod, who works under the *nom d'huil de serpent* Adam Dreamhealer and has the honor of a testimonial from paranormal investigator and ex-astronaut Edgar Mitchell, who reckons Dreamhealer remotely cured him of his kidney cancer. Dreamhealer uses a technique he calls quantum holography to visualize people's cancers and make them go away. If this doesn't work the first time, he's willing to have another try—for a repeat fee, of course. Like a surprising number of other woo therapists, Dreamhealer doesn't need a physical consultation for diagnosis, nor even for treatment: he can tweak your quantum hologram just fine while looking at a photo.

He apparently discovered his psychic powers as a teenager, when he experienced various involuntary telekinetic events—a better excuse than most teens can produce to explain how the car got scratched. After he'd cured his mother of trigeminal neuralgia he had a dream in which a huge black bird told him to go to Nootka—which proved to be an island in British Columbia. Off went the family to Nootka, and Dreamhealer met the huge bird in real life. It responded by downloading all the information in the universe into his brain. Oh, my.

Park's involvement began when ABC *Primetime* wanted him to consult on a program about Dreamhealer. Had the makers ensured, he asked, through medical records or biopsies, that Dreamhealer's "patients" had had cancer to begin with? The producers were startled; this hadn't occurred to them. Because they had no time to carry out the review Park suggested, they ran the documentary without, although they did craft it as an exposé—as well they might, since Dreamhealer failed to deliver the goods in any of the test cases they offered him.

As Park points out, Dreamhealer must have cried all the way to the bank. If only a tiny percentage of *Primetime*'s audience dissented from the documentary's debunking conclusions, this still represented many thousands of potential clients who would otherwise never have heard of him.

Of all the scientific research that's been carried out into woo medicine, one paper stands out. Called "A Close Look at Therapeutic Touch" and published in 1998 in the *Journal of the American Medical Association*, it detailed how its author, Emily Rosa, had devised an experiment to test the ability of healers to detect the energy field surrounding her hand. She found that ability lacking in the twenty-one she tested. The paper is still regarded as science's definitive verdict on therapeutic touch. The researcher, Rosa, devised and executed the experiment at its heart two years earlier, in 1996 . . . for her school's science fair, when she was nine years old.

4

PUFFING THE PRODUCT

It was in the campaigns by the tobacco industry to confuse the public about the links between smoking and life-threatening diseases that a pattern emerged which has since been followed in numerous other industry-funded campaigns to deny scientific work that might threaten short-term profits. The first step, which the tobacco industry deployed following the release in 1964 of a devastating report on smoking and lung cancer, involved denying entirely the possibility that smoking could be linked to *any* illness. However, this strategy was never very plausible—long before the 1960s people were aware of, for example, smoker's cough. Next the tobacco industry tried what might be seen as a more moderate approach—to muddle the science, or at least to muddle the public by claiming the science was controversial and by giving the impression there was a red-hot debate among scientists themselves over the health effects of smoking.

This was where all the covertly funded think tanks, astroturf[1] organizations, and activist groups of impressive but specifically unqualified scientists came in. While obviously such a tactic can't resist the inevitable forever, it can successfully delay action for not just years but decades—in this instance, plenty of time for the companies to restructure their business models so as to sell their products in developing countries, where, for example, the restrictions on marketing cigarettes to children are far laxer, if they exist at all. In the case of smoking, the period of delay ran from the 1950s until nearly the end of the twentieth century, and there are *still* people who deny smoking is a health hazard or who believe a scientific argument continues. Take this, accessed in December 2010 on a website called *Smoking Aloud*:

> *Smoking Conspiracy: Second Hand Smoke.* In yet another bogus report, this time by Surgeon General Richard Carmona, he claimes [*sic*] there is no safe level of secondhand smoke and calls for a workplace ban on smoking.
>
> "The scientific evidence is now indisputable: second-hand smoke is not a mere annoyance," Carmona said. "It is a serious health hazard that can lead to disease and premature death in children and nonsmoking adults."
>
> "The scientific evidence is now indisputable," he says.
>
> The truth is every study used by the anti-smoking group on second hand smoke has been proven to be flawed and the data manipulated.[2]

The writer then goes on to cite Steven J. Milloy (see page 272), as if Milloy were a health expert, and to make the claim: "In studies conducted by the World Health Organization, statistically significant evidence was found that childhood exposure to cigarette smoke cuts the risk of lung cancer by 22 per cent." The links attached to this claim lead one to articles from other denialist sources rather than to, say, a World Health Organization study.

In the case of smoking there were really two cycles of this pattern. At the end of the first cycle, there was a compromise, in that the tobacco industry accepted, finally, that smoking was harmful and placed prominent warnings on packs to this effect, while society accepted that, if grown-up people wanted to kill themselves, that was their own decision. But the subsequent cycle was more problematic for the industry, in that it concerned secondhand smoking. The people being affected were not making a free decision to risk their health. This time the compromise has been, in essence, that people can smoke in the open air or in the privacy of their own homes/cars, but not in enclosed public places where their smoke could harm others.

Even that may soon change, however. A 2010 report done by Karen Wilson et al. found that children living in apartment blocks showed elevated blood levels of cotinine, a nicotine product that's a highly sensitive indicator of exposure to tobacco smoke.[3] The supposition is that the smoke makes its way through ventilation systems, under doors, even through permeable walls, until, in coauthor Jonathan P. Winickoff's words, "Smoke contaminates the whole building."[4]

So how did the tobacco industry's great master plan to malign science come about?

The link between smoking and lung cancer was nothing new: German scientists had demonstrated it in the 1930s. The Nazis, who were actually quite sound on issues like public health and preventive medicine even though disastrous on many other scientific issues, campaigned hard against smoking. It's ironic that, after World War II, it took some while for the German science on the subject to be "rediscovered," purely because it was "Nazi science"—ironic because the Nazis themselves had blindly refused to acknowledge Relativity and much else on the grounds that it was "Jew science." In the US it wasn't until the early 1950s that it became clear the writing was on the wall for the idea of smoking as a harmless, pleasant habit. Accordingly, in December 1953 the presidents of American Tobacco, Benson & Hedges, Philip Morris, and US Tobacco met with John Hill, CEO of the public relations firm Hill & Knowlton, to hammer out a strategy to defend their companies from the findings of science.

From this meeting sprang the Tobacco Industry Research Committee (TIRC), which would fund research that might cast doubt on the links that were becoming established between smoking and lung cancer. TIRC also arranged meetings with media principals, made its members and their claims readily available to journalists, sent letters to editors whenever a story appeared about tobacco and health that "lacked balance," bombarded physicians with misleading literature, and so on. This is roughly the way the industry-funded campaign of AGW denial acts today, all based on the realization that "scientists" who aren't actually doing any science have a lot more time and energy available to them for campaigning on issues than do those engaged in active research.

Those involved created doubt in large part by "raising questions"—even if the answers were already known. One such was this: The number of female smokers had increased very rapidly since World War II. Why wasn't there a corresponding increase in the number of female cancer victims? The answer's pretty obvious—it can take decades for a case of cancer to become evident—but far too few in the media bothered to think this through. It was a remarkable achievement, as smoke-and-mirrors goes, because the press must have been aware that lung cancer, once rare, had become almost commonplace, in accord with the twentieth-century craze for cigarette smoking.

The method worked well all through the 1950s and into the 1960s, to the extent that, when Surgeon General Luther Terry set up the Advisory Committee on Smoking and Health in 1962, he invited industry representatives to participate, for the sake of "balance." On its release on January 11, 1964, *Smoking and Health: Report of the Advisory Committee to the Surgeon General of the United States* caused a firestorm. The evidence linking smoking to cancer was indisputable. What has only much more recently come to light is that the conclusions of the report were no surprise to the tobacco industry itself, whose scientists had earlier—sometimes years earlier—made similar inferences. The industry scientists had even gone beyond the report, concluding, which the report signally did not, that nicotine was an addictive drug, another fact the industry outwardly denied. Even so, and even despite the surgeon general's report, the industry continued to maintain in public that the science was doubtful.

One move was to rename TIRC the Council for Tobacco Research (CTR), thus removing from overt display the fact that it was an industry front group. A certain amount of research went into trying to create a safer cigarette—this even as the CTR was insisting cigarettes *were* safe. And in 1964, when the Federal Trade Commission decided that, from January and June 1965 respectively, cigarette packs and cigarette advertisements must display health warnings, the industry was still able to parade before Congress a due number of scientific "skeptics."

A great many of the tobacco industry's internal memos and other documentation from that era and later have since become public, and can be found online in such archives as the Legacy Tobacco Documents Library, maintained by the University of California, San Francisco.[5] Here, for example, are some of the minutes taken of a brainstorming session at Brown & Williamson in May 1977 about the launch of a new low-tar brand (one of the proposed names for which was, confusingly, Trojan):

- How can [we] blend in a satisfier like cocaine?
- Wish [we] had a name that was directionally totally masculine
- How to market an addictive product in an ethical manner
- Wish we had another name for nicotine. Get away from negative association[6]

There was also some speculation about "free nicotine," a variant that, having been treated with ammonia, affected the brain and central nervous system more rapidly than the ordinary variety—gave a quicker rush, in other words.

- Wish we could address ourselves to free nicotine as opposed to nicotine as we know it

This desire to make the product *even more addictive* does not, of course, reflect a very responsible attitude.

The most frequently cited smoking gun occurs in a document titled "Smoking and Health Proposal," which was produced at Brown & Williamson in 1969: "Doubt is our product since it is the best means of competing with the 'body of fact' that exists in the mind of the general public."[7]

The stratagem the tobacco industry used over and over was to pretend the natural uncertainty that must forever be a part of science means the science is shaky. A similar rathole—and an escape route the courts repeatedly permitted—was that, since it can never be demonstrated conclusively that *any specific instance* of lung cancer (or emphysema, or heart failure, or whatever) was caused by cigarette smoke, the tobacco industry couldn't be held accountable for a particular individual's plight. Naomi Oreskes and Erik M. Conway, in their *Merchants of Doubt* (2010), cite this maneuver being used as late as 1997.

Back in the 1980s, Horace R. Kornegay, chairman of the Tobacco Institute, had summarized the situation vis à.vis tobacco and science in a June 25, 1981, memo to his executive committee:

> *Science* remains the fundamental problem confronting the industry. Bad press, unwarranted regulation, and poor opinion are its *symptoms*. Symptoms cannot effectively be treated without attacking their underlying cause.

He went on to advocate renewed efforts to commission scientific research that would either counter the negative reports from independent research or at the very least muddy the water:

> Our scientific positions must be less defensive and more constructive. Our research support objectives must be reordered and our research support results must be publicized. Industry communications must contain less of "But *that* study is wrong" and more of "Look what *this* study shows."[8]

Of course, if scientific research were to be commissioned, ideally those doing the commissioning should be scientists themselves. Physicist Frederick Seitz was recruited by R. J. Reynolds in 1978 or 1979; we shall meet him again in the context of global warming (see pages 284ff.). Over the next decade or so Seitz doled out research grants. There's no doubt that some of the research done at Reynolds's expense was of genuine scientific value, such as Stanley B. Prusiner's discovery of prions, which brought him the 1997 Nobel Prize in Physiology or Medicine. Much, though, was simply chaff. But chaff was all the tobacco industry wanted—enough to cloud the public perception of the state of the science. Moreover, the industry was establishing a reservoir of scientists who might well be relied upon to give favorable testimony in future court cases.

Industry-friendly scientists were only one part—albeit a very large part—of the story. As usual, there were plenty of politicians open to the approaches of tobacco lobbyists and their generous campaign contributions. Perhaps an even more insidious move was the attempt to establish a sort of tobacco-friendly fifth column within the journalism industry, through, for example, the financial support by Philip Morris for a journalism school, the National Journalism Center (NJC). At one point the company played with the notion of buying a major news outlet. Overall: "The company [Philip Morris] planned to 'design innovative strategies to communicate [its] position on ETS [Environmental Tobacco Smoke, the term the tobacco industry preferred] through education programs targeting policy makers and the media' via the NJC. Finally, journalists associated with think tanks that were financially supported by Philip Morris wrote numerous articles critical of the EPA."[9]

Looking at the plethora of counterfactual articles in the world's press today on the topic of AGW, it's hard to escape the conclusion that a similar exercise is currently being mounted by the fossil-fuels industry—that, in essence, some journalists are being covertly funded to express the opinions they do.

Brown & Williamson had led the way in 1969: "Doubt is our product." The PR company APCO similarly advised Philip Morris in 1992 to start getting into the business of merchandizing doubt. While the US public certainly wasn't going to believe a tobacco executive in any argument with scientists over the health hazards of smoking, a network of amenable scientists and astroturf organizations could spread the notion that the conclusions being presented by the scientists were not carved in stone.

Looked at through any rational pair of eyes, this strategy might seem doomed—who'd get aboard a plane that the vast majority of aeronautical engineers declared wasn't airworthy? But APCO had the psychology right.

Cigarettes were so popular because people enjoyed smoking and it was addictive, so many and probably most smokers were, even if only unconsciously, desperately seeking an excuse to avoid having to quit. The first organization APCO created, The Advancement of Sound Science Coalition (TASSC), spread misinformation about other areas of science as well: nuclear waste disposal, biotechnology, and, in the long term most damagingly, global warming.

The APCO advice came in the aftermath of the devastating 525-page EPA report *Respiratory Health Effects of Passive Smoking: Lung Cancer and Other Disorders*,[10] whose appearance in December 1992 signaled a whole new round of tobacco-industry resistance to scientific reality. Earlier, the Department of Health and Human Services had produced an equally alarming report, *The Health Consequences of Involuntary Smoking* (1986), but the industry had weathered this. The new report was immediately recognized as a far more serious threat. Rather than address the science, the industry and its apologists chose to attack the integrity of the EPA.

Respiratory Health Effects drew upon the most expert scientists in various federal agencies. For peer review, the EPA's Science Advisory Board created a panel of nine scientists with relevant expertise, plus nine highly qualified consultants, plus advisers/assistants drawn from the Science Advisory Board itself; by way of comparison, the average peer-reviewed scientific paper is vetted by just three peers. This panel concluded: "The Committee concurs with the judgment of EPA that environmental tobacco smoke should be classified as a Class A carcinogen."[11]

The purportedly scientific counterblast, funded circuitously by Philip Morris, was *The EPA and the Science of Environmental Tobacco Smoke* (1994), published by the Tobacco Institute–funded Alexis de Tocqueville Institution and written by S. Fred Singer and Kent Jeffreys. Singer, as a physicist, had no relevant expertise to judge epidemiological matters. Jeffreys, as a lawyer, had no relevant expertise to judge scientific matters. Their report was published sans peer review.

Compare and contrast the attributes of the two reports.

The document was launched at a press conference mounted by Rep. Peter Geren (D-TX) and Rep. John Mica (R-FL). Authors Singer and Jeffreys were there too. The latter produced this gem: "I can't prove that ETS is not a risk of lung cancer, but EPA can't prove that it is." The EPA likewise couldn't prove with 100 percent certainty that getting run over by a steamroller was a bad plan, but . . .

Returning to the peer review of the EPA's *Respiratory Health Effects*: When the discussions were later analyzed, it was revealed that the reviewers' criticisms were not that the scientists involved had overstepped but that they had been too tentative. There seemed a good correlation between passive smoking and cardiovascular disease, between passive smoking and the incidence of respiratory diseases among teens in smoking households, and so on. There was also a near-certain relationship between passive smoking and sudden infant death syndrome; this was omitted from the executive summary because the researchers couldn't be certain if prenatal or postnatal smoking was to blame.[12] Such hesitancy does not seem quite consonant with the habitual portrayal by denialists of scientists as arrogant purveyors of barely supported hypotheses.

The EPA's caution was all the more remarkable in that the science concerning the dangers of passive smoking was nothing new. The industry's own scientists had realized as far back as the 1970s that there was a particular problem with the smoke people inhaled from cigarettes left smoldering in ashtrays (when cigarettes smolder the temperatures involved are lower than in active smoking, and a different raft of toxins is generated). The same comments apply to the smoke a smoker is likely to inhale from those parts of the cigarette that are not actually at the glowing tip. Long before the EPA arrived on the passive-smoking scene, the tobacco industry was nervously trying either to reduce the smoker's inhalation of such smoke (by redesigning filters, for example) or at least to make it less obvious what was going on.

A 2006 Gallup poll found that 12 percent of Americans still think passive smoking is either "not too harmful" or "not harmful at all."[13]

Finally, after half a century of obfuscation, wriggling, and delay, the tobacco industry was brought to book. In 1994 the attorneys general of Mississippi, Minnesota, Florida, and Texas sued the tobacco industry claiming reimbursement for healthcare costs those states had incurred because of the industry's products. While this litigation was proceeding, forty-two other states made similar claims. Eventually, in 1998, a Master Settlement Agreement was agreed whereby the American Tobacco Company, Brown & Williamson, Philip Morris, and R. J. Reynolds would make annual payments to the forty-six states involved and accept restrictions on the future advertising and marketing of tobacco products.

And in a separate action, begun in September 1999, the federal government filed a lawsuit against the major cigarette manufacturers, the Tobacco Institute, and TIRC/CTR under the Racketeer Influenced and Corrupt Organizations (RICO) Act. The Amended Final Opinion of the US District Court for the District of Columbia,[14] handed down in 2006, is a colossal document, running to 1,652 pages. In a way it provides one of the best available histories of the industry's deceptions—which amounted, the court under Judge Gladys Kessler concluded, to a criminal conspiracy.

The industry must have celebrated the fact that the court felt unable to exact the massive penalties the feds had been demanding, but it's likely their rejoicing began too soon: Judge Kessler's ruling opened the floodgates to a succession of successful *civil* suits against the tobacco companies. A Massachusetts jury in January 2011 awarded a total of over $150 million in a suit brought against Lorillard Tobacco, maker of Newports, in consequence of the death of Marie Evans, who died of lung cancer at age fifty-four after smoking since childhood. The reason for the enormous damages award to Evans's family—of the total, $81 million were punitive—is that Evans became addicted to cigarettes as a consequence of a Lorillard campaign in the 1950s to go around black schools giving away free cigarettes to the kids in hopes of addicting them.

Such legal cases have had the effect of curbing at least some of the worst of the tobacco industry's abuses; but there are still plenty of nations whose unfortunate inhabitants are ripe and ready for the industry's parasitic bite. And the industry is using legal muscle and bullying economic clout where need be. In 2010 Philip Morris International sued the government of Uruguay on the grounds that its tobacco regulations are overly restrictive; the company's annual turnover is of the order of $66 billion while Uruguay's GDP is only about half that. Philip Morris International has sued Brazil over the images which that country insists on incorporating into pack warnings, including one that shows a fetus with the caution that smoking can cause spontaneous abortion. Along with British American Tobacco and Imperial Tobacco, Philip Morris International attempted to sway the 2010 Australian election by financing a media campaign, supposedly mounted by the owners of mom and pop stores, against the incumbent Labour government's plans to inhibit pack design; and in June 2011, it was reported that the company intended to sue the Austrailian government over this issue. And so it goes on. Cigarette sales may be plummeting in many nations as smokers either quit or die, but worldwide they're rising at a steady rate. The good news is that over 170 nations, in conjunction with the World Health Organization, are working together to devise a global antismoking treaty.[15]

And about time, too. The WHO has estimated that smoking is responsible, worldwide, for about five million premature deaths a year. Approximately 50 percent of smokers die because of their habit. A 2010 WHO study, funded by the Swedish National Board of Health and Bloomberg Philanthropies, suggested that an additional six hundred thousand premature deaths worldwide are in consequence of secondhand smoke; a distressing number of these casualties are children.[16]

5

PAYING WITH THEIR LIVES

Much has been written about Bernarr Macfadden, the twentieth-century physical-culture guru who built up a media empire around his theories; his company published such diverse titles as *True Detective*, *Photoplay*, and the *New York Graphic*. He rejected much of medical science, preferring such regimes as rigorous fasting to drugs: "Drugs never cured anything, unless you call death a cure" was his axiom. Another of his adages was: "Weakness is a crime—don't be a criminal"; hence the title of Robert Ernst's biography *Weakness Is a Crime: The Life of Bernarr Macfadden* (1991).

In a tell-all book, *Dumbbells and Carrot Strips* (1953; with Emile Gauvreau) MacFadden's ex-wife, Mary Williamson, revealed the grim regime inflicted upon their children, as all their youthful illnesses—like measles and whooping cough—were treated by week-long fasts and other draconian methods. One son, Byron, died in infancy when Macfadden tried to treat his convulsions by dunking him in hot water. Such an event might make most of us pause to reconsider our attitudes toward more orthodox medical treatments. Not Macfadden. It was more important to maintain his theory than to amend it in the light of evidence.

The Byron Macfadden tragedy is, alas, far from unique. Children all over the globe pay regularly with their lives for their parents' denial of medical science. That this should be so, not just in the developing world but in such scientifically advanced nations as those of North America and Europe, is where the true bitterness of these tragedies lies.

When Kent Schaible fell ill in January 2009 with sore throat, congested lungs, diarrhea, and other symptoms, his parents Herbert and Catherine, of the First Century Gospel Church, Philadelphia, obeyed their church's precepts and, rather than take him to a doctor, prayed over him for the ten days it took him to die of what proved to be bacterial pneumonia. Philadelphia's assistant medical examiner, Edwin Lieberman, ruled the death a homicide, but when the parents were arrested in April 2009 they were charged instead with involuntary manslaughter; in February 2011 they were sentenced to ten years' probation with the requirement they seek routine and emergency medical care for their seven remaining children.

The First Century Gospel Church is run by the Reverend Nelson Ambrose Clark, who explained, "Our teaching is to trust Almighty God for everything in life: for health, for healing, for protection, for provisions, for avenging of wrongs."[1] So far as he was concerned, the Schaibles were being persecuted for their faith. He points to *Death by Medicine* (2010) by Gary Null and Martin Feldman (see page 83) as justification for his belief that people are better off with faith healing or just plain prayer—this despite the fact that in 1991 nearly five hundred children at his church and the nearby Faith Tabernacle Congregation, whose views on medicine are similar, fell sick with measles, six of them dying.

The organization Children's Healthcare is a Legal Duty (CHILD) was founded in 1983 to attempt to protect children from harm brought about by their elders' religious and cult beliefs, notably from medical neglect imposed by pious parents. It's in the vanguard of campaigns to abolish the so-called religion defense—whereby parents who harm their children can claim in court that religious imperatives overruled the dictates of common humanity. Alarmingly, nineteen states still permit this defense in cases of felony crimes against children, which would include instances where parents have allowed their children to die of medically treatable ailments. These nineteen are Arkansas, Idaho, Indiana, Iowa, Kansas, Louisiana, Minnesota, New Jersey, Ohio, Oklahoma, Oregon, Rhode Island, Tennessee, Texas, Utah, Virginia, Washington, West Virginia, and Wisconsin. Almost all states legitimize faith healing as healthcare, which offers another legal excuse to negligent parents.

CHILD's website contains a number of horrific case studies. In several, the religion concerned is Christian Science, like that of eleven-year-old Ian Lundman, who died of diabetes in 1989 in Minneapolis. As he became visibly more and more gravely ill, his mother and stepfather, rather than take him to a doctor, hired a "Christian Science practitioner" to pray for the child and an unlicensed "Christian Science nurse," Quinna Lamb Giebelhaus, to watch over him as he died; the latter's "concept of care was to give him drops of water through a straw and to tie a sandwich bag and washcloth around his scrotum. She did not call for medical help or ask his mother to obtain it." At the subsequent trial, Giebelhaus was asked "what training she had received specific to the care of sick children. Her only answer was that she had been taught to cut sandwiches in interesting shapes."[2]

The parents of two-year-old Harrison Johnson, who died in Tampa, Florida, in 1998, were members of The Fellowship, a religious group that refuses medical care in the belief that the practice of medicine is witchcraft. Young Harrison was attacked by a swarm of wasps, receiving 432 stings; he suffered for over six hours before, half an hour after he had become completely unresponsive and had probably died, his parents called the emergency services. His treatment during this excruciating time? Enthusiastic prayers by his parents and their church friends and the attentions of Fellowship "nurse" Carol Balizet, author of a number of Born Again–style religious novels and the home-birthing polemic *Born in Zion* (1992). According to the CHILD account of the Johnson case, "The judge . . . instructed the jury that the state must prove that the parents willfully or intentionally caused the harm to the child. The jury acquitted."

In 2008, in Weston, Wisconsin, the parents of eleven-year-old Madeline Neumann watched for a period of weeks as their daughter died from what proved to be diabetic ketoacidosis, all the while choosing to pray for her rather than call for medical help. The local sheriff commented: "They believed up to the time she stopped breathing she was going to get better. They just thought it was a spiritual attack. They believed if they prayed enough she would get through it."[3] The reason their daughter died, apparently, was that their "faith wasn't strong enough."

Another case is that of Dennis Lindberg, who died of leukemia in Seattle in 2007 because he and his aunt (his guardian), as Jehovah's Witnesses, refused to allow him blood transfusions. When Dennis's parents and the Child Protection Services went to court to try to save the boy's life, their case was rejected by Skagit County Superior Court Judge John Meyer on the grounds that "I don't believe Dennis' decision is the result of any coercion. He is mature and understands the consequences of his decision. This isn't something Dennis just came upon, and he believes with the transfusion he would be unclean and unworthy." Meyer also, according to CHILD's account, said, "He knows very well by stating the position he is, he's basically giving himself a death sentence." The individual whom Meyer was insisting had a legal right to commit suicide was, we remember, a fourteen-year-old.

In the case of autism, parents who resort to religious methods at least have the excuse to offer that nothing else works either. That doesn't make the tragedy any less. In his *Autism's False Prophets* (2008) Paul Offit cites the instance of the autistic eight-year-old Terrence Cottrell, who died in the course of a two-hour exorcism at the Faith Temple Church, Milwaukee, probably because the preacher in charge of the exorcism kept his knee firmly planted in the child's chest throughout the proceedings.

Although the strength of the religious motivation in most of these cases is evident even from the title of Shawn Francis Peters's heartwrenching book *When Prayer Fails: Faith Healing, Children, and the Law* (2008), other reasons can be offered by parents—as per Bernarr Macfadden. Some parents' belief in the efficacy of CAM comes close to an ideology; other parents are misled by the notion that CAM remedies must at the very least be harmless because, after all, they're "natural." The trouble with this sort of reasoning, as we've noted, is that even a harmless "remedy" may kill or severely harm someone if taken in place of an effective treatment. For "Adverse Events Associated with the Use of Complementary and Alternative Medicine in Children,"[4] Alissa Lim, Noel Cranswick, and Michael South of the Royal-Children's Hospital in Melbourne analyzed thirty-nine cases of CAM-associated "adverse events" reported to the Australian Paediatric Surveillance Unit during the period between January 2001 and December 2003. In seventeen of these cases, including all four in which the children died, it was reported that the damage was done not byz any harmful properties of the particular CAM "remedy" but by the failure to use conventional medicine.

Whatever the parents' motives—ideological or just misguided—it can't be denied that in these thirty-nine instances the children suffered because of a bad decision about which they had no say. As the authors conclude, the "most worrying feature" is "the significant proportion of life-threatening and fatal reports, particularly in families using CAM to the exclusion of conventional medicine. . . . The diversity of reports received demonstrates the difficulty of monitoring this area given the range of CAM therapies and different adverse outcomes."

In December 2010 the Roman Catholic Diocese of Phoenix stripped St. Joseph's Hospital and Medical Center, an establishment with an international reputation in the fields of neurology and neurosurgery, of its Catholic affiliation. The reason was that staff at the hospital had in 2009 performed an emergency abortion on a mother who was in imminent danger of death unless they did. Said Linda Hunt, the hospital's president:

> If we are presented with a situation in which a pregnancy threatens a woman's life, our first priority is to save both patients. If that is not possible, we will always save the life we can save, and that is what we did in this case. Morally, ethically, and legally, we simply cannot stand by and let someone die whose life we might be able to save.

Bishop Thomas Olmsted disagreed on ideological grounds: "In the decision to abort, the equal dignity of mother and her baby were not both upheld."[5] It's an odd notion of "equal dignity" that dictates the doctors should have allowed both to die, and one extraordinarily lacking in compassion.

At least they performed the necessary operation. In 2009 the Center for Reproductive Rights brought before the UN Committee on the Elimination of Discrimination against Women the case of a girl called "L.C." in Peru:

> L.C. was 13 years old when she was repeatedly raped by a 34-year-old man. . . . Her ordeal started in 2006, and by 2007 she learned that she was pregnant.
>
> Desperate, L.C. attempted to commit suicide by jumping off the roof of a building next door to her house. Neighbors discovered her and rushed her to the hospital. But even though doctors concluded her spine needed to be realigned immediately—and even though abortion in Peru is legal where the mother's health and life are at risk—they refused to operate on L.C. because she was pregnant.
>
> L.C. eventually suffered a miscarriage because of the severity of her injuries. Several weeks after the miscarriage, four months after she was told she needed surgery, L.C. underwent the spinal procedure.[6]

Because of the delay, the surgery was ineffective. L.C. is now confined to a wheelchair and has minimal use of her arms—her life destroyed by antiabortion ideology. Despite harrowing tales like these from nations with strict antiabortion laws, in the US midterm elections of 2010 no fewer than sixty-three house candidates, describing themselves as "pro-life," pledged a without-exceptions opposition to abortion.

6

THE ANTIVAXERS

We are now an organized and united group, thanks mainly to the power of the Internet. Our message has severely eroded confidence in the cornerstone of health care: THE CHILDHOOD VACCINE PROGRAM.

—Anne Dachel[1]

In late 2010 there was an outcry over a cinema "public awareness" advertisement created by the antivax[2] organization SafeMinds and the website Age of Autism. The ad claimed the Thimerosal in that year's flu shot was a hazard. In response to representations from the rational community, the AMC cinema chain canceled its plans to screen the item. Aghast, one of the principals of SafeMinds, Kim Stagliano, promised to retaliate by boycotting, with her family, AMC theaters all through the holiday season. Good. The last person you want to sit beside in a crowded movie theater is someone who hasn't been vaccinated.

Today's rejection of vaccination is not a new phenomenon. From the outset the practice was denounced by the Christian churches, which for centuries had maintained that diseases were caused by demons and/or were expressions of divine displeasure. If a disease had been visited upon us by an angry God, the reasoning went, surely any effort to counter or ameliorate it was an attempt to thwart God's will, and could have no effect but to make him even crosser. Of course, such theological deduction was faced with the unfortunate evidence that those who were inoculated tended to survive while the devout tended not to; and thus, gradually, vaccination was accepted as suiting God's purpose after all. Even so, suspicion lingers—and not just among evangelicals or the vaccination/autism crowd. The US comedian and self-proclaimed rationalist Bill Maher has frequently maintained on his TV shows that vaccination must "obviously" be dangerous because you're introducing a disease into a body that was hitherto disease-free. Hm. I carry around nitroglycerin pills to stick under my tongue in case of emergency. Is this a cunning plan by my physician to blow my head off?

The origin of the modern antivax movement, spanning much of the globe, is largely the responsibility of UK medic Andrew Wakefield, who in the late 1990s was a nonclinical researcher at the Royal Free Hospital's medical school, London. He originated the notion of a correlation between administration of the MMR (measles, mumps, rubella) vaccine and regressive autism—the form of autism that first manifests in infancy, unlike classical autism, whose symptoms are present from the start.

It was Wakefield's work on Crohn's disease that led him down the dark alley toward antivaxism. His research had established that the cause of Crohn's disease is reduced blood flow to the intestines because of blockage of the arteries leading there. So far, so good. The next question was: What blocks those blood vessels? The answer he came up with was: the measles virus, persisting in patients long after the disease has subsided. But then—in a 1995 paper called "Is Measles Vaccination a Risk Factor for Inflammatory Bowel Disease?"[3]—he and coauthors took the hypothesis a stage further, claiming Crohn's disease could be caused not just by the virus but also by the vaccine against it.

This was startling enough that researchers all over the world attempted to replicate his results—without success. Moreover, they found they couldn't replicate his results concerning the measles virus, either. In the face of this fallout, in 1998 Wakefield and his colleagues published another paper, "Measles Virus RNA Is Not Detected in Inflammatory Bowel Disease Using Hybrid Capture and Reverse Transcription Followed by the Polymerase Chain Reaction,"[4] which in essence admitted the earlier findings were wrong.

That paper appeared six months or so after a controversial press conference Wakefield held in London to announce his new hypothesis: that *autism* could be caused by the MMR vaccine. In this scenario, the dead virus in the vaccine, on reaching the intestine, creates physical damage there, which allows malignant proteins to escape from the intestine into the bloodstream and reach the brain. It was just a question of identifying the malignant proteins. In the meantime, there was good reason to take the precautionary measure of administering vaccines against the three diseases one by one, rather than in the MMR combo.

The practice of announcing scientific results via press conference does not have a distinguished history—think of the cold fusion fiasco, for example. In this instance, Wakefield could at least promise a supporting paper would be appearing in *Lancet*, which very soon it did.[5] The UK press, even the quality sheets and the BBC, went wild with scare stories about the dangers of the MMR vaccine. What failed to make the headlines was that a senior coauthor on the *Lancet* paper, Simon Murch, was among those cautioning against overreaction to Wakefield's claims: "This link is unproven and measles is a killing infection. If this precipitates a scare and immunization rates go down, as sure as night follows day, measles will return and children will die."[6]

Murch's fears could not have been more prescient. Others of Wakefield's coauthors were likewise quick to assert that their results were at best suggestive. Such caveats were still being glossed over in 2002 when Lorraine Fraser of the *Daily Telegraph* was named British Press Awards Health Writer of the Year. Her feat of journalistic prowess? Writing a string of articles promoting Wakefield and his claims that the MMR vaccine caused autism. (The *Telegraph* and the *Mail* were particularly strident in support of Wakefield, seemingly on ideological grounds—in the same way that these two papers have been the UK's most virulent in their AGW denialism.)[7]

Perhaps most irresponsible of all was the December 2003 TV movie screened by the BBC, *Hear the Silence*, which starred Hugh Bonneville as a heroic Wakefield, persevering in the teeth of the powerful bureaucratic Forces of Darkness because he had heard the parents of autistic children crying out in anguish . . . It's a caricature to be found duplicated endlessly in the comments sections of any web post on the subject of either autism or vaccination or both. Even after the *British Medical Journal* published an article in January 2011 exposing Wakefield's results as fraudulent (see below), the caricature continued. As an experiment, I went to the first such posting that attracted my attention[8] and checked the topmost comment on the stack. Sure enough:

> *Kathleen D.:* I find it appalling that doctors like Wakefield would lose their right to practice medicine because of their concern for the children and their parents. Think about it; he has little to gain and far more at stake here by divulging and writing what his investigations into the 5 studies conducted by different institutions and countries have determined. Unfortunately, the pharmas may "have an offer these institutions CAN'T refuse," and have backed down original claims of their own findings through extensive investigative studies.

The general sensationalism of the UK and Irish media led—understandably—to a fairly dramatic decline in the number of parents allowing their children to be vaccinated. Soon the levels of immunization in parts of both countries fell below that required for herd immunity. (Some individuals cannot sensibly be vaccinated because of allergies or the like; in others, the vaccine doesn't "take." If most of the "herd" are immunized, the fact that a small minority aren't doesn't matter much, because the disease isn't doing the rounds anyway.) The loss of herd immunity in turn led to the reappearance of outbreaks of measles for the first time in many years. As a result of a measles epidemic in Dublin during the winter of 1999–2000, 111 children were hospitalized, of whom 12 had to be treated in intensive care and 3 died.[9]

The antivax scare in the UK was compounded in late 2001 by the coy refusal of Prime Minister Tony Blair and his wife Cherie to say whether or not their newborn, Leo, had been vaccinated. Much later the Blairs announced that of course the babe had been vaccinated, but by then the damage had been done. Another politician to comment stupidly on the issue was Ken Livingstone, Mayor of London; in consequence, London's kids suffered worse than those in many other parts of the country from falling vaccination levels and a consequent rise in the incidence of measles.

The net result of the media frenzy and foolish utterances by public figures was that, by 2008, measles was declared once again endemic in England and Wales.

Not all UK journalists were as credulous as their peers. In particular, Brian Deer, a reporter at the *Sunday Times*, became increasingly suspicious of both Andrew Wakefield and the 1998 *Lancet* paper heralded by that press conference. Over the years, he chipped away at the story. His research revealed some alarming conflicts of interest, conflicts of interest about which Wakefield's coauthors and the editors of *Lancet* had been kept in the dark.

Most serious was that, some while before, Wakefield had been hired by an East Anglian lawyer, Richard Barr, to devise the scientific underpinning to help obtain government-financed Legal Aid for a class-action case against the MMR manufacturers. Ideal for the purpose would be a "bowel–brain syndrome," and indeed Wakefield worked up just such a scheme even before the production of the *Lancet* paper—i.e., prior to doing the research that supposedly indicated the syndrome's existence! For this work, Wakefield would over time be paid some £435,643 (then about $800,000), plus expenses.

Several of the twelve children used as subjects for the *Lancet* paper's research were referred to Wakefield by Barr; they were, in fact, children of Barr's clients. Others were referred to Wakefield by the antivax organization Justice Awareness and Basic Support (JABS). Thus the study's subjects were not randomly drawn from cases Wakefield had encountered while performing his duties at the Royal Free, as was the general assumption, but were in effect self-selected. There was another very serious methodological flaw. In order to demonstrate a direct causal link between MMR vaccination and regressive autism, it was essential to show that symptoms began to develop fairly shortly after the injection; if the gap were longer, the child could have been exposed to all sorts of other influences in the interim, meaning there'd be no unequivocal connection to the MMR shot. To readers of the paper it might have seemed that a short-term cause-and-effect link had indeed been established. What was not evident was that the estimations of the time between the child being vaccinated and the onset of the symptoms were derived from interviews conducted with the parents years later. The researchers had no sensible means of evaluating the accuracy of the parents' recollections. Add in that it was in the financial interests of those parents who were Barr's clients to remember a pretty close correlation,[10] and the information becomes yet more suspect. A useful study in this context is "Recall Bias, MMR, and Autism" by N. Andrews et al.[11]

Deer unearthed other problems. Wakefield had stated that his investigations of the children—involving highly invasive methods such as colonoscopies, barium meals, spinal taps, and biopsies—had been approved in advance by the relevant ethics boards. According to the much-later (May 2010) judgment of a General Medical Council (GMC) investigation that led to Wakefield and one of his coauthors, John Walker-Smith, being struck off the medical register, this was not so in the case of ten of the children. In seven instances, moreover, the "invasive practices" were found to be contrary to the child's clinical interests. For example, the colonoscopies administered to eight of the children were not clinically indicated. One child suffered perforation of the colon in a colonoscopy gone wrong.

In 2004, largely as a consequence of Deer's work, ten of Wakefield's twelve coauthors on the 1998 paper withdrew their names from it, and *Lancet*—having been warned by Deer of the revelations he was about to make in the *Sunday Times*—retracted it.

It took some years for the consequent GMC investigation to declare its conclusions—which it did, as noted, in May 2010—and during this interval Deer was beavering away at the story behind the 1998 paper. He was by now convinced, not just that the methods had been unethical and the results tailored, but that the study as a whole was fraudulent. At the start of 2011 he published a series of articles in the *British Medical Journal* laying out the devastating evidence he'd unearthed that this was indeed the case.[12] It was an astonishing demolition, and made headlines all round the world.

True to form, Russell L. Blaylock, the cancer-curing physician whose scholarly journal of choice seems to be *Newsmax*, perhaps sensing a kindred "maverick" spirit in Wakefield, chipped in with his dissent:

> Abundant evidence has shown that these very same people destroy the reputations of anyone producing evidence, no matter how well researched and of the highest ethical standards, if it in any way endangers [the] vaccine program. It is ironic that these accusers speak of "blatant fraud," when virtually all of the vaccine safety evidence they use abundantly is fraudulent by careful design. . . .
>
> They do studies that use as placebo controls people injected with a vaccine adjuvant. Placebos are supposed to be completely inert. The evidence shows that the greatest danger from vaccines is from the vaccine adjuvant—so, how can they use adjuvant-injected people as controls? Yet, all of their studies used such vaccinated controls—this is blatantly manipulated, and they know it.[13]

Wakefield's paper said nothing about the dangers of adjuvants, focusing instead on the vaccines themselves. As for Blaylock's "The evidence shows" phrase, the obvious response is: *What* evidence?

Jenny McCarthy, the ex-*Playboy* model turned actress who has become the antivax movement's celebrity figurehead, took a different approach, preferring—as is her wont—to draw on emotion rather than reason:

> Dr. Wakefield did something I wish all doctors would do: he listened to parents and reported what they said. His paper also said that, "Onset of behavioral symptoms was associated, by the parents, with measles, mumps and rubella vaccination in 8 of the 12 children," and that, "further investigations are needed to examine this syndrome [autism with gut disease] and its possible relation to this vaccine."
>
> Since when is repeating the words of parents and recommending further investigation a crime? As I've learned, the answer is whenever someone questions the safety of any vaccines.
>
> For some reason, parents aren't being told that this "new" information about Dr. Wakefield isn't a medical report, but merely the allegations of a single British journalist named Brian Deer. Why does one journalist's accusations against Dr. Wakefield now mean the vaccine–autism debate is over?[14]

The "allegations" can largely be found in the report of the GMC, which took some years over its deliberations. Further, as stated in the *British Medical Journal*, Deer's series was thoroughly peer-reviewed. McCarthy presumably knows these things but simply doesn't care. Further, no one in the world has said it's a crime for a physician to listen to and (issues of confidentiality aside) report what patients have said, or to recommend further investigation of an issue. McCarthy just made that bit up—as one might expect from someone who has boasted about being a graduate of "The University of Google."[15]

Long before being investigated by the GMC, Wakefield had to a great extent abandoned the UK for the greener grass of the US. There, his chief facilitator was Congressman Dan Burton (R-IN), then chair of the Committee on Government Reform, a born-again Christian, a conspiracy theorist (he led the anti-Clinton cavalry in their claims that Vince Foster had been murdered), and a long-time supporter of quack medicine. In 1977 he was behind the campaign to make the bogus cancer treatment Laetrile legal in Indiana despite the FDA's scientifically backed contentions that (a) it was ineffective as a cancer cure and (b) brought with it the serious risk of cyanide poisoning. Luckily for the citizens of Indiana, the FDA's regulations forbid Laetrile from being imported into the state, and likewise various of its ingredients—

so it cannot be manufactured there. As he's in the fortunate position of not having had a body count to prove the folly of his ways, Burton remembers this rather differently.

Similarly, in 1997, when the FDA moved to curb the use of the substance ephedra in dietary and stimulant supplements in the wake of medical research showing it to be dangerous even in small quantities, Burton leaped to the defense of the supplements industry. Along with the industry's own fake think tank, the Ephedra Education Council, Burton succeeded in delaying the banning of this dangerous substance until 2004. And those who attempted to thwart him did so at their peril, as noted (p. 130) by Dan Hurley in *Natural Causes: Death, Lies, and Politics in America's Vitamin and Herbal Supplement Industry* (2006).

It's hardly surprising, then, that Burton (whose Wikipedia page, when checked on January 8, 2011, mysteriously contained no mention of his campaigns on behalf of Laetrile and ephedra) should be so responsive to Wakefield's cause. More particularly, Burton suffered the horror of watching his grandson Christian succumb to regressive autism, an event he claimed happened shortly after the infant had received nine shots. He blamed this sad occurrence on the mercury-containing Thimerosal used as a preservative in vaccines.

A press release issued by Burton on October 26, 2000, is a classic example of antiscientific spin:

> On July 18, 2000 the Committee [on Government Reform] conducted a hearing entitled, "Mercury in Medicine: Are We Taking Unnecessary Risks?" During the hearing, the FDA admitted that children are being exposed to unsafe levels of mercury through vaccines containing Thimerosal. It was also determined that symptoms of mercury poisoning mimic symptoms of autism—a disease that has reached epidemic levels in the United States. However, the FDA has chosen to allow pharmaceutical companies to merely phase out their use of Thimerosal, leaving mercury-containing vaccines at public and private health facilities.[16]

It would be extraordinarily unlikely for the FDA to admit any such thing, because the Thimerosal in use in vaccines did *not* subject children to unsafe levels of mercury; there has never been any serious scientific suggestion that it did.

Although mercury is a toxin, it seems hardly as toxic as its reputation among people who don't care to probe science too much. Aside from the scare stories propagated about Thimerosal there have been plenty of panics, too, about the amalgam used for dental fillings, which contains about 40 to 50 percent mercury, traces of which slowly leach into the mouth and thence to the rest of the body. Various clinical trials have shown no adverse effects in children whose teeth have been filled using amalgam, in newborns whose mothers had amalgam-filled teeth, and in babies being breast-fed—the main groups perceived to be at risk—yet the fears live on. In 2006 and 2010 the FDA gave dental amalgams a clean bill of health, at least for nonpregnant people over the age of six. Yet, each time, lawmakers exercised their greater scientific expertise to delay full-scale approval.[17] (A separate issue concerns the adverse *environmental* effects of dental amalgam use: "In 1991, the World Health Organization confirmed that mercury contained in dental amalgam is the greatest source of mercury vapour in non-industrialized settings, exposing the concerned population to mercury levels significantly exceeding those set for food and for air."[18])

Despite Burton's efforts, in 2003 the Centers for Disease Control and Prevention, having scrutinized the scientific evidence, declined to ban the use of Thimerosal. In consequence, some CDC members received death threats. Where the manufacturers had removed Thimerosal from various vaccines it was for no scientific reason but simply in hopes of calming the hysteria Burton and others had stirred up, hysteria that was leading to a potentially catastrophic drop in the numbers of children being vaccinated. Rates of regressive autism were unaffected by the move.

But back to 2000. On April 6 Burton convened a hearing of his Committee on Government Reform, supposedly to investigate whether or not there was a relationship between the MMR vaccine and regressive autism, but in fact to press his own conviction that there was. The hearing saw a parade of scientists carefully picked as being among the tiny minority who thought such a connection might exist. The star turn was Wakefield; among others was Mary Megson of the Medical College of Virginia, who claimed to be getting great results treating autistic children with cod liver oil: "One child's IQ score went up 105 points."[19] Golly.

Standing against this torrent was Henry Waxman (D-CA), ranking minority member of the Reform Committee, who'd earlier tussled with Burton over ephedra. It was obvious to him that Burton was rigging the proceedings in favor of pseudoscience. Waxman's position was that it was up to scientists to determine whether vaccines were safe, not untutored congressmen: "The consequences of an unfair hearing on autism connected to vaccinations can cause people to die."[20] Like their UK counterparts not so many months earlier, however, the US press ignored the voice of reason and mounted a sensationalist blitzkrieg. The worldwide press behaved little better.

Burton's rigged hearing was as successful as he could have possibly hoped it to be—not in persuading Congress to legislate against vaccines scientifically proven safe but in scaring the pants off current and future parents of small children, so that rates of vaccination plunged. In due course, as had children in various UK communities before them, children in parts of the US lost their herd immunity. In late October 2010, CNN reported that California had suffered 5,978 known cases of whooping cough in the most severe epidemic the state had experienced in almost six decades; ten very young infants had died.[21] As Burton had said some years earlier, "I think every opportunity should be given to people so that they can survive and have a healthy life . . ."[22]

Probably more important than any other factor in spreading the belief that the Thimerosal in vaccines might be responsible for autism was an article called "Deadly Immunity" by Robert F. Kennedy Jr., which appeared in the June 20, 2005, issue of *Rolling Stone*. Although there were so many factual errors in Kennedy's article that *Rolling Stone* began rapidly issuing retractions,[23] these were lost amid the media storm, backed up by Kennedy's appearances on high-audience programs like Don Imus's radio show and Joe Scarborough's TV show. Quoted by Kennedy in his article were several scientists whose qualifications must have seemed to the public highly impressive, as indeed some of them actually were; less obvious was that the views of these scientists were far-flung outliers from the general scientific consensus. Among them was Boyd Haley, then a University of Kentucky chemistry professor (now retired) and at the time highly regarded. According to Haley:

> You couldn't even construct a study that shows thimerosal is safe. It's just too darn toxic. If you inject thimerosal into an animal, its brain will sicken and swell. If you apply it to living tissue, the cells die. If you put it in a Petri dish, the culture dies. Knowing these things, it would be shocking if one could inject it into an infant without causing damage.[24]

There was one major obstacle in the way of fingering Thimerosal as the guilty party: the properties of Thimerosal. Thimerosal is so safe that, in 1929, when Eli Lilly began experimenting with it in hopes it might prove to be an effective antibiotic (it was good at killing bacteria in labs but proved useless at killing them in people), they found that an injection of two full grams of the stuff caused no harm to the injectee at all. This dosage is about ten thousand times bigger than any to be found in a vaccine. Making lemonade out of a lemon, Eli Lilly realized its curious spectrum of properties meant Thimerosal would make an excellent preservative for vaccines: It would kill any organisms that might degrade the vaccine in storage, then metabolize away harmlessly inside the human body after the vaccine had been administered.

Nonetheless the myth was born that, since the product of Thimerosal's degradation in the body is ethylmercury, its use in vaccines could cause mercury poisoning. But ethylmercury does not have the same effects on the body as mercury—any more than sodium chloride affects the body in the same way as sodium. Ethylmercury does not hang about in the body long, being rapidly flushed away. Even so, the idea was born of chelation therapy as a treatment for autism, and anxious parents were relieved of untold millions of dollars.[25]

This is not to say that all advocates of chelation therapy as an autism treatment were crooks. The organization Generation Rescue was founded in 2005 by California financier J. B. Handley, whose son Jamie had been stricken by regressive autism a couple of years earlier. Convinced the cause was Thimerosal, the Handleys began daily rubbing a chelating agent on Jamie's limbs, and believed they could see a huge improvement—as perhaps they indeed could, because who knows the beneficial effect that frequent affectionate rubbing by Mommy and Daddy might have. We can understand why the stalwarts of Generation Rescue could regard this apparent empirical evidence as convincing, at least so long as Thimerosal was still used in vaccines.

Another powerful antivax organization, the National Vaccine Information Center (NVIC), founded in 1982 by Barbara Loe Fisher, Jeff Schwartz, and Kathi Williams, advised Congress in the years leading up to 1986's National Childhood Vaccine Injury Act, which had the seemingly benign effects of creating a compensation program for people injured through vaccination, requiring physicians to explain benefits and risks before vaccination, and so forth. The consequence of the act was, of course, to worry members of the public, who came to believe the risks of vaccination were far higher than they are. This seems to be the purpose as a whole of the NVIC, which capitalizes on its name and on its prior involvement with the government to give the impression it's a dispassionate, authoritative voice rather than an antivaccination advocacy group.

Fisher, who wrote the antivax tract *DPT: A Shot in the Dark* (1985; with Harris Coulter), was, as Barbara Loe Arthur, one of the signatories of a 2009 libel lawsuit brought against Paul Offit, Amy Wallace, and Conde Nast Publications over a brief remark Offit had made about her in a long interview with Wallace in *Wired* magazine:

> The November 2009 issue of *Wired* Magazine, published by Conde Nast Publications Inc., contains an article written by Amy Wallace that quotes Paul A. Offit . . . as saying that Plaintiff Fisher is a liar ("'she lies,' he said flatly"). That statement comes within the context of an article that portrays those like Fisher (who oppose mandatory vaccination) as unscientific, uneducated, and harmful to society.[26]

The complaint was thrown out in March 2010.

Generation Rescue's Handley had got the idea chelation therapy might be effective from research done by Mark and David Geier, father and son, respectively an obstetrician/gynecologist (with a PhD in genetics) and a biology graduate. In a 2003 study, they surveyed data from the national Vaccine Adverse Events Reporting System (VAERS), which collated reports from nurses, patients, and others who suspected a particular vaccination might have caused harm. The Geiers found, they believed, that Thimerosal-containing vaccines seemed to be causing a higher incidence of mental/behavioral problems than did others.[27]

One of their other recommendations was more drastic. It was already known boys were far more likely to develop autism than girls. Might this be because autistic boys were overproducing testosterone? With this in mind, and knowing testosterone tends to bind with mercury, the Geiers hit upon the idea that another autism treatment might be to cut off the boys' testosterone production—chemical castration, in other words. After using Lupron, a chemical that has this as a temporary effect, on over sixty autistic children for a period of weeks to impede their sexual development, while continuing chelation therapy, the Geiers believed, in 2006, they'd cracked the problem. It was, alas, yet another false dawn.

And then the Geiers' ascendant star was brutally extinguished. It emerged that Mark Geier had been yet another medical scientist to receive money from the UK lawyer Richard Barr. And the Geiers' 2003 paper came under assault by their scientific peers because of methodological and straightforwardly factual errors; for example, it claimed that infants were receiving increasing doses of Thimerosal when in fact doses had dropped off drastically in the preceding years as vaccine manufacturers abandoned its use; and it claimed instances of heart arrest (which is what happens when you die, whatever the cause) as examples of heart *attacks*. The data from VAERS were also suspect, since VAERS is not a scientific tool to provide the medical community with early warnings of possible problem vaccines; rather, it merely records the voluntarily submitted impressions of anyone involved, qualified or not, in a particular case—i.e., it has the same methodological problems as a website poll. In due course it was discovered that personal-injury lawyers were a principal source of information submitted to VAERS, thereby creating a perfect evidential circle for use in lawsuits: Lawyers could point at "scientific" data they had themselves created!

And what of the journal in which the 2003 paper appeared, the *Journal of American Physicians and Surgeons*? This is the official publication of an organization called the Association of American Physicians and Surgeons, which is a conservative think tank, an offshoot of the infamous Oregon Institute of Science and Medicine (see page 285). In a 2009 piece called "The Tea Party's Favorite Doctors" Stephanie Mencimer blew the AAPS's pretensions wide open:

> Yet despite the lab coats and the official-sounding name, the docs of the AAPS are hardly part of mainstream medical society. Think Glenn Beck with an MD. The group (which did not return calls for comment for this story) has been around since 1943. Some of its former leaders were John Birchers, and its political philosophy comes straight out of Ayn Rand. Its general counsel is Andrew Schlafly, son of the legendary conservative activist Phyllis. . . . Its website features claims that tobacco taxes harm public health and electronic medical records are a form of "data control" like that employed by the East German secret police. An article on the AAPS website speculated that Barack Obama may have won the presidency by hypnotizing voters, especially cohorts known to be susceptible to "neurolinguistic programming"—that is, according to the writer, young people, educated people, and possibly Jews.[28]

A recent craze in the antivax world is to claim that autism is caused by "mitochondrial disorders," or is a misdiagnosis of those. In real life, mitochondrial disorders are not unknown, but they're rare—unless of course they're far commoner than we think because they've been so frequently misdiagnosed as autism! Whatever, this new line of reasoning might seem to let vaccines off the hook, but no such luck. Vaccination apparently *exacerbates* the "mitochondrial disorders."

Another recent fad has been to point at the small amount of formaldehyde in some vaccines as a threat to health, specifically as a carcinogen; formaldehyde is, after all, a KNOWN TOXIN! Hm. It's a KNOWN TOXIN that your liver manufactures every day of your life. At any particular moment there's about ten times more formaldehyde swimming around in the body of even the smallest infant than in any vaccine you might administer.

Andrew Wakefield, meanwhile, was enjoying mixed fortunes. Having departed the Royal Free Hospital in 2001, he spent increasing time in the US, eventually becoming the mainspring of the Thoughtful House Center for Children, in Austin, Texas, an institution that claims both to conduct research into autism and to treat sufferers. In February 2010, as the GMC ruling loomed, Wakefield lost his position at Thoughtful House. In May 2010 he published his own account of the MMR brouhaha, *Callous Disregard: Autism and Vaccines—The Truth Behind a Tragedy*. It's perhaps unfair to mention that this has a foreword by Jenny McCarthy.

One of Wakefield's associates at Thoughtful House, Arthur Krigsman, came to wide public notice when, testifying before Congress's Government Reform Committee in June 22, 2002, he claimed to have independently confirmed Wakefield's results. Naturally the scientific world was eager for details. There was a long wait. Finally, on January 27, 2010, Krigsman published his supportive paper, "Clinical Presentation and Histologic Findings at Ileocolonoscopy in Children with Autistic Spectrum Disorder and Chronic Gastrointestinal Symptoms,"[29] in the first issue of an online journal called *Autism Insights*. Among this journal's editorial board were

Bryan Jepson, MD, Director of Medical Services, Medical Center, Thoughtful House Center for Children, Austin, TX, USA
Arthur Krigsman, MD, Director of Gastrointestinal Services, Pediatric Gastroenterology, Thoughtful House Center for Children, Austin, TX, USA
Carol Mary Stott, PhD, Senior Research Associate, Research Department, Thoughtful House Center for Children, Austin, TX, USA
Andrew Wakefield, MBBS, FRCS, FRCPath, Research Director, Thoughtful House Center for Children, Austin, TX, USA[30]

Unusually for learned journals, *Autism Insights* charges a "processing" fee of $1,699 to authors who submit papers. Discounts are available for authors making bulk submissions.

It can often seem, in discussions of the scientific case for a purported link between the MMR vaccine and autism, that it's a matter of Andrew Wakefield against the world. This isn't so. One staunch supporter of Wakefield has been the "nutritional therapist" Patrick Holford. It can be difficult for people outside the UK to understand the extent to which Holford, through a morass of websites and his own Institute for Optimum Nutrition, captures the public ear there. His support for Wakefield was thus no minor detail. Holford sells to parents, through various of his outlets, all they will need to treat their children's autism—dollops of dietary supplements plus a regime of expensive tests and consultations. Since there is zero evidence that autism will respond to vitamins or any other dietary enhancement, and a vast scientific consensus that it won't, it's something of a mystery as to why anyone should part with any money whatsoever to follow Holford's "action plan"—yet many do.

Wakefield's false conclusions were restricted to the MMR vaccine. Holford speaks out against vaccination in general, recommending parents do everything they can to build up their offspring's immune system. Mothers should therefore breastfeed their babies and then later ensure that—you've guessed it—the infants' diets are supplemented by the correct regime of vitamins, minerals, and the rest. He has suggested parents might investigate homeopathic alternatives to vaccination.

With all the hullabaloo being created by the antivaxers, it was only natural that opportunistic homeopaths would move in to offer "alternative vaccines." According to a BBC report in September 2010,[31] this was by no means universal among homeopaths; following up on claims that the three hundred–odd UK members of the Homeopathic Medical Association were offering the "replacement vaccines," their reporter interviewed six and found that three were more than happy to administer the orthodox MMR vaccine. But the Inverness-based homeopath Katie Jarvis said she offered "homeopathic prophylaxis" to parents unwilling to have their children vaccinated. On being asked if her remedies offered the same protection as the vaccine, Jarvis replied, "I'd like to say that they were safer, but I can't prove that." Too right. She also claimed she could protect patients homeopathically against other, even more serious diseases, including polio, tetanus, and diphtheria.

What is perhaps even more alarming is that the UK's National Health Service spends nearly £4 million ($6 million) annually on homeopathic treatments, money that could instead be spent on medicine.

Some years earlier, in 2002, Edzard Ernst and Katja Schmidt of the University of Exeter did an informal survey, e-mailing 168 registered homeopaths in the guise of a mother seeking guidance as to whether or not to let her child be given the MMR vaccine. Of the 168, 104 replied; of these, 27 dropped out on being informed (as was ethically necessary) that this was a survey rather than a genuine request. Of the remaining 77, 75 advised the mother against vaccination.[32]

To date the biggest study done in an attempt to find a link—any link—between the MMR vaccine and autism has been the one reported in the 2002 paper "A Population-Based Study of Measles, Mumps, and Rubella Vaccination and Autism."[33] Because of the efficient records kept in Denmark, the authors were able to analyze the health histories of almost all the children born in that country over a period of eight years, from the start of 1991 to the end of 1998. Of the over 530,000 children concerned, more than 440,000 had been vaccinated; the remainder had not. No difference was discovered in the rates of autism's incidence between vaccinated and unvaccinated children; further, among those autistic children who had been vaccinated, there was no discernible relationship between age at vaccination and the onset of autism.

One of the most gruesome ironies of the antivax campaign is that there *is* a known link between rubella and autism: If the mother catches rubella while pregnant, there's an increased chance (for reasons no one properly understands) that her child will suffer one of the autistic spectrum disorders. The irony arises because, of course, the more kids there are running around without immunity to rubella, the more likely it is pregnant mothers will catch the disease; in other words, far from reducing instances of autism, antivax campaigners are building up future cases of it.

I've spent most of this chapter talking about antipathy to the MMR vaccine, but all the other vaccines have their opponents as well—including surprises like the polio vaccine, which one might assume would be universally approved. Far from it; in fact, one could make a case that the modern antivax hysteria owes some of its primary memes to scares over the oral polio vaccine during the 1960s.

A consequence of the creation of the first polio vaccine, by Jonas Salk in the late 1950s—even before the oral vaccine came along—was that between 1954 and 1961 the annual number of cases of polio in the US dropped from 38,476 to 1,312, or by about 96.5 percent; the Surgeon General estimated that, of those residual polio cases, some 90 percent represented people who had refused vaccination.

There's a modern conspiracy theory claiming that polio was already well on its way out when, with the introduction of Salk's vaccine, it was given a new lease on life. Like all the best conspiracy theories, this one takes a few uncontested facts and heftily misinterprets them. It's true that polio rates had been in decline in the developed nations since the late nineteenth century; this was not because the disease was somehow losing its potency but simply—as polio spreads through ingesting feces—a result of better water purification and better personal hygiene. Since the introduction of the vaccines, polio rates have dropped to the point where the disease has virtually (although, because of increased globalization, not entirely) disappeared from the developed world and is fast on the wane in all other countries that have accepted the vaccine. But, just as individual areas experience cold snaps during global warming, so there can be temporary, small-scale, localized upturns in polio rates in the midst of an overall rapid decline. These give rise to anecdotal "evidence" of polio rates increasing after the introduction of the vaccine. The global statistics tell a different story, but statistics don't sell sensationalist books.

Salk's vaccine used dead viral fragments. The oral polio vaccine (OPV) devised by Albert Sabin in the 1960s used live virus—much attenuated, but nonetheless live. Parents ignorant of the underlying science quite understandably had fits at the thought of their little darlings consuming live viruses. The OPV *is* very slightly more risky than its injected counterpart—figures vary, but perhaps one in 750,000 recipients of all three doses of an OPV treatment can develop full-scale polio. (Many countries now use an injected vaccine that is 100 percent safe from this, although of course, as with any medical treatment, there's still a tiny risk of allergic or other adverse reactions.) This is a minuscule figure when set alongside the chances of catching polio in a society where herd immunity hasn't been established. Accordingly, in most developed countries polio vaccination was made a legal requirement . . . and the antivaxers of the era had a field day blaming the accursed politicians for putting the lives of children at risk. Since it took some time, despite the legal mandate, for herd immunity to be established, there were still for a while occasional polio outbreaks; rather than realize these were fewer and smaller than in the past, the antivaxers chose to blame the vaccine—"obviously" it was spreading rather than defeating the disease.

In 1988, when a global effort to eradicate polio through vaccination began, the annual incidence of polio worldwide was an estimated 350,000 cases. By 2001, this had dropped to just 483. That was an unusually good year for the "war on polio"—a more realistic average today is 1,000 cases annually. Hard to explain these figures in terms of the polio vaccine spreading the disease, isn't it?

Polio was well on its way to extinction, in fact, thanks to vaccination, until Muslim fundamentalists in the Nigerian state of Kano decided in 2003 that Western Christians were using the vaccine to spread AIDS and/or sterilize Muslim women. State governor Ibrahim Shekarau banned the procedure. Ten months later, in July 2004, the ban was lifted and an urgent campaign was begun to vaccinate four million of Kano's children under the age of five against the disease. By then, though, there had been polio outbreaks not just in Kano and other parts of Nigeria but in neighboring countries.[34]

The example of the polio vaccine is not an isolated one. The Measles Initiative was launched in Africa in 2001, its aim being to vaccinate as many children as possible. The results exceeded the most optimistic expectations. In the first five years of the program, annual measles deaths in Africa dropped from four hundred thousand to thirty-six thousand—by 91 percent.[35]

But in Zimbabwe some years later, members of a sect called the Vapositori refused vaccination for their children on religious grounds; the sect's leaders claim prayer is the best therapy. Between September 2009 and March 2010 there were 110 known deaths of children from measles, and in most the Vapositori were implicated. In May 2010 the Vapositori climbed down rather than incur hostile legislation.[36]

There are various examples of nations foolishly assuming that, just because rates of incidence of a particular disease have dropped to minuscule levels, there's no need to continue with universal vaccination. One example occurred in the UK in the early 1970s, when the government of the day decided to economize by cutting back on immunization against whooping cough. The result was the 1978 whooping cough epidemic in which a hundred thousand were infected and thirty-six people died. At roughly the same time there was a fall in the vaccination rates in Japan for whooping cough from 70 percent to perhaps 30 percent (estimates vary); in consequence, 1974's figures of 393 cases with zero deaths leaped to 1979's figures of 13,000 cases with 41 deaths. In the general political chaos following the breakup of the USSR, the practice of universal diphtheria vaccination in the Eastern Bloc countries was a casualty. Between 1989 and 1994, rates of diphtheria infection rose from a total of just 839 to nearly 50,000, with some 1,700 people dying of the disease in the latter year. And this massive diphtheria outbreak wasn't confined to Eastern Europe; the disease spread into the rest of Europe and thence all over the world.[37]

But one can go back much further than the late twentieth century. The March 9, 1888, issue of *Science* had this item:

> Vaccination Statistics. — The following extract from *The Sanitarian* would seem to indicate that a compulsory vaccination law has its advantages: "The success of the anti-vaccinationists is aptly shown by the results in Zurich, Switzerland, where for a number of years, until 1883, a compulsory vaccination law obtained, and small-pox was wholly prevented (not a single case occurred in 1882). This result was seized upon in the following year by the anti-vaccinationists, and used against the necessity for any such law, and it seems they had sufficient influence to cause its repeal. The death returns for that year (1883) showed that for every thousand deaths two were caused by small-pox; in 1884, there were three; in 1885, seventeen and in the first quarter of 1886, eighty-five."

It's worth spelling out a couple of things:

- By summer 2001, Thimerosal had largely been removed from vaccines in the US. The rate of reported autism continued to rise.
- Because of the scares begun by Andrew Wakefield's 1998 *Lancet* paper and perpetuated by media hysteria, politicians (like Dan Burton, John Kerry, Chris Dodd, and Joe Lieberman), and the various antivaxer groups, through most of the 2000s vaccination rates fell in most US demographies. Autism rates continued to rise.

Is there not perhaps a pattern detectable here? While it might be very possible to make a case that, all other things being equal, the cause of regressive autism could well be environmental, it seems patently obvious that the culprit can be neither Thimerosal nor the MMR vaccine.

If vaccination isn't to blame, why are autism rates rising? There's no easy answer except, just possibly: They aren't. Certainly part of any perceived increase is that doctors are becoming much more alert to autism, and thus more ready to diagnose the ailment in infants who're suffering the relevant developmental problems. Another factor is that, in the early 1990s, the definition of what is meant by autism was, for theoretical reasons, expanded to include other conditions—Asperger syndrome and PDD–NOS (Pervasive Developmental Disorder–Not Otherwise Specified)—that had previously been considered distinct. It's unclear as to whether these two trends, taken together, might not entirely explain the supposed increase.

Whatever the truth of the matter, bellowing false answers from bully pulpits, far from helping autistic children and those who love them, is merely perpetuating their anguish through distracting from the quest for the *true* answers. The celebrity antivaccination campaigners, from Jenny McCarthy on down, may believe they're spreading light and hope. In fact, through their shrill, self-absorbed efforts to drown every voice but their own, what they're doing is spreading sickness, and misery, and death.

7

THE AIDS "CONTROVERSY"

We know more about the diversity and origins of Human Immunodeficiency Virus (HIV) than perhaps about any other human pathogen.

—Nathan D. Wolfe and Tony Goldberg,
"HIV-1 Origins: What We Don't Know"[1]

Rush Limbaugh headlined his radio show for April 28, 2009, with the extraordinary claim that "AIDS was going to get really bad, but it didn't"[2]—a medical judgment that's on a par with his April 24, 2008, claim: "Polio is not a virus."[3]

Limbaugh is part of a larger syndrome. It seems that in the US conservatives largely dismissed the AIDS epidemic as yet another liberal conspiracy. Ronald Reagan was nearly five years in office before he thought this enormous threat to US public health worth mentioning (September 17, 1985), and even then only in answer to a direct question at a press conference. Five months passed before he alluded to the subject in public again, in his address to Congress on February 6, 1986. His first substantive public discussion of AIDS did not come until May 31, 1987, when he finally admitted there was an epidemic underway. By then, thirty thousand Americans had been diagnosed with the condition and an untold but very much larger number had been infected by the HIV retrovirus but didn't yet know it.

A lot of this reluctance on the part of the US Right to face the reality of AIDS may have been due to the fact that, in the early days, AIDS was assumed to be a "gay plague"; not only were gays a minority but the "gay lifestyle" was regarded as immoral and debased by the fundamentalists upon whose votes the GOP depended. Over a decade later, faced with the unpalatable facts that the number living with HIV/AIDS in the US was approaching ten million and that the single most effective way of curtailing the disease's spread among the young—the promotion of sex education and in particular the use of condoms—was anathema to the Christian Right, the George W. Bush administration opted instead to put all its efforts behind abstinence-only sex education, even though this approach had been repeatedly demonstrated not to work. Through its reluctance to support any endeavor, at home or abroad, that was connected to the promotion of condom use or needle exchange, and through its dissemination of misinformation on these subjects and indeed most matters sexual, the Bush administration grossly weakened the fight against AIDS worldwide. Although the denialism of the Mbeki administration in South Africa (see pages 135ff.) was indubitably more extreme, at least it was based on genuinely held—albeit nonsensical—convictions about medical science. The Bush administration's denialism on the subject of AIDS was ideological—or, perhaps even worse, not ideological at all but tailored to ensure the continued electoral support of the Christian Right.

In the opening of their "Statement on National HIV Vaccine Awareness Day" (May 18, 2010) Anthony S. Fauci, Margaret I. Johnston, and Gary J. Nabel wrote:

> More people today have access to life-saving antiretroviral therapy for HIV/AIDS than ever before. Yet for every person who begins treatment for HIV infection, two to three others become newly infected. Treatment alone will not curtail the HIV/AIDS pandemic. To control and ultimately end this pandemic, we need a powerful array of proven HIV prevention tools that are widely accessible to all who would benefit from them.
>
> Vaccines historically have been the most effective means to prevent and even eradicate infectious diseases. They safely and cost-effectively prevent illness, disability and death. We at the National Institute of Allergy and Infectious Diseases (NIAID), part of the National Institutes of Health, have been working for more than two decades with our colleagues worldwide to develop an HIV vaccine, and this research continues to rank among our top priorities.[4]

They pointed to one encouraging piece of research done during the preceding year. In a major clinical trial in Thailand, a prime-boost regimen of the two particular vaccines had been shown to have a 31 percent effectiveness rate in preventing HIV infection. At less than one in three, this may not seem much of an improvement; yet it's a striking advance on anything achieved before. The three authors were nonetheless pragmatic as they assessed the situation:

> As we recognize recent progress in HIV vaccine research and hope for continued advances, we must remember that a vaccine alone will not end the HIV/AIDS pandemic. If an HIV vaccine is developed, it will need to be used in concert with multiple other scientifically proven HIV prevention tools . . . including pre-exposure prophylaxis with antiretroviral drugs, microbicides, and expanded HIV testing and treatment with linkage to care.

Naturally, such hopes for the future will be dismissed by many AIDS denialists as yet another example of Big Pharma aiming to cash in on the vaccine and drug market. As in the campaigns of the antivaxers, it's a recurring theme in AIDS denialism that Big Pharma has corrupted governments, the NIH, and indeed the entirety of the biomedical establishment. There is little doubt that Big Pharma will corrupt where it can, but the sort of conspiracy the deniers are here claiming beggars belief.

Even without the paranoia about Big Pharma, the lethal recommendation of the AIDS pseudoscientists is that those diagnosed with HIV infection refuse antiretroviral drugs like AZT,[5] which are used in combating the development of the disease. They claim that

- AIDS is not caused by the HIV retrovirus so the drugs are unnecessary, or
- the drugs are dangerous, or even
- it is the drugs themselves that *cause* AIDS!

Saddest are the cases of HIV-infected pregnant mothers who are deceived by this nonsense into rejecting medication, with the consequence that their children are born with, and very likely die painfully and young from, HIV infection. A tragic example concerned the HIV-denying activist Christine Joy Maggiore, who founded the denialist organization Alive & Well AIDS Alternatives.[6] Maggiore herself died in 2008 in her midthirties; but the real horror was earlier. Although Maggiore was HIV-positive at the time of daughter

Eliza Jane's conception, she refused to take anti-HIV medication during the pregnancy and to allow Eliza Jane, during the child's brief life, to be tested for HIV. The consequence was that Eliza Jane died of untreated AIDS in 2005 aged just three and a half. Even then, Maggiore denied what had happened, hiring an AIDS-denialist vet to produce a rival autopsy report.

The deniers do not reserve their venom solely for the drugs used to mitigate the effects of AIDS. Another myth concerns HIV testing. The reality is that HIV testing is approximately 99.99 percent accurate—that is, it can throw up a false positive or a false negative just once in ten thousand instances. This makes it one of the most accurate tests in modern medicine. Yet some deniers produce figures that "prove" the inaccuracy of the test. It's worth looking for the dates on the sources they cite for those figures. When medical science first became aware of the HIV retrovirus, tests were a bit chancy. Obviously, these were then progressively refined until today's standard of accuracy was achieved; that's the way science functions. Presumably those who keep trotting out the 1980s figures are perfectly well aware they're outmoded.

Despite the strength of the science concerning AIDS and the efforts to combat it, a few contrarians are qualified scientists although, as we see in other instances of scientists bucking the consensus, their qualifications are almost always in unrelated fields. Casper Schmidt, for example, was a psychiatrist. In his oft-quoted 1984 paper "The Group-Fantasy Origins of AIDS"[7] he claimed the already burgeoning AIDS epidemic was merely a sort of social hysteria, a reaction to conservative moral strictures regarding homosexuality. A decade after publication of his paper, Schmidt died of social hysteria. Similarly lacking in relevant qualifications are the primary members of the Perth Group, a set of Australian scientists who claim the HIV retrovirus does not exist: biophysicist Eleni Papadopulos-Eleopulos, pathologist John Papadimitriou, and ER physician Valendar Turner. According to Michael Specter in his 2007 essay "The Denialists," the members of the Perth Group "insist that AIDS in gay men results from drug abuse and repeated exposure to semen."[8]

And sometimes the matter of scientific qualifications is a little cloudy, as in the case of David Rasnick, a member of South African president Thabo

Mbeki's AIDS Advisory Panel. In a 2006 op-ed in the South African newspaper *The Citizen*, Rasnick claimed to be a visiting scholar in the department of molecular and cell biology at Berkeley. Alas for Rasnick, Richard Harland, professor and chair of that department, publicly disputed this claim.[9]

By far the most influential scientist to deny the HIV–AIDS connection is US molecular and cell biologist Peter Duesberg. His early work on cancer, done in the 1970s, rightly gained him international acclaim. What he has never done is any research on HIV/AIDS; his views on the subject are theoretical. Similarly, the Group for the Scientific Reappraisal of the HIV/AIDS Hypothesis, set up by Duesberg and other denialists in 1991, has never soiled its hands by doing any hands-on AIDS research.

In *Denying AIDS* (2009; p. 39) Seth Kalichman sums up Duesberg's hypotheses concerning the causes of AIDS thus:

> Duesberg's explanation for AIDS is actually far more complex and convoluted than how HIV actually causes AIDS. . . . For gay men, drug use causes AIDS. For gay men who do not use drugs, HIV medications cause AIDS. In Africa, malnutrition causes AIDS. If you are a wealthy African, AZT causes AIDS. If you are a hemophiliac, treatments for hemophilia cause AIDS.

The frequent claim that Duesberg lost NIH funding because of his AIDS theorizing seems as ill founded as his denialist views; he lost the funding long *before* he began to pontificate on AIDS because, having done sterling work on the role of retroviruses in and the genetic underpinning of cancer, he suddenly changed his mind, deciding that instead cancer was caused by the cellular condition aneuploidy (excess chromosomes) and that retroviruses were harmless. Since this flew in the face of the vast body of cancer research, not least his own, it seemed—and seems—a less than promising line to pursue; not unnaturally, NIH declined to put its precious dollars into it. And, since NIH has never funded Duesberg to do AIDS research, another oft-repeated allegation, that NIH cancelled his funding for AIDS research, is obviously likewise false.

Similarly, much has been made in denialist circles of the rejection by NIDA (National Institute on Drug Abuse) in 1993 of a grant application made by Duesberg for research into the possibility of AIDS-like reactions in mice subjected to nitrites inhalation. In reality, this was because *another* study of the effects of nitrites inhalation on the immune system was already being funded by NIDA.

Some of Duesberg's claims go beyond the outrageous to the bizarre. One is that there's no evidence of an HIV/AIDS epidemic among female prostitutes—despite the fact that numerous studies show a high rate of HIV/AIDS among sex workers. Clearly needle-sharing plays a large part in the HIV epidemic among the world's sex workers, but to say the epidemic doesn't exist is crazy—and promoting the denial seems cruelly counterproductive.

Another strange claim of Duesberg's is that there's no real AIDS problem among children. To the contrary, the WHO estimated for 2008[10] that about 430,000 children worldwide were born with HIV in that year, bringing the total for children under fifteen living with HIV to about 2.1 million, while some 270,000 died of HIV-related illness. It is difficult, faced with such data, not to describe this as an epidemic. Of course, because Duesberg believes the HIV virus isn't responsible for AIDS, he can argue that these figures are irrelevant. Yet that's a quibble: the WHO may call it "HIV-related illness" as an acknowledgement of the role of the retrovirus, but the rest of us call it AIDS.

After Duesberg's 1988 presentation in Washington, DC, at the scientific forum sponsored by the American Foundation for AIDS Research, Anthony Fauci, the US government's top AIDS expert, had had enough. "This is murder!" he exploded. "It's really just that simple."[11]

Several other scientific and quasiscientific memes that do the rounds of AIDS-denialist circles are worth noting. One challenge often hurled at orthodox medicine is that it must explain its claim that such an array of diseases could be caused by a single virus. This reveals a fundamental ignorance of the workings of the HIV retrovirus. The virus's assault does indeed cause a single condition: the collapse of the immune system. Thereafter, the body is vulnerable to an array of diseases, one or other of which is likely to prove fatal. It's almost impossible to credit that the pushers of HIV/AIDS denialism don't know this.

Another denialist challenge is to produce a "pure" form of the HIV retrovirus—one "uncontaminated" by cell proteins. Without the ability to do this, they say, HIV testing is meaningless. They're quite right to say science has never achieved the feat of isolating the retrovirus without those proteins . . . for the very good reason that the way viruses *function* involves hijacking cell proteins to use as their own. If you found a virus "uncontaminated" by cell proteins it wouldn't be a virus.

Perhaps the most startling notion is that HIV cannot be transmitted via heterosexual vaginal sex. Many such diehards refer to a single landmark paper of AIDS research. The paper's misrepresentation in the denialist literature typically takes the form of stating that this decade-long study of heterosexual couples found only a 0.1 percent rate of transmission through heterosexual sex—a rate so low it could well be explained as a statistical anomaly or by, say, an unadmitted act of gay infidelity. In fact, as the paper's chief author, Nancy Padian, explained in the official statement on the subject she eventually felt impelled to issue, the conclusions of the study were that safe sex and condom use worked, *not* that heterosexual HIV transmission was a myth.[12]

Hank Barnes's rabid denialist blog *You Bet Your Life* produced one of the more moronic retorts to Padian's statement: "Then, these jokers trot out ditzy Nancy to pontificate a bit. . . . Well, that's because infectivity for HIV is low. You estimated 1/1000 sex acts for a woman to get it from an infected man, and 1/10,000 for a man to get it from an infected woman."[13]

"Ditzy"? Barnes and his commenters then go into an extensive exercise in cherry picking to demonstrate they know better than Padian the meaning of her research.[14] This reaction was, alas, fairly typical of the denialists.

Among other odd claims about AIDS is that it's a consequence of polio vaccination. According to this notion, certain chimpanzees in the Democratic Republic of Congo suffered from simian immunodeficiency virus (SIV); when tissues from these chimps were used in oral polio vaccine (OPV), the SIV mutated in its new hosts to become HIV. It's a clever hypothesis. The idea fails because "the circulating [SIV] virus is phylogenetically distinct from all strains of HIV-1, providing direct evidence that these chimpanzees were not the source of the human AIDS pandemic."[15] As the medical blogger Orac sums up, "Basically, testimony by eyewitnesses, documents from the time, epidemiological analysis, as well as phylogenetic, virologic and PCR [polymerase chain reaction] data all converge to reject as false the hypothesis that HIV/AIDS was derived from this polio vaccine."[16]

Even despite its having been repeatedly proven false over a period of many years, the hypothesis still turns up—and not only among AIDS cranks.

Orac's comment was made in connection with an article that had appeared on the antivaxer site Age of Autism a few days earlier, "How Vaccine Damage Deniers Threaten Us All" by Jake Crosby.[17]

During the first part of the last century the US suffered a plague of quacks who, like Ruth Drown and John R. Brinkley, sold "cures" for cancer and other diseases. Through persuading people to forgo the options offered by conventional medicine, they caused an untold but likely enormous toll of premature deaths. Their spiritual descendants today are all too eager to sell you means of protecting yourself from HIV/AIDS, or of curing you of the disease should you be unfortunate enough to catch it. Once again, the bodies have been piling up, not because the "cures" are in themselves dangerous but because gullible people are persuaded to reject effective treatments. Further casualties arise because these individuals, in their willful denial of their own toxicity, infect others before they die. Through their success in persuading political leaders like President Thabo Mbeki of South Africa that AIDS is not caused by the HIV retrovirus and that the alternative "cures" are effective, these AIDS opportunists (some of whom are quite genuine in their beliefs, others less so) are responsible for hundreds of thousands of deaths, maybe more.

The "cures" can be bizarre. Writing in 2003, Meera Nanda told of the Indian government, then controlled by the nationalist Bharatiya Janata Party (BJP), "investing in research, development and sale of cow urine, sold as a cure for all ailments from the Acquired Immune Deficiency Syndrome (AIDS) to tuberculosis (TB)."[18] In South Africa, the AIDS "cure" known as *ubhejane*, marketed from his "clinic" in downtown Durban by Zeblon Gwala, was invented not by the ex-trucker himself but by his long-dead grandfather, who appeared to him in a dream and gave him instructions on how to produce two herbal stews that would cure AIDS; Gwala does a roaring trade, and was supported by prominent members of the Mbeki administration.[19] And in the US the lab technologist Roberto Giraldo not only claims the HIV retrovirus doesn't exist but offers "cures" based on this rationale, including acupuncture, aromatherapy, color therapy, digitopuncture, homeopathy, hyperthermia, orthomolecular medicine, and even yoga—approaching the full quack spectrum.[20]

Some of these "therapists" have a bully pulpit. Becoming president of

The Gambia after a bloodless coup in July 1994, Yahya Jammeh has been persistently linked with human rights abuses and press muzzling. Although his declaration on homosexuality was overshadowed by the international outcry from 2009 onward over Uganda's proposed new legal crackdown on gays, the policy Jammeh announced in The Gambia on May 15, 2008, was even more draconian.[21] In January 2007 he stunned the world by claiming his ancestors had revealed to him in a dream the true, nonviral cause of AIDS and a means of curing it, an herbal concoction comprising "a green herbal paste, a bitter yellow liquid and eating bananas"[22] that would effect a cure in a mere three to ten days, the herbal medicines being both taken orally and smeared on the body. Side effects were noticeable but not serious: "[T]hey should be kept at a place that has adequate toilets facilities because they can be going to toilet every five minutes."

The UN development program's representative in The Gambia, Fadzai Gwaradzimba, disagreed strongly, pointing out the therapy was nonsense and that the promise of such an easy and relatively mild "cure" would surely encourage people to engage in behavior likely to increase their chances of catching and spreading HIV/AIDS. Her reward was to be expelled from The Gambia.[23]

In the UK one of the most prominent denialists of the science concerning HIV/AIDS is the popular "nutritional therapist" Patrick Holford (see page 119), who in 1984 founded the Institute for Optimum Nutrition, a "not for profit educational charity whose purposes are to advance education of the public and health professionals in all matters relating to nutrition [and] to preserve and protect the health of the general public by giving advice, assistance and where necessary treatment through nutritional therapy" (from its website[24]). The institute dishes out a diploma, the DipION, to would-be nutritionists who complete a moderately expensive three-year course. The DipION is not recognized as an academic qualification; in the UK any amateur, with or without a DipION, can hang his or her shingle as a nutritionist. In fact, Holford himself is one such. Although he has a BSc in experimental psychology from the University of York, his sole qualification in the field of nutrition is a DipION awarded by his own institute.

His claim in *The Optimum Nutrition Bible* (1997) that "AZT, the first prescribable anti-HIV drug, is potentially harmful, and proving less effective than Vitamin C" (p. 208) was checked out by the journalist Ben Goldacre, whose "Bad Science" column for the *Guardian* has become required reading for many in the field.[25] The claim was based on a paper called "Suppression of Human Immunodeficiency Virus Replication by Ascorbate in Chronically and Acutely Infected Cells" by S. Harakeh, R. J. Jariwalla and L. Pauling, published in the September 1990 issue of the *Proceedings of the National Academy of Sciences*. The paper (in fact a laboratory report rather than a clinical study), which nowhere mentions AZT, describes experiments dosing HIV-infected cells with large quantities of vitamin C; while it seems to indicate the vitamin might inhibit various HIV-related activities in isolated cells, it tells us nothing about whether vitamin C doses would have any effect on HIV-infected humans.

Although his specialty is in the denial of climate change, Christopher Monckton (see pages 293ff.) has not been silent on the subject of AIDS. His early contribution to the field is characterized by a 1987 article for the *American Spectator*'s January 1987 issue called "AIDS: A British View." Its oft-quoted paragraph (p. 30) reads:

> For there is only one way to stop AIDS. That is to screen the entire population regularly and to quarantine all carriers of the disease for life to halt the transmission of the disease to those who are uninfected. Every member of the population should be blood-tested every month to detect the presence of antibodies against the disease, and all those found to be infected with the virus, even if only as carriers, should be isolated compulsorily, immediately, and permanently.

More recently Monckton has disavowed these views, although characteristically he has done so while declining to admit there might be anything . . . well, *wrong* with them. Because his advice wasn't heeded, you see, there are now so many cases of AIDS in the world that quarantine is no longer a viable option. At the moment, according to an e-mail posted on November 13, 2009, he is "working on a cure for infections including HIV"[26]—so we await the published results with eagerness. In fact, we've been waiting for some while now . . .

Although there is not a country in the world unaffected by the scourge of AIDS, in a very real sense the AIDS tragedy is an African tragedy. All the records for rates of HIV infection, and for AIDS deaths, are held by African countries. International assistance has been hampered by, in some instances, doctrinal shackles on the donor countries and in others by governmental corruption, stupidity, or ideological obstinacy on the part of potential recipient nations. Nowhere on earth, however, has the tragedy been greater—because it could so easily have been ameliorated—than in South Africa during the years 1999 to 2008, when the country was governed by the administrations of President Thabo Mbeki.[27]

In Pretoria on May 6, 2000, Mbeki addressed the panel he had assembled to advise him on AIDS. Among its members were a number of prominent AIDS denialists, including Peter Duesberg, Harvey Bialy, Roberto Giraldo, Eleni Papadopulos-Eleopulos, and David Rasnick. This is somewhat as if Congress had called an advisory panel on climate change and included among its "experts" people like Christopher Monckton and Michael Crichton. What a ridiculous thought. Oh, wait a minute . . .

When Mbeki spoke before this motley band he was, of course, addressing the broader public:

> There is an approach which asks why is this President of South Africa trying to give legitimacy to discredited scientists, because after all, all the questions of science concerning this matter had been resolved by the year 1984. I don't know of any science that gets resolved in that manner with a cut-off year beyond which science does not develop any further. It sounds like a biblical absolute truth and I do not imagine that science consists of biblical absolute truths.[28]

This plays well to the gallery, of course, making it sound as if hidebound establishment scientists are attempting to stifle dissent—and, besides, science certainly *isn't* in the business of "absolute truths." Science does, however, declare some things to be established beyond all sensible doubt in the light of current knowledge, and one of these is that AIDS is not the result of aneuploidy, biochemical imbalance, social hysteria, malnutrition, the displeasure of God, or any of a hundred other crank explanations . . . That these hypotheses are false is as near to an "absolute truth" as you're ever likely to find. It has also been established beyond all reasonable doubt that AIDS is a consequence of HIV infection. Even so, according to widespread news reports, Mbeki was in that same year, 2000, openly speculating that the claim of a link between HIV and AIDS might be a dastardly plot by the CIA, and that the medicines being offered to South Africa by the West might be designed not to counter HIV but for a far different purpose: harming innocent Africans.[29]

A few weeks later, in July 2000, an International AIDS Conference was held in Durban. Alarmed by Mbeki's support for AIDS denialists, and that of his health minister, Manto Tshabalala-Msimang, over five thousand relevant scientists, including eleven Nobel laureates, signed a statement that was published in the scholarly journal *Nature* to coincide with the event. To ensure there could be no accusations of conflict of interest (they came anyway, of course), the organizers requested that scientists who worked for pharmaceutical and other relevant commercial companies refrain from participating. In part the document reads:

> *A Declaration by Scientists and Physicians Affirming HIV Is the Cause of AIDS*. . . . The evidence that AIDS is caused by HIV-1 or HIV-2 is clear-cut, exhaustive and unambiguous, meeting the highest standards of science. The data fulfil exactly the same criteria as for other viral diseases, such as polio, measles and smallpox. . . .[30]

The Mbeki administration was upset, and made disparaging and dismissive comments in the media. The most ludicrous reaction was Tshabalala-Msimang's: She called it "elitist"—an accusation that's the last refuge of the scoundrel.

A denialist document of import in South Africa was *Castro Hlongwane, Caravans, Cats, Geese, Foot & Mouth and Statistics*, a long tract circulated at the 51st National Conference of the African National Congress in 2002; it is rumored that Mbeki, even if he had no hand in its writing, was in contact with its author while it was being written—and certainly he has said he agrees with its stance. The Castro Hlongwane Document, as it's often called for short, claimed that a number of prominent South Africans who'd died of AIDS had, rather, been poisoned by the antiretroviral drugs they'd been taking. Its primary claim was that AIDS is really a cluster of traditional African diseases exacerbated by poverty and malnutrition.

In due course the sheer logical unsustainability of the Castro Hlongwane Document would backfire, forcing Mbeki to moderate his stance; but in the shorter term the president and his political allies took full advantage of the situation. Resources were withdrawn from the relevant public health programs, so the antiretroviral drugs that might have controlled their illness became effectively unobtainable to South African AIDS victims. In their place appeared all sorts of "traditional remedies," usually herbal, as opportunists, presumably unable to believe their luck, cashed in. The death toll, and infection rates, inexorably mounted.

Not content with making it difficult for South African AIDS sufferers to gain access to the antiretroviral drugs they desperately needed, Mbeki threw his political clout behind the development of a drug called Virodene, whose main active ingredient is the industrial solvent dimethylformamide (DMF). Virodene was the brainchild of Olga Visser, who claimed once to have been head of the department of new technologies in medicine at Moderna University in Lisbon, Portugal (no such department exists at the university), and who, while more recently working as a medical technician at Pretoria Hospital, discovered in 1995 that dimethylformamide had antibacterial properties. Since AIDS is not a bacterial disease, this might not seem relevant; however, Visser and her entrepreneur ex-husband, Jacques Siegfried "Zigi" Visser, conducted unauthorized small-scale trials in South Africa. When the regulatory clamps came down in South Africa, they shifted their experimental locale to Tanzania; trials of Virodene were declared illegal there, too, but the Tanzanian army was prepared to allow the experiments to go ahead in military hospitals.

Visser reported to the world that tests of Virodene had produced "magnificent results," and she gained the ear of Mbeki. He pushed for further investigation of the drug, even though as early as 1998 South Africa's Medicines Control Council had declared the substance potentially harmful and likely useless. "The agency also noted a test-tube study published in 1997 in the journal *AIDS Research and Human Retroviruses* by University of Washington researchers that suggested that DMF could actually inflame HIV."[31] And there has been little evidence since that Virodene is anything more than a quack nostrum. Its cruelest effect was to divert South African resources and energies away from a realistic approach to the country's AIDS crisis, at the cost of untold lives.

Perhaps the most important figure on the South African AIDS scene after President Mbeki and his health minister, Manto Tshabalala-Msimang, has been the German-born vitamin seller Matthias Rath, who claims huge doses of his products can be substituted for antiretroviral drugs. He has achieved his eminence in South African society through a campaign not so much of justification for his pseudoscientific claims as of vilification of orthodox medicine, and in particular Big Pharma. To audiences away from his home territory some of his rhetoric may seem immediately risible; but then we're not living in communities terrorized beyond the realms of reason by a plague killing nearly a thousand people a day. Here's a sample entry, dated May 13, 2005, from Rath's website, *Dr. Rath Health Foundation Africa*:

> THE PHARMACEUTICAL DRUG CARTEL
> LAUNCHES WORLD WAR III TO PREVENT
> THE CONSTRUCTION OF A HEALTHY WORLD
>
> **Never before** in the history of mankind was a greater crime committed than the genocide organized by the pharmaceutical drug cartel in the interest of the multibillion-dollar investment business with disease. Hundreds of millions of people have died unnecessarily from AIDS, cancer, heart disease and other preventable diseases and the only reason that these epidemics are still haunting mankind is that they are the multibillion-dollar marketplace for the pharmaceutical drug cartel.
>
> **Never before** in history was there a larger, more profitable and bloodier fraud than that perpetrated by the pharmaceutical drug cartel. . . .
>
> **Never before** were the accomplices to this global drug-genocide more desperate about hiding this fraud and their crimes against humanity. . . .
>
> **Never before** were the drug cartel and its stakeholders in medicine, the media and their storm troopers in the streets more eager to silence those who expose this global fraud and genocide. That is the background of the attacks against the Dr. Rath Health Foundation, against traditional medicine and against the government of South Africa.
>
> **Now, we, the people, must build a world without diseases!**

This is not exactly the sort of marketing spiel you'd expect to find on the website of a Merck or a Pfizer. The approach obviously works for Rath, though, because for many years he's enjoyed the kind of market dominance in South Africa that the average corporation can only dream about. It hardly needs pointing out that Rath, a hugely wealthy globetrotting Westerner, is not, so far as the poor South African communities most ravaged by AIDS are concerned, in any communal sense a member of "we, the people."

A good short account of Rath's activities can be found in chapter 10 of Ben Goldacre's *Bad Science* (2009; see pages 70–71).

The success of Mbeki's policies can be gauged from the fact that in 1990 the rate of HIV infection in South Africa was approximately zero. By 2004 the number of the infected had risen to about one person in three. According to an estimate by the country's Medical Research Council, in 2005 and 2006 no fewer than 336,000 South Africans died of AIDS.

8

SELFISH HELP

Undoubtedly the most notorious of all New Age gurus—although it wasn't his gurudom that brought him the notoriety—was one Dragan Dabič, who practiced alternative medicine and wrote successful magazine articles on this and related subjects in Belgrade for a number of years until 2008; his specialty was human quantum energy, which means . . . well, your guess is as good as mine. Dragan Dabič was the pseudonym of onetime Bosnian Serb president and accused criminal against humanity Radovan Karadzič during his years on the run.

According to a BBC report following his arrest in July 2008, Karadzič, "[b]illed as Dabič, Spiritual Explorer . . . gave lectures comparing spiritual meditation and silent techniques practiced by orthodox monks. . . . In one lecture programme he was billed as a 'researcher in the fields of psychology and bio-energy.'"[1] In fact, Karadzič did have qualifications in psychology—his profession was psychiatry before he became a nationalist politician. As for the rest, he appears to have been making it up as he went along—a dramatic demonstration of how the possession of minimal or zero expertise need not in any way adversely affect the lucrative career of a self-help guru: All you need is the gift of the gab, and the gullibility of too many members of the public will do the rest for you.

In *Not in Kansas Anymore* (2005) Christine Wicker portrays various members of modern US society who believe themselves vampires or werewolves, who practice witchcraft or hoodoo, and so forth. Amid these voyeuristic entertainments, she intermittently conveys a much more serious picture, one of a society riddled with magical thinking—even though its members are rarely aware this is the case. As an example, consider the practice of grabbing the family Bible and stabbing a thumb down onto a random verse to get God's advice. Doubtless the people who do this believe they're acting in a rightly Christian fashion; in fact they're engaging in the ancient dark art of bibliomancy. In 2008 the US Christian group Pray at the Pump claimed credit for having brought gas prices down through intercessory prayer, and reacted skeptically to "just theories" by everyone else that the price drop was owing not to God but to economic factors—after all, they *had* prayed, and gas prices *had* come down. Again, magical thinking, here confusing correlation with causation.

As Wicker points out, where magical thinking becomes most obvious in modern society is in the plethora of self-help books that habitually clutter the bestseller lists. Most of us know that "Use the Force, Luke" is just fantasy. Alas, even the small minority who don't still translates into a heck of a lot of people, and it's at this vulnerability that self-help gurus strike.

The Secret (2006), by Rhonda Byrne, an expanded, illustrated, and highly designed catalogue of New Age gurus to whom the gullible might wish to send money, is only the latest and most egregious of these books. It would probably have gone nowhere had it not been for its enthusiastic endorsement by Oprah Winfrey.[2] Its overall thesis, expressed in a foggy and often nonsensical manner, is an old, old one: If you *wish* something hard enough, the universe will "hear," and will respond by giving you the thing you desire. If, perversely, the universe doesn't deliver, then obviously this must be *because you got the spell wrong*. The universe, you see, doesn't differentiate between good and bad; more importantly, it doesn't differentiate between the things you're thinking about because you want them and those you're thinking about because you *don't* want them—it doesn't understand negations. Thus, your thought "I don't want to get run over by a truck" enters the process exactly as if it were "I want to get run over by a truck." This leads the reader to an urgent speculation: Just how dimwitted can the universe possibly be?

It's difficult to understand how the underlying physics of the Secret could actually work. How *could* your brain radiate thoughts that would instantly permeate all corners of the universe? Leaving aside Byrne's scattered claims that everything in her book is explicable through quantum mechanics, there's her strong argument that you shouldn't concern your pretty little head with how it works. Her analogy here is with TV. Few, she says, understand how their television set works, yet we all know how to use it. The analogy seems fine at first glance, but not at second. If we want to learn how our TV works, we can read a book to find out. If we don't understand how the Secret works, the book we buy to find out—Byrne's *The Secret*—makes a point of not telling us.

There are other, peripheral problems with the physics. Thoughts behave like magnets, the book tells us: Positive thoughts attract positive results, negative thoughts negative ones. Confusingly, then, in behaving like magnets it seems thoughts don't behave like magnets at all, since the rule of magnetic poles is—as any schoolchild knows—that opposites attract and likes repel.

As many commenters have pointed out, *The Secret* is essentially a gospel of greed. Over and over again the focus is on the acquisition of wealth—more specifically, on getting money for nothing, or at least for minimal effort. Thus Byrne's contributor Bob Proctor, talking in the movie version:

> Wise people have always known this. . . . Why do you think that 1% of the population earns around 96% of all the money that's being earned? Do you think that's an accident? It's designed that way. They understand something. They understand *The Secret.*

Which is to say Mahatma Gandhi was an idiot because he could surely have manipulated his fame into a nice little earner. Curiously, Jesus Christ, according to *The Secret*, was a millionaire.

Most earlier promoters of the power of positive thinking were less than enthusiastic about "something for nothing" ideas, acknowledging that an important component of the whole process was the individual's directed hard work. Self-belief would *help* you get what you wanted, but wouldn't actually do it all for you. And many of the self-help gurus involved in *The Secret* have been tacitly backing away from this "something for nothing" aspect ever since the book's release, recognizing on their own websites that a bit of input from the wisher might be a good plan. But only tacitly: An explicit retraction would of course undercut the guru's supposed infallibility.

Thus Jack Canfield, responsible for the *Chicken Soup for the Soul* books, tells the audience of *The Secret* that, way back when, he gained the $100,000 he wanted through visualizing it—so much so that he mocked up a $100,000 bill and pinned it to the ceiling over his bed. And this did the trick! A few weeks later he suddenly realized that if a book he'd written were published, he might earn as much as $100,000 from it. And so it came to pass. He worked hard and he fulfilled his dream. Yet Canfield and Byrne present this as an example of the Secret performing as expected.

It's not just the raw greed that leaves an unpleasant taste in the reader's mouth. While there are occasional clumsy attempts to get around one obvious consequence of *The Secret*'s central thesis, it's inescapable: Our misfortunes are *all our own fault*. Thus, for example, the Holocaust was all the fault of the European Jews for harboring negative thoughts. Byrne makes various clumsy attempts to dodge this implication.

She is not modest in her claims about some of *The Secret*'s contributors—for example (p. 192), that John Hagelin is "regarded by many as one of the greatest scientists on the planet today." In fact Hagelin is known not as a scientist at all but primarily as the three-time Natural Law Party candidate for the US presidency. Another of Byrne's expert contributors, investment guru David Schirmer, may have found that practicing the Secret brought wealth to him rather too efficiently. In June 2010 the Australian Securities Investment Commission announced on its website that "Mr David Gary Schirmer . . . has been permanently banned from providing financial services following an ASIC investigation."[3] There have been further little difficulties among the roster of *The Secret*'s contributors. Two distinct DVD versions of *The Secret* were released in 2006, half a year apart, because of squabbling between Byrne and contributor Esther Hicks—distinctly nonspiritual, non-New Agey arguments over money, in fact.

Byrne and her contributors are not short of pithy observations that might help YOU improve your life, or at the very least enlighten you. A few from Byrne:

- Quantum physicists tell us that the entire Universe emerged from thought!
- To know what you're thinking, ask yourself how you are *feeling*. Emotions are valuable tools that instantly tell us what we are thinking.
- Every thought has a frequency. We can measure a thought.

And a couple from others:

- You can cook a man's dinner with electricity, and you can also cook the man!

—Bob Proctor

- . . . [I]t has been scientifically proven that an affirmative thought is hundreds of times more powerful than a negative thought.

—Michael Bernard Beckwith

In 2008, *The Secret* received the ultimate reverse accolade: It was listed by the UK economy hotel chain Travelodge as one of their top ten books most left behind by summer guests.

Byrne's initial inspiration was supposedly the book *The Science of Getting Rich* (1910) by Wallace D. Wattles. Wattles's thesis (p. 14) was that it's normal to want to be wealthy, and that if you don't achieve this you "are derelict in your duty to God, yourself and humanity"—in other words, if you're poor you're morally deficient. Ironically, when Wattles died in 1911, aged forty-one, just a year after publication of his book . . . you got there ahead of me, didn't you?

A more obvious ancestor of *The Secret* is one of the great classics of the field (still in print today), *The Power of Positive Thinking* (1952) by Norman Vincent Peale. Peale's book was full of the same pablum—imagine hard enough that you have something and it'll surely come to you; make yourself think happy thoughts and you'll find you *are* happier; and so on—but had a few sinister undertones. Peale had some uncustomarily harsh words for independent-minded women, for example, whose determination, he averred, made them ugly, whereas unquestioning faith in God (and in God's agents, their husbands) would soften their features, make them glow from within, and all the rest. When I suggest as much to my wife, the universe is only too likely to think I have a yearning for a black eye.

Still, in those prefeminist days perhaps that didn't seem so bad. What might have been more concerning was Peale's subtext that the way to make yourself happier—male or female—was to abdicate responsibility for yourself, to discard your notions of independence. Whether you were surrendering your *self* to God, gods, or simply the magisterial dicta of Norman Vincent Peale was not the point; what was important was that you were, in effect, giving up control of your life to some other entity. Doubtless that made a lot of people feel good in the shortish term, but sooner or later the arrival of the electricity bill must have been a rude reminder that there's more to happiness than a permanent plastic grin.

This idea of surrendering responsibility to another lies at the heart of the Recovery movement even today. Many of the Recovery courses, whoever might be the shaman offering them, follow the model of the Alcoholics Anonymous (AA) 12-Step Plan, an integral part of which is the profession of faith: Individuals aren't capable of effecting their own cures; they need to abandon their identity and let God, aided by the group, do the job for them.

AA claims a success rate of about 40 percent, which seems estimable—nearly half—until one looks at the figures for non-AA approaches. A 1967 survey published in the *American Journal of Psychiatry* found that, of 301 surveyed people arrested in San Diego for public drunkenness and instructed to partake in one of (a) the AA, (b) professional counseling, or (c) toughing it out on their own, the people in category (c)—judged by the rate of rearrest during the next twelve months—did by far the best, the people in category (a) by far the worst. In his book *The Natural History of Alcoholism* (1983) psychiatrist George Vaillant could find no evidence, after extensive studies, that any antialcohol therapy *at all* was useful; the best method seemed to be willpower.

Steve Salerno, author of the exposé *SHAM: How the Self-Help Movement Made America Helpless* (2005), in an earlier incarnation worked for the huge self-help publisher Rodale, and he lifts the curtain a little to give us a glimpse of practices there. Of particular interest is the Eighteen-Month Rule (*SHAM*, p. 6): "The most likely customer for a book on any given topic was someone who had bought a similar book within the preceding eighteen months." Further, Rodale was happy to splice and dice content around between different books, reusing chunks of text several times over, so the books bought by repeat customers weren't necessarily entirely new to them.

Obviously people buy multiple books on subjects that interest them; but, as Salerno points out forcefully, this habit shouldn't apply in the case of self-help books. But it would seem consumers happily accept that, if a self-help program doesn't work, it's not the program's fault but the person's. In a certain sense this is true—if you buy the Brooklyn Bridge on eBay, you have only yourself to blame when the bridge doesn't arrive in the mail as promised. But in that instance it's accepted that the *moral* (and legal) blame lies elsewhere: with the crook who gulled you. In the case of self-help books and programs, however, the purveyors claim self-righteously that they retain the moral high ground even as they're in effect selling everyone Brooklyn Bridges. The other difference is, of course, that most of us don't buy the Brooklyn Bridge more than once.

In short, it's not in the interests of the self-help gurus that more than a few of their clients should ever be *fixed*; should this happen the gurus might, horror of horrors, lose business. Hence the importance of clients being brainwashed into accepting that any failure of the program is owing to the client's inadequacy. After this brainwashing has been effected, the guru is set for life.

A major reason self-help books are bought in such vast numbers is that, precisely because of the huge readership, there are plenty of people around who for one or other reason became successful after reading a self-help book—and these people are naturally featured on TV chat shows by the likes of Oprah Winfrey. The countless people who read self-help books and whose lives continue unchanged are of course never featured anywhere; when was the last time you heard someone say, "Jim Smith read *The Secret* and it made no difference at all"? By contrast, had Jim Smith read *The Secret* and the following week won the lottery, *everyone*—Jim Smith included—would sure as eggs be talking about it . . . and giving the credit to *The Secret*. It's to a great extent the lack of negative stories that gives people the idea that self-help gurus have something valuable to say.

9

DISSENT ABOUT DESCENT

Colleen Thomas, who claims to have arrived on earth as mother of a host of "Pleadians,"[1] explains how the Creation story in Genesis is really, really scientific:

> The serpent of the Bible is in reality the subtle transverse wave of phonon magnetic fields (photon magnetic fields are far more excited and energetic and as such hotter) which are cold dark matter with cool energy (1,000 degrees Kelvin at most), the mediators of darkness and our contracted awareness of reality. Phonons emit radio frequencies that sound like a white noise hiss, that is why the etymology of the serpent in Hebrew is "hisser." Snakes move in transverse waves as well. It is clear that the story of creation is in reality the story of the physics of creation and [an] explanation of the causes of human conditions experienced on Earth as result of dark matter imposed intellectual and spiritual blindness. In and of themselves phonons are essential for creation to have contrast and form and as such no evil really exists. It is merely our ability to imagine evil that imposes evil conditions where none otherwise would exists, our minds have that kind of creative power!!![2]

She is apparently writing a book.

Although Darwin's is the name irrevocably linked in the public mind to the concept of human evolution, he was very far from the first to have evolutionary ideas; even among the ancients there were those—not excluding the author of Genesis, with its ordering of the emergence of living creatures—who showed glimmerings of evolutionary concepts, however wide of the mark. In the useful historical sketch included in the third (1861) and later editions of *Origin of Species*, Darwin mentions Aristotle's *Physicae Auscultationes*. Lucretius, in the first century BCE, spoke in *De Rerum Natura* of the concept of past extinctions, an important element of the mechanism of natural selection. But not all of the relevant speculations of the ancients seem with hindsight quite so enlightened: In the sixth century BCE Anaximander believed the first humans had been born, full-grown, from the wombs of fish, and it was attributed to Empedocles, a century or more later, that he believed the precursors of the first creatures were their separately emergent limbs and organs, which eventually came together rather as if forming colonial organisms.

In the eighteenth century the Comte de Buffon argued that God had created the universe but had thereafter left it on its own to evolve, living things included, according to the natural laws he had created as part of the package. Buffon's notions can be seen as a precursor to Lamarckism. In the early nineteenth century the French naturalist Jean Baptiste de Lamarck put forward a more substantive major evolutionary theory, one which still has a few adherents today. Where Darwin would invoke natural selection as an evolutionary mechanism, Lamarckism hypothesized that characteristics developed by individuals during their lifetimes could be transmitted to their offspring: If a weightlifter developed huge biceps, his children might expect to inherit that feature. Lamarck's theory could be summed up as: Characteristics acquired during an individual's lifetime can be inherited by the individual's offspring.

What's puzzling about Lamarck's scheme is that it seems to focus only on *useful* characteristics—longer necks to help giraffes reach higher leaves, and so on. But most of the changes we acquire in our lifetimes are *dis*advantageous. Wouldn't the children of elderly parents have bad backs, fallible memories, and an inborn urge to tell people to get off their lawn? More seriously, a major problem for the Lamarckian notion of heredity, and one that exercised many fine minds, was the matter of circumcision: If a culture has for centuries or millennia been circumcising its boy children as a matter of course, how come those pesky foreskins keep reappearing in each new generation?

Lamarck continued to have followers even after the publication of *Origin of Species* in 1859. In the US in the latter part of the nineteenth century the paleontologist Edward Drinker Cope, enthusiastically promoting evolution to a not always receptive populace, had the Lamarckian notion that evolutionary changes could be effected in one's offspring by willpower alone. Somewhat later, in 1920s Austria, Paul Kammerer produced a string of experiments that seemed to show acquired characteristics could indeed be inherited; he committed suicide when one of these was exposed as fakery.[3] In the USSR the science of genetics was crippled until the mid-1960s because genetics ran counter to the crank biological ideas of Trofim D. Lysenko, which included an extreme form of Lamarckism.

Returning to Darwin's precursors, another of interest was his grandfather, Erasmus Darwin, who expressed his evolutionary ideas primarily in *Zoönomia, or The Laws of Organic Life* (1794). These notions were not dissimilar from those of Buffon; however, Erasmus Darwin acknowledged instead the ideas of Lord Monboddo, the Scottish polymath and eccentric. The great difference between Monboddo's evolutionary scheme and Buffon's was that, where Buffon refused to contemplate that mankind might be cast from the same mold as the other primates, Monboddo was insistent this was the case, to the point of describing the anthropoid apes as Man's Brothers.

Arguably a far more important precursor than any of these was the anonymous 1844 bestseller *Vestiges of the Natural History of Creation* (after his death in 1871 the author was revealed as the Scottish writer, publisher, and scientific dabbler Robert Chambers). At least until the end of the nineteenth century this continued to outsell Darwin's *Origin*. Tennyson's famous lines from canto 56 of his poem *In Memoriam A.H.H.* (1849)—

> Who trusted God was love indeed
> And love Creation's final law
> Tho' Nature, red in tooth and claw
> With ravine, shriek'd against his creed

—often cited as a commentary on Darwinism, in fact predated the publication of *Origin of Species* (1859) by a decade, and was likely inspired by *Vestiges*. Chambers's book was a sort of compendium of those modern scientific theories that concerned transmutation—such as the evolution of stars and the changes within species over time. (Chambers was, bizarrely, inspired to write the book by conversations he'd had with phrenologists.) It's widely believed the popularity of *Vestiges* helped Darwin's theory gain acceptance ten years later.[4]

It is very obvious, then, that evolutionary ideas were much in the air for most of the first half of the nineteenth century.[5] Why then was *Origin* such a sensation? Richard Holmes, in *The Age of Wonder* (2008; p. 451), points to the answer:

> [W]ith the growing public knowledge of geology and astronomy, and the recognition of "deep space" and "deep time," fewer and fewer men or women of education can have believed in a literal, Biblical six days of creation. However, science itself had yet to produce its own theory (or myth) of creation, and there was no alternative Newtonian Book of Genesis—as yet. That is why Darwin's *On the Origin of Species* appeared so devastating when it was finally published in 1859. It was not that it reduced the six days of Biblical creation to myth: this had already been largely done by Lyell and the geologists. What it demonstrated was that there was no need for a divine creation at all. . . . The process of evolution by "natural selection" replaced any need for "intelligent design" in nature.

A major difference between Darwin's scheme of evolution and the earlier ideas along similar lines is that he saw no purposeful *directedness* in evolution, no teleology. There was no force, supernatural or otherwise, guiding evolution toward higher and higher forms—in fact, the notion of higher forms was itself misleading: a snail is as much a marvel of nature as is a human being. It's tempting to say that *this* was his great conceptual breakthrough: the recognition that evolution at its core relied on randomness, that it wasn't aiming toward some future goal.

This lack of directedness is a point missed by many of the less educated opponents of evolutionary concepts, who often cry, "How come monkeys aren't still evolving into human beings?"[6] Leaving aside the fact that no scientist has ever said we came from monkeys, the real answer is that there's no reason for monkeys to evolve in this direction. What monkeys are likely to evolve into is different monkeys.

In *Darwin's Gift* (2007) Francisco J. Ayala maintains that Darwin's great idea wasn't evolution, which is really just a byproduct, but the mechanism by which organisms and their organs can be so well "designed" for their niches: natural selection. In such a light, the nineteenth-century chorus of objections to Darwinism can be seen as coming not so much from Creationists, although these certainly added their voices, as from those who perfectly well accepted evolution but couldn't credit natural selection as a workable mechanism for it. And their objections weren't necessarily stupid: Challenged to explain how natural selection might function—how those favorable characteristics might be transmitted from one generation to the next—Darwin was stumped, veering sometimes toward Lamarckism and elsewhere toward his own, wrong hypothesis, pangenesis.[7] It was only with the rediscovery in 1900 of Mendel's theory of heredity that natural selection began to make perfect sense.

In large part, though, the rejections of evolution, whether from theists or scientists, were expressions of a gut reaction—a repugnance toward the notion that glorious, near-godly mankind could be, not just kin to the loathly beasts, but a product of randomness. (The process of natural selection is not random, in that undesirable mutations die out quickly while desirable ones *may* persist, but the trivial mutations that are the grist to natural selection's mill, while not completely random, are largely so within the limits of what genetic changes can accomplish.) This started long before Darwin formulated his and Wallace's theory. At the beginning of the nineteenth century the French naturalist Georges Cuvier argued vehemently against the protoevolutionary notions then being put forward, proposing instead that each species was the product of a separate Creation; he even echoed Empedocles (and foreshadowed the modern IDers) by claiming that each *organ* was the result of a special intervention by the Designer.

Like Buffon, the German naturalist Karl Ernst von Baer, writing in the same year that *Origin* was published, could conceive the evolution of other life forms but not of humans. This is still a prevalent view, even though it's hard to see how any pet owner, witnessing the countless similarities, could sustain it. (A relative once explained the difference to me: "We don't lick our butts clean.") Sir Richard Owen, the supposed pillar of British paleontology, also attacked Darwinian notions, though this seems to have been more because he feared being eclipsed.

A famous antievolutionary idea was advanced by Philip Henry Gosse in *Omphalos: An Attempt to Untie the Geological Knot* (1857). Gosse believed the earth had been created in 4004 BCE, and he set out to reconcile this with the evidence being produced by geologists, astronomers, and paleontologists for events that had taken place long, long before. Moreover, he had to consider the weighty matter of the navel (*omphalos* is Greek for "navel"). There had been major theological debates as to whether or not Adam had one. Gosse was on the side of those claiming he did, and went further: Adam had been born fully grown and with nails and hair, both of which would normally require prior growth.

Gosse solved his dilemma by proposing that God created the world, the universe, and Adam *as if* they'd had a previous history. He had planted the fossils in the ground as a part of his scheme. The evidence of the astronomers and geologists, far from demonstrating the immense antiquity of things, was instead a remarkable exposition of the masterful completeness of God's artifice: Each new piece of evidence they unearthed was a tribute to the Creator's conscientious craftsmanship. Nothing the scientists found implied that events had occurred prior to the Creation, in 4004 BCE.

Clearly, a theory like this can never be either proved or disproved—indeed, it could be used to support a claim that the Creation occurred just five minutes ago, with God having carefully constructed all our memories to mislead us. The theory begins to stumble, of course, as soon as we start to wonder why God should trouble to play this prank, as even Gosse's contemporaries pointed out; Charles Kingsley wrote to him that "I cannot . . . believe that God has written on the rocks one enormous and superfluous lie for all mankind."[8]

Among those who superficially accepted Darwinism but sought to "improve" it by reintroducing strong Lamarckian elements was the French philosopher Henri Bergson, who abhorred natural selection's overtly mechanistic thrust. In books like *L'Evolution Créatrice* (1907; *Creative Evolution*) he attempted to marry Darwinism with his own theory of consciousness. To the biological side of this coupling he brought the ancient concept of the *élan vital*, an immaterial force that directs the progress of life. This innovation was not much appreciated by biologists; as Julian Huxley pointed out, it was rather like telling a railway engineer the reason his engine worked was an undetectable *élan locomotif*. But it seemingly went down well among nonscientists. In the preface to *Back to Methuselah* (1921) George Bernard Shaw expressed at length his support for the notion of Creative Evolution, the *élan vital*, and Lamarckism in general by contrast with natural selection and its nasty mechanistic functionality.

It's hard to take seriously the rival evolutionary theory offered by the US zoologist Henry Fairfield Osborn. In his scheme, new types of life forms had emerged periodically through some form of spontaneous generation. When they first appeared they were confined to a single locality; they had very generalist characteristics and possessed a mysterious attribute called "race plasm." As they spread away from their original territory, specialist adaptations to the template occurred in response to the new environments the creatures found themselves in, giving rise to species; the effectiveness of the adaptations depended on the quality of the "race plasm." Osborn, a racist of the Aryan school, invented the concept of Dawn Man, an entirely unevidenced primordial species from which Nordics had descended, with all the other races owing their ancestry instead to shambling "apemen" like the Neanderthals. The idea that the different races might have arisen from separate ancestors was not a new one; called polygeny, it had been long discredited. The form embraced by Osborn and others, orthogenesis, for a while had some minor influence.

It is often not realized that for a while around the cusp of the nineteenth and twentieth centuries it looked as if Darwin's theory would be rejected by science: Its acceptance was certainly on the ebb. What would eventually turn the table back in favor of Darwinism was the discovery in 1900 of Mendel's theory of heredity, his work on which had been lurking forgotten in dusty journal pages for three and a half decades. Once genetics[9] was married to natural selection, Darwin's theory seemed inescapable; but it was some time before that happened. For much of the first half of the twentieth century it wasn't evident to biologists how natural selection and genetics meshed together—how they were the two essential complements in explaining how evolution functioned. It took the pioneering work of a dozen or more great geneticists—notably including J. B. S. Haldane—to achieve the fusion, as summarized in Julian Huxley's *Evolution: The Modern Synthesis* (1942).

A rival theory of evolution that sprang up as the nineteenth century passed and the twentieth century began was likewise given additional impetus by the rediscovery of Mendel's work on heredity. This was mutationism, the idea that evolution proceeds not via the slow accumulation of minuscule changes, as Darwin insisted, but through abrupt discontinuities—i.e., rare individuals that, through mutation, differ radically and advantageously from their parents. Obviously, these mutations would have to be capable of transmitting the new characteristics to their offspring.

Mutationism's supporters were by no means fringe scientists. Foremost was Hugo de Vries, whose *Die Mutationstheorie* (1901) popularized the idea. He backed up his case with his own studies of the evening primrose, maintaining that among these flowers true-breeding mutations appeared frequently. Later studies demonstrated that de Vries had been somewhat (not entirely) overstating his claim. Even as the theory fell from favor, the word "mutation" was adopted into genetics, where it soon gained its modern meaning.

Something shared by Mendel's theory of heredity and the theory of evolution by natural selection is that, despite the fact that between them they were to revolutionize our understanding, neither were appreciated at the time as anything beyond humdrum. Mendel's work outlining the laws of heredity was left to rot for over thirty years. And the reaction of the Linnaean Society to the presentation in 1858 of two crucial papers by Darwin on the origin of species and one by Wallace can be summarized by the comment made by the society's president, Thomas Bell, when giving his review of that year:

> The year which has passed . . . has not been unproductive in contributions of interest and value, in those sciences to which we are professedly more particularly addicted, as well as in every other walk of scientific research. It has not, indeed, been marked by any of those striking discoveries which at once revolutionize, so as to speak, the department of science on which they bear.[10]

Not all theologians had difficulty accepting that humans had evolved from lower forms of life. It was perfectly reasonable to some that God could have worked indirectly to concoct his finest creation. This was put with peculiar elegance by Augustus Hopkins Strong of the Rochester Theological Seminary in his *Christ in Creation and Ethical Monism* (1899; p. 169): "The wine in the miracle was not water because water had been used in the making of it, nor is man a brute because the brute has made some contributions to its creation." This attitude of compromise between religion and evolutionary science is still prevalent, in various forms, today.

Many old-earth Creationists—and obviously all IDers, too—have adopted what they regard as a sort of halfway stage between Creationism and Darwinian evolution. Confronted by the profuse evidence of evolutionary adaptation going on in the world around us, they concede that *micro*evolution is a reality—that adaptation occurs within species—while rejecting the possibility of new species arising through adaptation, or *macro*evolution. In reality, the acceptance of microevolution is, ipso facto, an acceptance of macroevolution, too. There's no abrupt transition line between the two processes. A further difficulty for the stance is that macroevolution—the production of new species—has been observed in action in the real world.

Perhaps remembering how ill its attempts at denialism had fared during a previous scientific revolution, the Roman Catholic Church adapted rather more readily than some of the other denominations to the notion of natural selection. A decree issued in 1909 relaxed the doctrinal necessity for Catholics to believe the days of the Creation, as described in Genesis, were literally that: days. The interpretation "period of time" became acceptable, thereby allowing for the vast eras required by evolution. For decades the Catholic Church and its officers tacitly assumed the validity of evolution while making no statement one way or the other. On October 23, 1996, though, Pope John Paul II came out of the closet on the subject. This wasn't an admission that God played no part in the process. He cited Pius XII's 1950 encyclical *Humani Generis* to stress God's role: "If the human body take its origin from pre-existent living matter, the spiritual soul is immediately created by God."[11]

Although there was some initial nitpicking among the faithful as to the exact reading of John Paul's statement, it became clear he meant exactly what the Creationists dreaded: The Roman Catholic Church fully accepted natural selection as the mechanism for human evolution.

On July 7, 2005, however, Cardinal Christoph Schönborn published in the *New York Times* an op-ed called "Finding Design in Nature" that appeared to support ID.[12] Since Schönborn was close to Benedict XVI, who'd been elected pope just three months earlier, this article was widely interpreted as signaling a radical change of tack by the Catholic Church. According to Damian Thompson in his 2008 book *Counterknowledge* (p. 32), it was fairly soon discovered that Schönborn had been assisted in writing the op-ed by Mark Ryland, a senior fellow of the heftily misnomered Center for Science and Culture, the Discovery Institute's prime disseminator of ID propaganda (see page 180). It seems all too likely Schönborn allowed himself to become a victim of said propaganda; and indeed he was soon backtracking.

Far more definitive was an article carried in the Vatican newspaper *L'Osservatore Romano* on January 17, 2006, by Fiorenzo Facchini, an evolutionary biologist at the University of Bologna. Facchini spelled out the principles of Darwinian evolution and didn't mince his words about ID: "It doesn't belong to science and the pretext that it be taught as a scientific theory alongside Darwin's explanation is unjustified." (At the same time, he insisted there was a role for a designer in our *spiritual* evolution.) On January 31, 2006, addressing the Palm Beach Atlantic University in Florida on the subject "Science Does Not Need God, or Does It? A Catholic Scientist

Looks at Evolution," the Vatican Observatory Director, Father George V. Coyne, heavily underlined the Church's position:

> I would essentially like to share with you two convictions in this presentation: (1) that the Intelligent Design (ID) movement, while evoking a God of power and might, a designer God, actually belittles God, makes her/him too small and paltry; (2) that our scientific understanding of the universe, untainted by religious considerations, provides for those who believe in God a marvelous opportunity to reflect upon their beliefs. Please note carefully that I distinguish . . . that science and religion are totally separate human pursuits.[13]

Naturally, Conservapedia® is full of articles deriding evolution, which it regards as still "just another theory" that must seek to prove itself against the established default explanation, Creationism. Since this is a perspective that doesn't play too well in the greater world, Conservapedia's editors have had to resort to some sophisticated arguments in support of their position . . . such as that old kindergarten favorite: "You're fat!"

Hence the appearance of articles like "Evolutionists Who Have Had Problems with Being Overweight and/or Obese"[14] and the special feature "Atheism and Obesity,"[15] with sections like "Lesbianism, Atheism, and Obesity" and "Picture of an Overweight Atheist Christopher Hitchens."

As if that weren't intellectual ammunition enough, there's "Essay: Does Richard Dawkins Have Machismo?"[16] which presents the burning question: "Is the atheist and evolutionist Richard Dawkins a man filled with courage, truth, and conviction or a man who is a cowardly pseudo intellectual pantywaist?"

Arguing with Creationists is not always easy. A Facebook exchange that exemplifies the problem went viral at the end of 2010. The original post:

> ** Fact—If the earth was 10 ft closer to the sun we would all burn up and if it was 10 ft further we would freeze to death . . . God is amazing!!

One of the commenters tried to introduce some science to the conversation:

> to anyone wondering, that's not true. 1) Earth's orbit is elliptical and the distance from the sun varies from around 147 million kilometers to 152 million kilometers on any given year. 2) Every star has a habitable zone that is affected by the size of the star and its intensity. The sun's habitable zone

> is about 0.95 AU to 1.37 AU. . . . Earth's orbit could be decreased by 4,500,000 miles or increase by 34,000,000 miles and still be in the habitable zone. 3) if your claim was true any moderately sized earthquake could have taken us out of the habitable zone. sorry.

To which the original poster replied:

> Okay thats cool and alll but don't ever comment on my status telling me that i am wrong everrr again. I didn't ask you did i? Answer: NO[17]

Michael Swanwick, in his time-travel novel *Bones of the Earth* (2002), brilliantly captures the goalposts-shifting rhetorical technique of the modern Creationist. "If time travel is real," writes the author of the fictitious tract *Darwin Antichrist*, "then why haven't we found human footprints among the fossil dinosaur tracks?" (p. 122).

Arguing online or in print is one thing, arguing face-to-face another. In his entertaining essay "Fighting for Our Sanity in Tennessee: Life on the Front Lines" (2001), evolutionary biologist Niall Shanks tells the tale of being invited to debate the young-earth Creationist Duane Gish. Shanks's suspicions that he was in some way being set up were aroused by Gish's insistence that he, Shanks, go first in the debate.

Accordingly, Shanks borrowed videos of some of Gish's earlier debates to try to get some tips. He discovered that Gish's presentation never changed: Whatever his debating rival might have said, Gish's "response" was identical. Shanks therefore prepared a presentation in which he introduced each point he knew Gish was going to make, and demolished it.

But he went further. He stole all of Gish's jokes and one-liners, too . . .

One assumes Gish was presenting his case as a scientific one, not as a matter of theology or faith. In his essay "Evolution, Thermodynamics, and Entropy" (1973), Henry Morris, one of Creationism's great figures, likewise invoked some hard science:

> Not only is there no evidence that evolution ever *has* taken place, but there is also firm evidence that evolution never *could* take place. The *law of increasing entropy* is an impenetrable barrier which no evolutionary mechanism yet suggested has ever been able to overcome. Evolution and entropy are opposing and mutually exclusive concepts. If the entropy principle is really a universal law, then evolution must be impossible.[18]

Using exactly this argument, we can demonstrate that it's impossible for a baby to arise from the fusion of an undifferentiated sperm cell with an undifferentiated egg cell, and for an adult to develop from a baby. Since both processes indubitably *do* happen, there must be something wrong with the argument. This leads us to an interesting difference in modes of thought between real science and Creation Science. The Creationist is here claiming evolution of any kind—whether of life forms or of stars—is a violation of the second law of thermodynamics (the "law of increasing entropy"). Of course, evolution *isn't*; but, were it shown this was so, the approach of real science would be to question the second law of thermodynamics—just as, for example, Einstein questioned Newtonian gravitation. On the one hand, we can see stellar and organic evolution happening; on the other, we have a theory, an authority. Science seeks a resolution and, if one can't be found, re-examines the authority. By contrast, the Creationists' approach is an appeal to authority—with the second law standing in for the authority they're actually invoking.

For more hard science, here's the explanation of the Noachian Flood, as offered by Carl E. Baugh of the Creation Evidence Museum in Glen Rose, Texas:

> The voice of God (whether by direct vocal intervention or by indirect vibrational disruption) at microwave energy level penetrated the great water reservoir beneath the earth's granite crust. With microwave's unique effect on water, this agitated medium rapidly disrupted the planet's subterranean structure which housed the designed nuclear reactors and internal foundations.
>
> The violent heated waters ruptured the granite crust and sent hot jets of steam upward through the thin firmament suspended above the earth. This action opened channel windows in the crystalline canopy and caused its collapse. The mass fell as liquid rain in the temperate zones and dropped as ice at the poles. Subsequent expulsion of water and chemical elements from earth's disrupted interior saturated the surface floods and trapped living organisms as fossils in sedimentary deposits.
>
> Gravitational attraction of the moon brought the global body of floodwaters into resonance. Cyclical tidal action deposited organic and inorganic materials into conformable sedimentary layers. The vast majority of the "geologic column" is adequately explained by a single year of global flood activity.[19]

An interesting feature of the museum is the pair of hyperbaric chambers in whose interiors Baugh has recreated what he believes to be a replica of the harmonious environment of the primordial earth before Eve arrived, bringing sin and death with her. Baugh expects that dinosaurs will soon begin spontaneously to emerge in these biospheres.

Another tourist magnet, Dinosaur Adventure Land, in Pensacola, Florida, was founded in 2001 by Kent Hovind, one of the more colorful recent figures in young-earth Creationism. As Jim Gardner has summarized, Hovind "is the *cause célèbre* of the monumentally ignorant creationism movement, because he has the nerve to stand up and say publicly what normal people seek medical attention for thinking in private."[20]

That dinosaurs and humans must have coexisted is a major plank of Hovind's quaint cosmology, because he recognizes that six thousand years simply doesn't allow enough time for the dinosaurs to have been extinguished in the Flood before *Homo sapiens* came on the scene. Besides, it says in Genesis that God created the first humans just a few days into the whole tableau. Moreover, we know humans were around before the Flood because otherwise Noah wouldn't have been there to build the Ark.

In Hovind's version, the Flood happened when God sent a giant ice meteor hurtling toward the earth; much of it broke off to form the craters of the moon, Mars, etc., but much came crashing to earth in the polar regions in the form of ultracold snow, which froze the mammoths so quickly they didn't have a chance to run away. Speaking of the Ark, since Noah saved two of every creature aboard it, that doughty vessel must have included all varieties of dinosaurs, as well, so there could very well still be dinosaurs alive today, only we don't know about them. Or do we? Just to cover all asses, as it were, Dinosaur Adventure Land contained a presentation featuring the Loch Ness Monster.

Unsurprisingly, Hovind's views have been derided—when noticed at all—by scientists and many others. What *is* surprising is that among the latter have been quite a few of his fellow young-earth Creationists. Clearly there are different degrees of crazy.

In January 2007 Hovind was sentenced to ten years in prison, having been found guilty of fifty-eight federal crimes concerned with tax evasion and illicit bank transactions.

The Creationist group Answers in Genesis (AiG) opened its own Creation Museum at Petersburg, Kentucky, in May 2007. In December 2010 the organization announced it would open its theme park, Ark Encounter, in Kentucky in 2014—an enterprise in part funded by taxpayer dollars. Most excitingly, AiG officials appear to have promised that *their* Ark will contain *real dinosaurs*!

A recurring problem for people like Hovind and AiG is to explain how Noah managed to get sample pairs of dinosaurs onto the Ark—after all, brachiosaurs were *big*. AiG's Ken Ham has offered a solution: The wily mariner got round the problem by choosing *young* dinosaurs! Being only partially grown, they took up less space.

The Australian-born Ham is another of young-earth Creationism's luminaries. He has the frequent debating tactic of asking the rhetorical question: "Were you there?" If you weren't there to witness the beginnings of all things, according to Ham, you have no way of knowing the Genesis account is untrue . . . which means it *is* true.

In May 2007, the Australian Creationist organization Creation Ministries International (CMI) asked Ham and his AiG colleagues the pertinent question "Were you there?"—or some variant thereof—before the Queensland Supreme Court as they sued to try to find out why so much of their funds had been devoted to paying Ham's and others' expenses. The suit was eventually settled "amicably."

But perhaps most colorful of all the current crop of young-earth evolution deniers is the New Zealand-born evangelist Ray Comfort who, frequently in conjunction with ex-child actor Kirk Cameron, has demonstrated a publicity flair most of his rivals must envy. Together, the pair released a hugely popular YouTube video of themselves discussing a banana, their point being that surely this fruit cannot have evolved but must have been designed by God, for otherwise how could it accommodate so conformably to the human hand and unzip so niftily for eating? Primatologists pointed out that humans habitually open bananas from the wrong end, the far *better* technique being the one employed by our close primate relatives: pinch the other end hard so the tough bit comes off, then squeeze out the banana's contents like toothpaste from a tube. Obviously it was with chimps, not us, in mind that God designed the banana.

A further antic by Comfort (whose recent books are published by WND Books, the publishing arm of Far-Right organization *WorldNetDaily*) was the production in late 2009 of a new edition of Darwin's *Origin of Species* as a free giveaway for schools and colleges. This edition found room for a rambling fifty-page foreword by Comfort in which he explained why silly old Darwin had got it all wrong, but (despite the claim on Comfort's *Living Waters* website that "[n]othing has been removed from Darwin's original work"), Darwin's preface, glossary, and four chapters of his original text, which, as biologists were swift to point out, just happen to contain arguments that Creationists have traditionally had difficulty countering, were omitted.

Comfort's blog *Atheist Central* revealed, in his November 6, 2010, posting, this strange claim:

> The name of Jesus Christ is the most despised name in history. If you disagree, then name one other historical figure who has his name used in place of a cuss word. How about wicked men like Hitler, Judas, or Rasputin? None were despised enough to use their name in such a way.

That's because, when you yell "Jesus!" on stubbing your toe, you're invoking the deity. Most of us don't invoke Hitler.

> Heres a problem you can't answer. Why did evolution stop? Why don't human people have wings? I think your badly misguided. You should look for real answers in the Bible, not evolutionist propaganda. Jesus wasn't a monkey, he was the Son of God. A God that also made you in his image. That image is sacred and should not be mocked with evolutionist fantasy. If you won't pray for your salvation, I will.
>
> —"Saturday hate mail-a-palooza," *Daily Kos*, October 30, 2010[21]

On October 24, 1996, David W. Cloud of the Fundamental Baptist Information Service released a document rebutting Pope John Paul II's then-recent declaration of support for Darwinism. At first Cloud's rebuttal—called simply "Pope Supports Evolution"[22]—seems the usual trite, muddleheaded stuff, but the devil (or whoever) lies in the details. For example:

> Genesis says the world was created perfect, then fell under sin and God's curse. This is consistent with everything we can observe. Everything is winding down. Everything is proceeding from order to chaos. Everything is corrupting. Evolution would require the exact opposite.

Perhaps this argument was what Concerned Women for America's spokeswoman Christine O'Donnell was thinking of when she said: "Well, creationism, in essence, is believing that the world began as the Bible in Genesis says, that God created the Earth in six days, six 24-hour periods. And there is just as much, if not more, evidence supporting that."[23]

But back to Cloud:

> [I]f evolution were true . . . [w]e would have a world filled with monsters and unpredictable madness, part one thing and part another, a fish becoming a bird, a frog becoming a rat, a lizard becoming a bird, partially formed beaks which do not yet have a purpose, partially formed feet, partially formed wings, partially formed eyes, partially formed brains.

We *would*? Is it just a matter of subterfuge that those wacky evolutionists have omitted all this good stuff from their textbooks? Cloud's ideology seems to have rendered him less able than the average ten year old to understand what the principle of evolution by natural selection actually means.

Cloud, like other Christians who deny evolution for fundamentalist reasons, does so essentially because God explained in Genesis how humans came into being. Unfortunately, God told two different, mutually incompatible Creation stories in Genesis. Since both can't be right, we have to conclude God's an unreliable witness.

But the choice is broader than between these two. The second edition of David A. Leeming's *Encyclopedia of Creation Myths* (2009) runs to two volumes and discusses over two hundred different versions of the Creation. In the Creation myth of the Bakuba people of central Africa, for example, in the beginning there was just darkness, water, and a giant called Mbombo. The sun and stars came about when Mbombo got a stomach upset and vomited them. The sun evaporated a lot of the water, revealing dry land. Later, Mbombo upchucked again, his vomit this time being all of the earth's living things, us included. The Masai have a myth that the first humans emerged from the severed leg of some other creature. The leg swelled as it decomposed, until finally it burst, releasing a man on one side and a woman on the other. The Wapangwa people of Tanzania have the myth that the earth originated as a vast ball of feces produced by white ants. How *dare* you insult someone's religion by rejecting any of these!

The Church of Jesus Christ of Latter-day Saints, while accepting the

Genesis account, has a different interpretation of it. In the Mormon view, the matter of which we and everything else are made is eternal. However, despite being eternal, it has not always had its current forms; in order to gain these it has had to be organized. When God created the world, then, he didn't create it out of nothing but instead organized matter that was already in existence. Likewise, when creating Adam, he organized matter to form a container for Adam's soul, which already existed. Souls originated long before the earth did, when Heavenly Mother and Heavenly Father procreated to generate untold numbers of spirit offspring—i.e., souls. Depending on how good or bad we were at this stage in our existence, we have a good or a bad time of it in our earthly incarnation. Thus, the Emperor Nero was obviously a virtuous fellow because . . . no, that can't be right.

It would be a mistake to think the only breed of fundamentalists to bitterly oppose evolution are Christians. There are strong schools of Creationism among adherents of the other major religions too.

> THEREFORE *DARWIN*;
> IS THE WORST *FASCIST* THERE HAS EVER BEEN,
> AND THE WORST *RACIST* HISTORY HAS EVER WITNESSED.
> —"Harun Yahya," "Darwinism Is the Main Source Of Racism," 2009[24]

In a sense, Creationism has always been rife in the Islamic world. Since the Qur'an is accepted as the indisputable word of God, and since the Qur'an contains a (very sketchy) account of the Creation, the discussion ends there. At the same time, Creationism has not until relatively recently been an issue in Islam except perhaps among scholars. It would be a simplification to say that, in essence, Muslim scholars ignore evolution while the general Islamic populace remains ignorant of it, but only just.

The group almost single-handedly responsible for changing the Muslim world's indifference to evolution is the Bilim Arastirma Vakfi (BAV; the name translates as "Science Research Foundation"), founded in Turkey in 1990 by the radical Islamist, virulent anti-atheist, and conspiracy theorist, Adnan Oktar. Although the BAV is internationally the more prominent of the two major organizations Oktar has created, the other, Millî Degerleri Koruma Vakfi ("Foundation for Protection of National Values"), founded in 1995, is significant domestically; it promotes a rightwing, conservative religious agenda analogous to that promoted by most US groups whose names feature "Family" or "Values." Oktar has a daily two-hour television show on which he promotes, Pat Robertson-style, his often highly offensive views.

Following publication of his first book, *Yahudilik ve Masonluk* (1986; "Judaism and Freemasonry"), Oktar served nineteen months in prison because of its inflammatory antisemitism; later he was to blame his imprisonment on a plot by Freemasons determined to stop him from writing about them. The Freemasonic plot didn't work. Oktar's *Global Masonluk* (2002; "Global Freemasonry") is a ferment of conspiracy theory. Not only are Freemasons mounting an attempt to impose an atheistical One World Government, they're also deceiving the rest of us into thinking Muslims do bad things. A sample from Oktar's associated website, globalfreemasonry.com:

> Freemasonry, the system of the antichrist[25] that holds the world under its sway and is leading the whole planet toward irreligion, wants to give a false impression of Islam by pretending to be Christian. In order to achieve that aim it seeks to train terrorists who appear to be Muslims.

Oktar's 1995 book *Soykirim Yalani* ("The Holocaust Lie") is a work of Holocaust denial that would be hilarious were its subject less grim. It seems odd he should mount an exoneration of the Nazis, because elsewhere he excoriates Darwin and evolution for being responsible for, among other things, Nazism. Perhaps, confronted by Jews, Darwin, and Nazism, he just couldn't decide which he hated the most.

Soykirim Yalani was published, like all of Oktar's books, under the pseudonym "Harun Yahya." It's hard to know quite how many books "Harun Yahya" has published, many of them cheap productions designed to be given away free to luckless citizens and college students; some estimates put the number around two hundred, and the "Harun Yahya" website claims three hundred. It's believed that for some years now "Harun Yahya" has been less Oktar's personal pseudonym than a collective house name for upward of thirty BAV members (although Oktar claims otherwise).

The collective's magnum opus to date has been *Yaratilis Atlasi* (2006; translated as *Atlas of Creation*, the title under which it is now more generally known), nearly eight hundred large-format pages packed with paired photographs comparing living creatures with fossils in an endeavor to show there's no difference between them—that claims of modern forms having evolved from earlier ones are a sham. (Two further massive volumes followed in 2007.) Thousands of copies were sent out free to educators worldwide. A few Western biologists, including Richard Dawkins, reviewed it, finding it full of elementary errors of identification, examples of captions declaring two images near identical when they quite clearly aren't, and, most worryingly, copious evidence that "Harun Yahya" does not understand (or chooses to lie about) the very basics of the scientific theory "he" is doing his best to expunge from human consciousness.[26]

Oktar presents himself and the BAV as the acceptable face of Creationism, but the authoritarian streak is broad and clearly visible. When in 1999 he was brought to trial on the grounds that the BAV was a criminal organization and that he personally had used bribery, threats, and intimidation, the case eventually had to be abandoned because so many of the witnesses inexplicably withdrew their testimony. Among other charges were that attractive young female BAV adherents were being used as sexual decoys to attract rich men, with the encounters being secretly filmed for purposes of blackmail—and to discourage the women themselves from leaving the BAV should they ever choose to do so. It is hard to escape the view that the BAV isn't so much an antiscientific movement or even an ideological/religious movement as a cult, with heavy Mob overtones.

Adnan Oktar is far from the only Islamic Creationist. Another of note is Mustafa Akyol, an individual with no scientific credentials who has nevertheless somehow managed to parlay his way into being called as an expert witness in Creationist trials worldwide, like the 2005 hearings in Kansas when the State Board of Education and its State Board Science Hearing Committee deliberated inserting ID into the school science standards along with a statement that evolution is "just a theory" rather than established fact. (The measures were passed, but in 2007 they were overruled when religious conservatives lost their majority on the State Board of Education.) Akyol thinks rejection of evolution on both sides of the east/west divide offers a unique opportunity to bring those two sides together, and is thus a good thing:

> [T]here are some westerners who say we know that materialism is a philosophy but science doesn't really support that philosophy; science can also support theism, the fact that there is a God. So when Muslims recognize this they can stop looking at the world in terms of west versus the east; so we can have a dialogue with the west.[27]

So if we simply unite in ignorance and falsehood we'll all live happily ever after—a policy that didn't work some centuries ago when Islamic science, once the splendor of the world, plummeted into a pit from which it has still not emerged.

What grounds are there for optimism about Turkey's fight against a retreat to the Dark Age? The European Union may hold the key. Turkey is eager to become a part of the EU and cannot understand why there is so much resistance within the EU to the idea. In large part, yes, it's because the EU is unlikely to admit any nation that seems so sluggish in cleaning up its human rights record. But another reason may well be the growing political desire, not just among average Turks but within government circles, to reject all the advances in human understanding and welfare associated with the Enlightenment in favor of a worldview that better belongs to the Middle Ages. Among the most significant players in bringing this sorry state of affairs to pass is the BAV. In 2007, Nicolas Birch reported the BAV's Tarkan Yavas as claiming the real reason Turkey's application to the EU was going so badly was that "Darwinism breeds immorality, and an immoral Turkey is of no use to the European Union at all."[28]

In March 2009, as the rest of the developed world geared up for the Darwin anniversary celebrations, the editor of Turkey's equivalent of *New Scientist*, *Bilim ve Teknik* ("Science and Technology"), was sacked for planning a cover feature about Darwin. Turkish teachers have been suspended for giving instruction on evolution, even though doing so (alongside Creationism) is part of the country's school standards. The activities of the BAV have the tacit and often not-so-tacit support of senior government figures. And, as Aykut Kence observes, "Turkey is the only secular state in the world that has creationism in its science textbooks."[29]

The most distressing aspect of Turkish Creationism is perhaps that the Turkish Republic was founded with such high ideals. Rightly or wrongly, it is claimed that Kemal Atatürk wrote the chapters on evolution in the school textbooks introduced after his successful revolt against theocracy; certainly that would have been consonant with what he hoped for from the nation he founded.

Although the classical rabbinical teachings offered a form of Creationism that sounds like its fundamentalist Christian counterpart—for example, their own calculations led the rabbis to think the world was about six thousand years old—in general, Jewry had little difficulty adapting itself to the idea of evolution and the long timescales it demanded. Today evolution is accepted by almost all Jewish denominations as compatible with Judaism, albeit often in a version that allows room for a Creator to have started the whole thing off. There is also a strand of Jewish thought that gives credence to what is confusingly called Intelligent Design; this is not the ID of Christian fundamentalists and the Discovery Institute but a bolting onto the back of evolution of ideas concerning evolution's purpose and goal. In a sense this is a denial of the theory of evolution by natural selection, one of whose major thrusts is that evolution has neither purpose nor goal. This Jewish compromise seems much like its Christian equivalents.

There are also those who, like Muslims who interpret passages in the Qur'an and claim they describe the latest discoveries of science (see page 46), make attempts to match up the Genesis account with what science has revealed to us about origins. A doyen of this approach is Gerald L. Schroeder, whose books on the topic are *Genesis and the Big Bang* (1990), *The Science of God* (1997), *The Hidden Face of God* (2002), and *God According to God* (2009). An extensive demolition of the first two of these (with a short note about the third) is Mark Perakh's "Not a Very Big Bang about Genesis" (1999; revised 2001).[30]

Schroeder's first concern is to match the Genesis story's six days of Creation with the fact that science has revealed the universe to be some 14 billion years old. In his first book he calls upon the Special Theory of Relativity, pointing out that the measure of time's passage depends very much on the frame of reference in which you happen to be: The faster your frame of reference is moving, the more quickly time will seem to be passing in a frame of reference that's "stationary." So, if God were moving around at very high velocity all the while he was creating the universe and guiding its evolution up until the creation of Adam, it's conceivable that each of the days he experienced was equivalent to billions of years undergone by the universe. At the moment he created Adam, according to Schroeder, God decided to leave his prior frame of reference and share Adam's—i.e., to experience time's flow at the same rate as we do now.

This piece of speculation in no way demonstrates that science and the Torah are essentially telling the same story, since there's no evidence to support, just for starters, the notion of a near-light-speed-traveling deity.

In *The Science of God* Schroeder tries a different Relativistic tack, abandoning the Special Theory in favor of the General Theory. Time-dilation effects occur not just through high velocity, as per the Special Theory, but also in gravitational fields, as per the General Theory. Schroeder posits that immediately after the Big Bang the universe consisted solely of energy; there was therefore no gravity. Soon, though, energy started being converted into matter (via $E = mc^2$), meaning there was progressively more mass in the universe and therefore increasing gravitational forces. The universe's clock has thus been slowing down ever since the Big Bang (at least until the creation of Adam); Schroeder carefully calculates that the universe's first "day" lasted 8 billion years, its second 4 billion years, and so on. Adding up the results of his calculations, he arrives at the same figure for the universe's age as the Torah's "six days"—what a surprise!

In both of these scenarios, on creating Adam, God abandons his previous frame of reference to adopt the same one as Adam's. A pertinent question is: Why? It's a question Schroeder fails to answer.

Other Jewish apologists for the Creation myth have included the fourteenth-century Rabbi Isaac of Akko. Like most of the old-earth Creationists seven centuries later, he took succor from the passage in Psalms that reads: "A thousand years in your [i.e., God's] eyes is like a day gone by." If we regard a God-day as being worth one thousand years of our time, then one of God's years will be 365,250 of our years. Rabbi Isaac also calculated the earth's age as 42,000 years. If you multiply 42,000 by 365,250 you get a figure that's a bit over 15 billion years, which is not too far adrift from science's reckoning.

As modern Jewish apologists like to point out, that's impressive. The only trouble is that the earth, whose creation is the one Rabbi Isaac was talking about, is only about 4.5 billion years old.

Hindu theologians don't have to bother about trying to equate evolution with a young earth, since it's a basic tenet of Hinduism that the world/universe and ourselves have been in existence forever. This presents different dilemmas, of course, since there's no room for evolution if humans have been present since the start and in the same form as they are today.

One way of denying scientific information is to pretend that it's telling you something else. The late-nineteenth-century Hindu philosopher Vivekananda (born Narendranath Dutta), who believed himself a scientific figure despite his resolute faith in reincarnation, attempted to perform this trick on the newly emergent Darwinism. There was, he claimed, a natural tendency for species to evolve toward higher ones: Just as apes had within them the potential to evolve toward human status, so were we in the process of evolving toward the Absolute. This ineffable natural principle obviated any notion of accommodation to the environment, so certainly wasn't Darwinism; and it certainly wasn't Lamarckism either, since any characteristics the newborn might have acquired came not from previous generations but from the individual's previous incarnation(s). Yet it suited Vivekananda to pretend his ideas were but a hairsbreadth from those of the European evolutionists.

A little later, the Bengali nationalist Aurobindo Ghose produced a variant evolutionary idea that once again bears little relation to either Darwin or nature. Evolution, Aurobindo proposed, was the process whereby the World Spirit worked toward fulfillment via "progressively higher levels of consciousness, from [inanimate] matter to man to the yet-to-come harmonious 'supermind' of a socialist collective," in the words of Meera Nanda.[31]

The modern face of Hindu Creationism is no product of the Indian subcontinent but, rather, of the US. At the forefront is the Hare Krishna movement, which has supported fundamentalist Christian groups in the latter's efforts to get Creationism or ID taught in schools. But the real buzz is Vedic Creationism, as promoted by Michael Cremo and Richard L. Thompson, authors of such extraordinarily huge books as *Forbidden Archeology: The Hidden History of the Human Race* (1994) and *Human Devolution: A Vedic Alternative to Darwin's Theory* (2003). These texts give the impression of complete authoritativeness as they dissect, one by one, countless pieces of evidence that archaeologists supposedly regard as integral to the story they have constructed of our prehistory; with the official version thereby demolished, the hole in our knowledge can be filled by the version of prehistory devised by Cremo and Thompson—which is that humans like ourselves, and human societies similar to the ones we know today, have been in existence for some two billion years. Obviously this must be true, because the establishment account isn't.

Archaeologists in droves have pointed out that the examples of archaeological evidence chosen as weak points by Cremo and Thompson are those very same items that archaeologists *don't* rely upon, because they know they're weak.

Central to the conceit of Vedic Creationism, which is derived from the Vedas and various other ancient writings, is the notion of *atman*, pure consciousness, which transmigrates through eternity in countless cycles of births and rebirths. All living creatures are devolved from *atman*; they are material forms adopted by *atman*. Physical evolution is not just rejected by the Vedic Creationists but regarded as an unimportant notion beside the truly significant matter of spiritual evolution. In this element we can see that, for all the pretensions to science of Cremo and Thompson, really the argument they're putting forward is a religious one.

The Hare Krishna movement has realized the raison d'être of the ID movement is primarily to try to get Creationism into US classrooms, and that, if ID opens the door, then Vedic Creationism has good grounds for claiming equal treatment. There's thus some degree of cooperation between the two movements, each declaring the "science" of the other has merit even though the two schools of thought are actually completely incompatible. Presumably the fact that both are anti-Darwinian is enough to satisfy believers that really they're saying the same thing.

10

WE'RE (BADLY) DESIGNED

Mano Singham sums it up succinctly in *God vs. Darwin* (2009, p. 101):

> "Intelligent design" (ID) can best be understood as a carefully crafted theory designed to eliminate those features that had led to the defeat (because of the Establishment Clause) of prior efforts to combat the teaching of evolution in public schools.

In short, ID is really just Creationism by stealth. Interviewed by the *Washington Post* in 2005, law professor (and born-again Christian) Phillip Johnson, the instigator of the modern recrudescence of ID theory, more or less admitted as much: "I realized that if the pure Darwinist account was accurate and life is all about an undirected material process, then Christian metaphysics and religious belief are fantasy."[1] Untold millions of dollars and billions of man-hours have been wasted in the US because Johnson didn't fully consider the latter option.

Even in the immediate aftermath of the publication of *Origin of Species* there were those who subscribed to some form of ID. One was the astronomer Sir John Herschel, who proposed that an Intelligence might control the course of evolution, although following scientific laws in so doing; the obvious counterargument would be that, since this situation is indistinguishable from the scientific laws operating *without* guidance, why invoke the Intelligence? The US botanist Asa Gray, a friend of Darwin's, accepted evolution but insisted on the directedness that Darwin had specifically rejected; in Gray's view, useful new variations were created by God, natural selection's role being merely to preserve these after their establishment.

Another important early IDer was Charles Hodge, who in books like *What Is Darwinism?* (1874) equated evolution with atheism, which meant it was obviously false.

The UK philosopher and theologian (and zealous crusader against slavery) William Paley is best remembered today for his book *Natural Theology, or Evidences of the Existence and Attributes of the Deity* (1802), in which he presented what is now technically known as the teleological argument from design—in other words, the contention that examination of the universe around us reveals countless examples of beauties, complexities, and concordances that can only be attributable to their having been designed; this implies the existence of a Designer. In his text, Paley showed all the observational traits one could desire in a naturalist, and he was not afraid to spell out fairly arguments that opposed his own. The book was a great bestseller all through the nineteenth century, even after the publication in 1859 of *Origin of Species*, whose author cheerfully admitted that on his own first encounter with *Natural Theology* he'd been convinced by Paley's arguments.

The most famous of these was the watchmaker argument homaged much more recently by Richard Dawkins in the title of his book *The Blind Watchmaker* (1986). Paley set this argument out right at the beginning of his book:

> In crossing a heath, suppose I pitched my foot against a *stone*, and were asked how the stone came to be there; I might possibly answer, that, for any thing I knew to the contrary, it had lain there for ever: nor would it perhaps be very easy to show the absurdity of this answer. But suppose I had found a *watch* upon the ground, and it should be inquired how the watch happened to be in that place; I should hardly think of the answer which I had before given, that, for any thing I knew, the watch might have always been there. Yet why should not this answer serve for the watch as well as for the stone? why is it not as admissible in the second case, as in the first? For this reason, and for no other, viz. that, when we come to inspect the watch, we perceive (what we could not discover in the stone) that its several parts are framed and put together for a purpose. . . . This mechanism being observed . . . , the inference, we think, is inevitable, that the watch must have had a maker.[2]

The reason Paley and his *Natural Theology* are anything more than a philosophical curio to us today is the rise of the ID movement. And it's more than a tad depressing that, even though the specifics may have changed, the arguments he produced in 1802 are essentially identical with those being advanced by the IDers in 2011. His watchmaker argument is, after all, little different from the argument from irreducible complexity.

Design ideas can be traced back far beyond Paley, to ancient Greek philosophers like Aristotle and Romans like Cicero. In the Christian tradition, the argument from design began in the writings of Thomas Aquinas in the thirteenth century; it was one of his celebrated five proofs of the existence of God.

Curiously, the argument from design was born out of the same mechanistic view of the universe and of life that modern IDers so decry. This is evident in the work of Sir Isaac Newton, the scientist who, despite his own spiritual explorations into alchemy and the like, more than any established our mechanistic view of the universe. Here he is with a piece of argument from design:

> Atheism is so senseless & odious to mankind that it never had many professors. Can it be by accident that all birds beasts & men have their right side & left side alike shaped (except in their bowells) & just two eyes & no more on either side the face & just two ears on either side the head & a nose with two holes & no more between the eyes & one mouth under the nose & either two fore leggs or two wings or two arms on the sholders & two leggs on the hipps one on either side & no more? Whence arises this uniformity in all their outward shapes but from the counsel & contrivance of an Author?[3]

It's but a short step from here to Paley's watchmaker. So here we have what we might call the IDer's Dilemma: Do we revile Newton as the *uber*materialist father of the mechanistic view of the universe, or do we praise him as one of the great early progenitors of ID theory?

Richard Dawkins's *The Blind Watchmaker*, mentioned above, effectively demolished—at length—any residues that might still have been lurking in the popular mind of the argument from design. What Dawkins couldn't have predicted is that his book would be responsible—along with Michael Denton's attack on Darwinism, *Evolution: A Theory in Crisis* (1985)—for kicking off the latest resurgence of the argument from design. In response to reading these two works, Phillip Johnson wrote his Creationist tract *Darwin on Trial* (1991), the book that launched the ID movement.

It seems likely Johnson didn't get to the end of Dawkins's book, because in its final chapter, "Doomed Rivals," Dawkins demonstrates not that natural selection happens to be the correct mechanism for evolution—he's already done so—but that "Darwinism is the only known theory that is in principle *capable* of explaining certain aspects of life" (p. 408). What he shows, in other words, is not just that the other hypotheses aren't true but that they *couldn't* be true. As part of this argument, he considers what the IDers would call irreducible complexity. Dawkins's refutation of this conceit is very lucid; as I say, it's hard to believe Johnson read it.

The idea of irreducible complexity is inextricably associated with the inventor of that term, US biochemistry professor Michael Behe. In his book *Darwin's Black Box: The Biochemical Challenge to Evolution* (1996) he defined an irreducibly complex system (p. 39) as a "single system composed of several well-matched, interacting parts that contribute to the basic function, wherein the removal of any one of the parts causes the system to effectively cease functioning."

Behe presented a number of examples of biological and biochemical systems that he claimed were irreducibly complex, like the bacterial flagellum (a feature that acts like a cute little propeller to drive bacteria around) and the human eye. Although subsequent editions of *Darwin's Black Box* mysteriously fail to bear corresponding amendments, Behe was forced in the witness box at the 2005 *Kitzmiller v. Dover Area School District* trial to admit that satisfactory evolutionary mechanisms have been worked out for many if not all of the systems he claimed as irreducibly complex. Even dedicated IDers are now displaying caution when advancing irreducible complexity as a mainstay of their pseudoscience.

It's Behe's claim that "[n]o one on earth has the vaguest idea how the [blood] coagulation cascade came to be."[4] We're reminded of Behe's claim at the Dover trial that science has "no answers to the origin of the immune system"; on being presented in court with a stack of peer-reviewed papers, books, and book chapters offering exactly those answers, he both admitted he hadn't read any of them and protested that, anyway, none of the evidence they offered was good enough. Not only is the evolution of mammalian blood clotting very well understood, Nature—or, rather, natural selection—has solved the clotting problem *more than once*, and in different ways. Lobsters, for example, enjoy the benefits of a different (and simpler) clotting mechanism than ours. If natural selection can find several different ways of solving a problem, this hardly argues in favor of one in particular being irreducibly complex. Quite the contrary: Any clotting mechanism more complex than the simplest on offer can be considered evidence favoring natural selection, whose structures and mechanisms often have a Rube Goldbergish quality through having evolved from pre-existing elements whose purposes were something else.

One of Behe's many precursors in calling attention to the "design" of the human eye was Paley. In fact, Paley's discussion of the subject was among several in which he came agonizingly close to evolutionary ideas before backing off.

The first eyes may have evolved as long ago as 530 million years, if traces in the fossil record are being read correctly. Even before that, there were species possessing light-sensitive cells—even if unable to see, they'd have had the advantage of knowing which direction light was coming from. There have been at least forty separate evolutionary emergences of a functional eye. Definitions vary, but it is frequently said that there are ten different basic "designs" of animal eyes, from the compound eyes found in the insect world to the various types of lensed eyes, such as ours. Lensed eyes are known to have evolved on at least seven different occasions. The "design" shared by humans and many other mammals (but not, for example, cats) is just one—and, for that matter, not necessarily the best "designed": The squid eye, though it cannot perceive color, has greater acuity than ours and no blind spot. Birds have the most acute vision of any.

It's a legitimate question as to why the Designer didn't give the pinnacle of his Creation the best possible eye available. There are various design failures in the human eye, not least that in effect the retina is back-to-front: the photosensitive cells are at the rear. In order to reach those cells, the light has to pass through a litter of processing cells on the retina's front surface. According to evolutionists, this back-to-front arrangement has come about because the eye developed, a very long time ago, from an optically sensitive outcrop of the cortex.

What then of the famous flagellum? There are all sorts of different "designs" of bacterial flagella, and some species swim around perfectly happily without any flagellum at all. Again, Nature has solved this propulsion problem several times and in different ways, some better than others—a clear sign of natural selection at work.

Another legitimate question to be asked of the IDers is why the Designer made such an incompetent job of us. We're poorly adapted to our upright stance, even after all these millions of years. We have "bad backs" and "dodgy knees" because of it. Even hemorrhoids can be traced to our poorly adapted posture—in fact, as with sea cucumbers, wasps, and so much else, it's hard to see why the Designer would have wanted to create hemorrhoids at all. Childbirth is abominably painful and even dangerous, to no apparent purpose. It's hard, too, to understand why we seem to be in so many ways poorly designed to withstand the rigors of the earth's gravitational field: We can kill ourselves by falling down a flight of stairs, and even tripping over our own feet can result in a broken bone or two. And then there's the matter of spontaneous abortion. The Designer terminates more pregnancies than does the abortion industry.

There are plenty of examples of bad "design." Behe is aware of these, and in *Darwin's Black Box* (p. 223) he tries to address them:

> Clearly, designers who have the ability to make better designs do not necessarily do so. For example, in manufacturing, "built-in obsolescence" is not uncommon—a product is intentionally made so it will not last as long as it might. . . . Most people throughout history have thought that life was designed despite sickness, death, and other obvious imperfections.

That last throwaway line is rather like claiming that, since most people throughout history believed the sun went round the earth, we should consider the claim seriously.

The main point he seems to be making in the passage above is that, where structures, organs, or organisms appear to have been "designed" extremely well for their function or niche, we are to accept this as evidence of the Designer. Where, by contrast, they're "designed" badly, we're supposed to take that as evidence of the Designer *as well*; we'd recognize the "design" as good if we knew the Designer's secret motives.

Behe's argument is strongly reminiscent of one made decades earlier, in 1914, by Sunni theologian Abu al-Majid Muhammad Rida al-Isfahani in *Naqd Falsafat Darwin* ("Critique of Darwin's Philosophy"), attacking the notion of vestigial organs:

> Can we assume that because the function of the unused organs is unknown to us, they are non-functional? By so doing, we close the door to scientific investigations of the nature of these organs. Darwin himself was not sure of the function of the organs.[5]

One of the messages to be taken away from Niall Shanks's *God, the Devil, and Darwin* (2006)—probably the best extant critique of ID—is that Behe has seen all the evidence in favor of natural selection and chosen to ignore it. Time after time Shanks goes to papers that Behe has cited and discovers their conclusions differ from those Behe claims to have drawn from them. It's depressing to find a biochemist of Behe's stature playing this game.

Another argument used by IDers is a probabilistic one. They point out that the odds against a particular evolutionary outcome (us, for example) are mindbogglingly huge. Surely it couldn't have happened without the intervention of the Designer. This is one of the approaches adopted by, among others, William Dembski, and it has two considerable flaws.

The first is a failure to understand that the odds against any *particular* outcome, evolutionary or otherwise, are often enormous. To point at a particular organ or feature and say the odds of it having developed are minuscule is to state a truth . . . but one that's meaningless. The fact that you, dear reader, exist at all is a statistical nightmare: Who could count all the chances and coincidences throughout human history that have led to the outcome of the person reading these words being *you*? Same for anyone you meet on the street. Yet the likelihood of your meeting *people* as you walk down the street is high. Likewise, while the odds of *that particular "design"* of bacterial flagellum having occurred through natural selection alone are small, the odds of bacteria having developed flagella are much higher, and the odds of bacteria having developed propulsion higher still. The argument can be extended indefinitely.

Dembski has shown some awareness of the criticism that the probability of a particular complex structure emerging varies considerably according to whether or not the structure has been specified in advance. He has therefore come up with the concept of "complex specified information" (CSI). If you could point at a bacterial flagellum, for example, and say, "That structure was specified in advance," you would be making a very much stronger argument for the intervention of a Designer than otherwise. The trouble with CSI would seem to be that, in the natural world, the only complex structures you can point at are ones that already exist, which means it's too late to say whether their designs were specified in advance. Since you can't tell if CSI even exists, it seems pointless to speculate further.

A related probabilistic issue is the so-called anthropic principle—or, rather, anthropic principles. The weak anthropic principle states merely that any theorizing about the universe should not ignore the fact that life—including us—is a part of it. Clearly a cosmological model in which the emergence of life in the universe was impossible would not be a particularly useful description of the universe we know. Far more controversial is the strong anthropic principle—which is the one that affects ID theories. This has it not just that our cosmological model must permit the emergence of life but that it *requires* it. The notion is controversial because it seems to add an aspect of directed purposefulness to the universe, whereas the message of science is surely that, while (obviously) things have turned out the way they have, this could very well not have been so. Even our own universe could very well have existed without life ever having appeared. Most certainly, the history of the universe could well have omitted *us*—we're not an essential component.

Allied are notions concerning the fact that the universe seems to have been rather precisely tailored to permit the formation of carbon-based (and conceivably other) life forms. There are some basic constants built into the universe—indeed, into the Big Bang—that, had they been just a trifle different, would have made for a very different kind of universe and one in which life as we know it would have been impossible. Martin Rees explores these constants in his book *Just Six Numbers* (1999). The six are not something so simple as the gravitational constant, as one might expect. Instead . . . well, here's a sample (p. 2):

> The cosmos is so vast because there is one crucially important huge number *N* in nature, equal to 1,000,000,000,000,000,000,000,000,000,000,000,000. This number measures the strength of the electrical forces that hold atoms together, divided by the force of gravity between them. If *N* had a few less zeroes, only a short-lived miniature universe could exist: . . . there would be no time for biological evolution.

The sixth of Rees's numbers is the one that's most easily accessible to commonsense: the number of spatial dimensions, D. As we know, $D = 3$. No matter how many other, *non*spatial dimensions might exist (e.g., time), and there may be many, "Life couldn't exist if D were two or four."

The odds against all six of Rees's numbers—and the combination of them—being just right for the emergence of life forms such as ourselves are clearly, well, astronomical. Could it possibly be a product of coincidence?

One school of thought regards that as a silly question: It's only because we're here to ask the question that the concept of coincidence arises; if the numbers were a bit different, we wouldn't be here to do the asking. There's no coincidence involved because that's just the way things have turned out.

To the Creationist, this problem doesn't arise, because the Creationist knows that isn't how the universe works, since it has had only a few thousand years of existence. The supposedly more moderate IDer, conceding that the universe is billions of years old, will, if willing to accept the Big Bang theory, claim the happy concordance of the six numbers as obvious evidence of the Designer. Surely it's inconceivable that things could have turned out this way simply by chance.

Many modern cosmologists agree, which is why the notion of the multiverse has emerged in recent years. For a long time we thought the earth was unique, and the rest of creation revolved around it; then we discovered differently. Next we thought the sun was the center of the universe. For a while we thought the contents of our galaxy constituted the entire universe. What if our universe, too, is only one among many? It's not difficult to contemplate the possibility of there being countless other universes, big and small (whatever those terms might in context mean), alongside our own in a giant structure of which we can as yet see nothing, the multiverse. Clearly, the multiverse is as yet a highly speculative concept, yet it would serve to explain quite a number of things, not least the coincidence of the basic constants of our own universe (if there are billions of universes, then every now and then one of them's going to turn out just right for folks like us), as well as the fact that our universe had a *beginning*: It's much easier to countenance a universe of (so far) finite lifetime existing within an eternal structure than it is to explain how such a universe could exist in isolation.

> Within this first model, to postulate a non-material cause—such as an *unevolved intelligence* or *vital force*—for any event is to depart altogether from science and enter the territory of *religion*. For scientific materialists, this is equivalent to departing from objective reality into subjective belief. What we call intelligent design in biology is by this definition inherently antithetical to science, and so there cannot conceivably be evidence for it.[6]

What's alarming is that the author of these words, Phillip Johnson, is not attempting to demolish the notion of ID. He's offering this argument as *justification*.

In a way, it's a mistake to think of ID as a scientific movement at all. In reality, it's a *political* movement, its agenda being the replacement, as a substrate of modern civilization, of rational—"materialistic"—thought by a sort of faith-based mélange. The prime mover behind the ID campaign is the conservative Discovery Institute, founded in 1996. The key to understanding the Discovery Institute's agenda is the Wedge Document, compiled by Creationists within the Center for the Renewal of Science & Culture sometime in the late 1990s and soon thereafter leaked to the web. As Niall Shanks sums it up in *God, the Devil, and Darwin* (p. 12):

> At the thinnest end of the wedge are questions about Darwinism. As the wedge thickens slightly, issues about the nature of intelligent causation are introduced. As the wedge thickens still further, the interest in intelligent causation evolves into an interest in supernatural intelligent causation. At the fat end of the wedge is a bloated evangelical theology.

Naturally, rationalists worldwide cried "Foul!" when this profoundly antidemocratic plan to march human knowledge back to the Middle Ages was revealed. And they were joined by many Christian theologians, horrified by it not just as a negation of truth but as bad theology. In the *Sydney Morning Herald* for November 15, 2005, Dr. Neil Ormerod, professor of theology at Australian Catholic University, summed up the feelings of many theologians in an article called "How Design Supporters Insult God's Intelligence," describing ID as "an unnecessary hypothesis which should be consigned to the dustbin of scientific and theological history."[7] Another is John F. Haught. In *God After Darwin* (2000; expanded 2008), Haught identifies the three attributes that distinguish religion from science: that religion claims to have as its territory (a) the explanation of ultimate causes, (b) the exploration of inaccessible mystery, and (c) the appeal to a personal deity (or deities). He then demonstrates that ID matches all three of these characteristics. Moreover (p. 199),

> [a] sure indication that ID is not science lies in the fact that its chief architects openly present ID as an alternative to naturalism or materialism rather than solely as an alternative to another scientific theory. In doing so they themselves rhetorically locate ID in the arena of belief systems rather than exclusively empirical science.

He adds that "Dembski, for example, explicitly states that ID is part of a program to defeat naturalism."

There's a curious sociological aspect to all this. Usually, when something like the Wedge Document is exposed, revealing the agenda of a particular pressure group to be quite different from the one its recruited supporters were led to believe, those supporters tend to back off, their ardor cooled by the realization they've been exploited as useful idiots. In the case of the ID movement, however, it seems to have made little difference.

In 2005 the Discovery Institute hired the public relations company CRC (Creative Response Concept). This company had come to prominence the year before through masterminding the Swift Boat Veterans for Truth campaign, which had perpetrated not truth but falsehood about the Vietnam War career of presidential candidate John Kerry. It seems the Designer, while capable of tweaking an eye here or a flagellum there, requires mundane assistance when it comes to tweaking the truth.

ID's status as science wasn't helped when, in 2010, William Dembski made a public declaration to the effect that he was a biblical literalist. Statements such as "I believe that Adam and Eve were real people, that as the initial pair of humans they were the progenitors of the whole human race, that they were specially created by God, and thus that they were not the result of an evolutionary process from primate or hominid ancestors" are not the way to guarantee the continuation of constructive dialogue with scientists.[8] Really, though, this was far from the first time Dembski had said something similar. In *Intelligent Design: The Bridge Between Science and Theology* (1999, p. 206), the book that brought his ID views to notice, he famously stated: "My thesis is that all disciplines find their completion in Christ and cannot be properly understood apart from Christ. . . . [A]ny view of the sciences that leaves Christ out of the picture must be seen as fundamentally deficient." This may or may not be good theology, but it most certainly isn't a valid scientific approach. It becomes increasingly hard to see why anyone takes Dembski's pronouncements on science seriously at all.

In an interview with Michelangelo D'Agostino in 2006, Phillip Johnson, the father of ID, seemed to admit the whole crusade had been based on, if not a hoax, then a false pretense: "I also don't think that there is really a theory of intelligent design at the present time to propose as a comparable alternative to the Darwinian theory."[9] Yet, five years later, the Discovery Institute is still churning out its pseudoscience on a daily basis, still claiming ID is a scientific theory.

11

NO SAFE CLASSROOM?

In December 2010 newly elected Oklahoma state senator Josh Brecheen announced his plan to introduce the teaching of Creationism into the state's schools. He did this in an op-ed in the *Durant Daily Democrat*,[1] regurgitating many of the usual Creationist misconceptions—"I specifically remember asking [a professor at college] how in 4,000 years of recorded history how we have yet to see the ongoing evidence of evolution (i.e. a monkey jumping out of a tree and putting on a business suit)." Unfortunately, much of the regurgitated matter was soon recognized as having come from the book *Case for a Creator* (2004) by Lee Strobel. Among the comments the op-ed's page attracted objecting to this plagiarism, though, Brecheen did have his supporters, like "annettejohnson":

> I am so glad that someone finally has the nerve to stand up for our creator "GOD"! I say BRAVO to you Mr. Brecheen!! Were our schools not founded in the begining [*sic*] to teach reading so that our people could read the Bible? The TRUTH! . . . Wouldn't you rather believe in God and find out he is real than not believe and find out he isn't?[2]

Brecheen is just the latest in a long line of elected officials who've sought to get Creationism/ID into US classrooms or, in earlier times, to ban the teaching of evolution from those same classrooms. The most famous example of the latter situation was the Scopes Trial in the 1920s, although it wasn't everything it might have seemed to be.

The story of the "Monkey Trial" began on March 21, 1925, when Governor Austin Peay of Tennessee signed into law the Butler Act:

> AN ACT prohibiting the teaching of the Evolution Theory in all the Universities, and all other public schools of Tennessee, which are supported in whole or in part by the public school funds of the State, and to provide penalties for the violations thereof.[3]

The civic leaders of the small town of Dayton realized there could be a financial boon to be gained by whichever community was the first to bring a prosecution under the act: Not only would there be an influx of reporters, commentators, and general sightseers for the trial itself, but the town concerned would for at least a few days be the focus of national attention—free publicity that could surely redound only to the town's good.[4]

So a prominent local businessman, the local school superintendent, and the chairman of the school board approached three local attorneys, two to prosecute this hypothetical case and one to act in the defense. Next they had to find a teacher who would break the law on their behalf. The chairman of the school board owned a drugstore, and one of the regular customers was a general sciences teacher called John T. Scopes; although he wasn't the regular biology teacher, he occasionally acted as cover. Asked by the connivers in the drugstore if he'd be willing to be prosecuted, he agreed; a warrant was promptly drawn up, a cop was called, and then Scopes, having been served with his warrant, went off for a game of tennis.

There were various reasons why Scopes was an ideal culprit: He accepted the truth of evolution by natural selection, so was only too willing to help demonstrate in public that the law was a ass; and, as a single man without local ties, it wouldn't be much of a blow when—as all concerned knew would be the result—he was found guilty. The only problem was that nobody, Scopes included, could remember him ever having discussed evolution in the classroom. Later, Scopes had to persuade a couple of cooperative students that they could "remember" him having done so, and to tutor them in what he had supposedly said.

From that point onward, everything got a little out of hand. The ACLU was only too willing to fund the defense; they were actually a little miffed when Dayton's civic leaders, eager for publicity reasons to headhunt as many big-name participants as possible for the trial (they even asked H. G. Wells to be a defense witness, but he had other commitments and regretfully declined), snagged Clarence Darrow to be Scopes's lawyer. The celebrated liberal politician William Jennings Bryan, a devout antievolutionist (and anti-Social Darwinist), agreed to appear for the prosecution, and also offered, if no one else would, to pay Scopes's fine. Darrow, who shared Bryan's progressive political views, had worked vigorously for Bryan's political campaigns.

In other words, the single most famous court case in the history of the warfare over the teaching of evolution in US schools was not so much a court case as a staged event—staged by participants whose motives, in many cases, had nothing to do with either the rightness or wrongness of the issue or even with the rightness or wrongness of the Butler Act: The idea, for most, was to generate business for Dayton. It's ironic that the Scopes Trial's most famous retelling was likewise staged: the highly fictionalized 1955 play by Jerome Lawrence and Robert Edwin Lee, *Inherit the Wind*, later the basis for the classic 1960 Stanley Kramer movie of the same name, with Spencer Tracy as Darrow and Fredric March as Bryan.[5]

While the trial itself might have been just a stunt, the same can't be said of Bryan's interrogation by Darrow. As H. L. Mencken concluded,[6] "Bryan went there in a hero's shining armor, bent deliberately upon a gross crime against sense. He came out a wrecked and preposterous charlatan, his tail between his legs. Few Americans have ever done so much for their country in a whole lifetime as Darrow did in two hours."

Mencken was of course overoptimistic in thinking this would signal an end to the teaching of Creationism. Many of the attempts have ended up in court and, since Scopes, the Creationists have been consistently the losers. Of the several big grandstanding cases, the most recent has been the *Kitzmiller v. Dover Area School District* trial of 2005, where this Pennsylvania school board, backed by the Discovery Institute, attempted to force teachers to discuss evolution as "just a theory" while giving the kids a fairly firm shove toward ID tracts, notably the book *Of Pandas and People* (1989). The result was a fiasco for the ID proponents, while those members of the school board who had passed the measure were summarily ejected by the voters even before judgment had been announced. In his ruling Judge John E. Jones III included a scathing demolition of ID's pretensions to scientific status:

> ID is not science. We find that ID fails on three different levels, any one of which is sufficient to preclude a determination that ID is science. They are: (1) ID violates the centuries-old ground rules of science by invoking and

> permitting supernatural causation; (2) the argument of irreducible complexity, central to ID, employs the same flawed and illogical contrived dualism that doomed creation science in the 1980s; and (3) ID's negative attacks on evolution have been refuted by the scientific community. . . .
>
> ID is predicated on supernatural causation, as we previously explained and as various expert testimony revealed . . . ID takes a natural phenomenon and, instead of accepting or seeking a natural explanation, argues that the explanation is supernatural.[7]

Yet Brecheen is hoping to introduce Creationism to the classrooms of Tennessee. And one consequence of the 2010 machinations of the Texas Board of Education (see pages 33ff.)—the tactic the ousted Don McLeroy is proudest of—is the introduction of a provision allowing the board to adopt supplemental materials in addition to the core textbooks. This might seem an innocuous measure; but its bite became apparent almost immediately. Because the $350 million required to replace Texas's science textbooks for 2011 with ones in line with the new standards was deemed budget-straining, the board elected to use supplemental materials as a stopgap. There is nothing to stop the introduction of a tract like *Of Pandas and People* to Texas schools as a supplementary text.

Amid all the noise made by the Christian Right, it's too easy to think US Christians are fairly united in denying Darwinism. A dramatic demonstration that this is not so was offered in 2004 when the school board in Grantsburg, Wisconsin, passed a series of antievolution policies dressed up as "balance" measures. The board cannot have anticipated the receipt, shortly afterward, of a strongly worded letter of protest sent, not by scientists (of course one arrived from those) or even by religious studies professors (one came in from those, too), but by 188 of the ordinary working Christian pastors of Wisconsin. By July 3, 2010, the online version of that letter had since been signed by 12,598 Christian clergy around the country. Here is perhaps the crux of it:

> We believe that the theory of evolution is a foundational scientific truth, one that has stood up to rigorous scrutiny and upon which much of human knowledge and achievement rests. To reject this truth or to treat it as "one

> theory among others" is to deliberately embrace scientific ignorance and transmit such ignorance to our children.[8]

Some of the campaigns against education in evolution focus on other areas than the classroom. The seven-part TV documentary series *Evolution*, produced in 2001 by PBS, was widely regarded in the science community as an excellent piece of educational popularization. The Christian Right, by contrast, vilified it as an attack on God—this despite the series' last episode making a case that evolution and faith are not incompatible. The Discovery Institute created a website to campaign against the series and mounted a petition called "A Scientific Dissent from Darwinism," which, it claimed, demonstrated there were plenty of scientists who likewise disputed evolution. The Intelligent Design and Evolution Awareness Center, based in California, produced various documents for fundamentalist biology teachers to use to counter any understanding of evolution's workings that their luckless students might have derived from the documentary; these tracts included "Ten questions to ask your Students about the PBS *Evolution* series." Among the ten questions are such imbecilities as "If induced mutations on fruit flies are deadly, weakening, or disadvantageous for survival (i.e. legs growing out of the head), what are the implications for the chances of advantageous mutations arising in the wild?" (Oddly enough, it was discovered a few years later that certain species of ants have, if not legs growing out of their heads, one of their pairs of legs growing upward rather than downward from the body. Having the upward-turned pair of legs means the ants, which habitually live in tunneled nests, have the advantage of being able to push against the tunnel's ceiling at the same time they're walking on its floor. Thus does real life imitate Creationist ridicule.)

One consequence was that the top-notch teaching aids PBS had produced were largely kept out of the nation's classrooms. Meanwhile, the Bush administration used the manufactured brouhaha over the series as an excuse to accuse PBS of liberal bias and cut its funding. Remember, it's a recurrent theme of the Discovery Institute and others of their breed that those plucky few who dare dissent from the prevailing acceptance of evolution are being persecuted by nasty, nasty scientists.

But there are shafts of good news. One is that, on June 18, 2010, the United States District Court for the Western District of Texas (Austin Division) upheld the decision by the Texas Higher Education Coordinating Board to refuse the Institute of Creation Research Graduate School a certificate of authority to offer master-of-science degrees.[9] "Essentially, the panel reasoned much of the course content was outside the realm of science and lacked potential to help students understand the nature of science and the history and nature of the natural world." The ICRGS had not long earlier made the move from California to Texas hoping for a more lenient attitude toward their claims to be teaching science. No such luck, even in Texas.

12

EVILUTION

[W]e keep marrying other species and other ethnics and other—I mean, the Swedes—See, the problem is, the Swedes have pure genes. . . . Because they marry other Swedes. Because that's the rule. Finland—Finns marry other Finns, so they have a pure society. In America, we marry everybody.

—Brian Kilmeade, *Fox & Friends*, July 8, 2009

As reported in *Time* for May 16, 1981, Braswell Deen Jr. a judge of the Georgia State Court of Appeals, stated in 1979 that "[t]his monkey mythology of Darwin is the cause of permissiveness, promiscuity, pills, prophylactics, perversions, pregnancies, abortions, pornotherapy, pollution, poisoning and proliferation of crimes of all types."[1]

Darwin gets blamed for a lot—and Deen's not the only one to lay (so to speak) teen promiscuity at his doorstep. Here's Brock Lee of *Young Earth Creation* writing in July 2010 in the *Owatanna People's Press*:

> What is a species? The evolutionary answer seems simple enough: a species is an interbreeding population.
>
> Most people walk away from biology classes with this definition, and it is this definition which causes problems. . . .
>
> Notice that by this definition, those that are not interbreeding are not part of the species. This means that a virgin is not, by this definition, a human. . . . Do you see how something so simple as a wrong definition can have devastating effects?
>
> Hormones, peers and the teaching of evolution work as a trifecta to push teens into being sexually active.[2]

And the habit of blaming all of the world's perceived ills on Darwin is nothing new. Here's the Zionist and social critic Max Nordau in an 1889 essay called "The Philosophy and Morals of War":

> The greatest authority of all the advocates of war is Darwin. Since the theory of evolution has been promulgated, they can cover their natural barbarism with the name of Darwin and proclaim the sanguinary instincts of their inmost hearts as the last word of science.[3]

The assertion, often repeated by evolution denialists, that Darwinism—or even Darwin himself—was responsible for the genocidal atrocities of Hitler and Stalin is a classic example of ignoring the gorilla in the room. The single best-known datum about Soviet sciences under Stalin is that natural selection and genetics were rejected in favor of the Lamarckian crackpottery of Trofim D. Lysenko. And in Nazi Germany science was basically under the thumb of Himmler, who detested the notion that human beings might have evolved from "lower" animals. In *The Master Plan* (2006) Heather Pringle adds: "Himmler . . . was also outraged by an idea proposed by another German researcher that the Cro-Magnon arose from the Neanderthal. To Himmler, both these hypotheses were 'scientifically totally false.' They were also 'quite insulting to humans'" (p. 134). John Cornwell, in his exhaustive *Hitler's Scientists* (2003), dismisses the Nazi version of evolutionary science—hugely perverted as it was by the need for conformity with Aryan supremacist mythology—as "pseudo-Darwinism" (p. 87), and that's perhaps the best term for it.

So any connection between the Holocaust and Darwinism is tenuous at best. One might be better trying to argue a link between Darwinism and the concept of American Exceptionalism, which led to the near-genocide of North America's indigenous population. And Christians who try to make political capital by claiming that all sorts of vile political theories—and even viler crimes against humanity—can be laid at the door of *Origin of Species* might pause to conduct a similar exercise substituting the Bible in place of Darwin's book. It is, after all, logically difficult—although I'm sure it's been done—to blame Darwin for the Crusades, the Inquisition, the witch hunts, pogroms . . . The truth is, of course, that both books—like all other books—have had evil readers.

Similarly, even if Darwinism *had* led to the Holocaust, that wouldn't mean that Darwinism was bad, wrong, or false science. To use an analogy, the hydrogen bomb is an extremely evil consequence of a lot of very *good* science—if the science hadn't been good, the bomb wouldn't have worked. The vileness of the bomb doesn't invalidate the science that led to it. Similarly, just because some racists and the like have (falsely) claimed their vile attitudes are justified by Darwinism doesn't invalidate natural selection as a fundamentally accurate explanation of how we and the rest of the living world came to be as we are.

The human animal has shown itself adept at finding ways to base disgusting beliefs on anything put in its way. The Social Darwinists (see pages 207ff.) were not the only ones to pervert Darwin's theory into something destructive; there were plenty of others, among them those white scientists and pseudoscientists who seized on evolution as a justification for ill-treating people of color, such as the slew of Southern scientists and philosophers who, in the wake of the Civil War, were eager to prove the inferiority of the Negro race.

Before the acceptance of Darwinism, many of these individuals had called upon the doctrine of Preadamism. The Preadamitic notion was that, since the different races of humankind all looked different, it was inconceivable they could all be descended from a single Adam and Eve. Obviously Adam and Eve must have been white, since God would not have had it otherwise. Thus, there must have been at least one and more likely several Adams created before God got it right. Descendants of these prototype Adams were clearly inferior to the descendants of the final model. When Adamitic and Preadamitic people had first begun to interbreed, God had been so miffed that he'd brought down the Flood upon the world; to avert his wrath, this must not be allowed to happen again. (It's difficult to understand how the philosophers who presented this latter view could do so with a straight face, since the South was widely populated with people of mixed race.) Typical of the racist Preadamite screeds is *Anthropology for the People: A Refutation of the Theory of the Adamic Origin of All Races* (1891) by the Reverend William H. Campbell, writing under the nom de plume "Caucasian."[4]

With the slowly spreading acceptance of the new Darwinian theory, these Southern thinkers, rather than admit Preadamism was nonsense and its conclusions likewise, decided the hypothesis required some tinkering. One tinkerer was Alexander Winchell, who claimed in his *Adamites and Preadamites* (1878) that evolution did happen but was divinely guided and that the observed mental and biological inferiority of the Negro was proof the real Adam must have had at least one inferior humanoid precursor from which the Negro could have evolved. Winchell spent his later career in Michigan, as state geologist and as professor of geology and paleontology at the University of Michigan. It might seem difficult for us now to reconcile the scientific prestige of these posts with the crankery of his racist evolutionary hypotheses.

A variant notion was proposed by the nineteenth-century German zoologist Carl (or Karl) Vogt and a number of others: that the different races of humankind were different *species*, descended from different ape species. Chimps were the ancestors of the whites, gorillas of the blacks, and orangutans of the Asiatics. That these different "species" of humanity could happily interbreed to produce fertile offspring was ignored.

Although Darwin has often been blamed for racism, as if that particular bigotry had not existed long before his time,[5] much of the early resistance to Darwinism was not theological but racist. The unwelcome message of Darwinism was that all varieties of human beings were, at the very least, extremely closely related. Hence a large part of the outcry—and hence, too, the almost immediate efforts to interpret evolution by natural selection as supportive of racist preconceptions.

It was a racist society into which *Origin of Species* was released. Even those Victorians we look back upon as enlightened liberals—including Darwin himself, with his vehement denunciations of slavery and at least one firm friendship, in his student days, with a black scientist—exhibited many attitudes that today we regard as appalling. Darwin's friend T. H. Huxley, another great progressive, famously thought white people had big skulls to house the big brains they needed to be as good at thinking as they were, while black people had big jaws because they were better at biting and chewing than they were at thinking. These are racist views (although Huxley expressed them in the midst of a ringing declaration in favor of Emancipation), yet in the context of the society in which they were expressed they were unexceptional. Huxley's views on the inferiority of women were likewise. In the most simplistic terms, Victorian society needed an explanation as to why the Industrial Revolution and all the developments thereafter had created not a land of milk and honey but one in which, if anything, the lot of the poor was even worse than before. That this might be because there was something inherently rotten in the structure of their own society was not conscionable, and so they fell back on the more amenable Calvinist notion that the people at the bottom of the heap were there because they deserved to be: God had made some people better than others. (This bizarre habit of blaming the underprivileged for their poverty is, alas, still with us.) It was easy enough to extend these notions to other races—and probably also to the French. If they lacked the will or the technological wherewithal to resist being trampled underfoot by the English, this must be because they were innately inferior. Although in 1865 Huxley did not realize the acceptance of evolution would progressively erode such smug assumptions, it did later begin to dawn on him.

As for the antievolutionists, their racism was far more rife and far less tolerant. While Huxley's paternalism—"our dusky cousins"—sets the teeth on edge, it was clear to Soapy Sam Wilberforce's audience, when he made his notorious remarks in the famous Oxford debate about not wishing to be descended from apes, that he wasn't talking solely about chimpanzees.

That Darwin should be blamed for racism is bitterly ironic, since everything we know of the man tells us that he was, for someone of his era and milieu, in the vanguard of antiracism. In their *Darwin's Sacred Cause* (2009) Adrian Desmond and James Moore argue eloquently and at length that the mainspring driving Darwin's determination to get his theory out into the public arena was his detestation of slavery—a detestation that was at the time by no means universal among the English middle classes. Desmond and Moore cite many examples of Darwin's feelings on the matter, extending to the contemporarily common mistreatment of colored people in general.

13

EUGENICALLY SPEAKING

We've noted that ID is not so much a science or a pseudoscience as a political scheme, and the same is in a sense true of eugenics. It's a belief system more than anything else. That the human species could be improved through positive eugenics (promotion of breeding between high-quality individuals) or negative eugenics (prevention of breeding between "undesirables") goes back as far as Plato and probably farther. It was only after publication of *Origin of Species* in 1859, though, that the devotees of eugenics felt they had a good sciencey underpinning for their beliefs.

The major flaw in the claimed scientific underpinning of eugenics is that it depends on an overly simplistic version of Mendelian heredity. For example, eugenics ignores entirely the role of environment in development; yet as early as 1915 the work of Thomas Hunt Morgan et al. showed how important environmental factors were in the equation. In the human context, it was recognized early that a host of influences other than the genes an individual is born with—and not just the influences but the interactions between those influences—can affect the outcome. It is simply not the case that "good genes will out," as the cliché has it: All other things being equal, they might; but all other things never are. A more efficient way of raising a population's intelligence than any heredity-based stratagem is to improve its diet, and that's even before we start to take education into consideration.

By far the most important figure in the early story of the eugenics movement is Francis Galton, a cousin of Charles Darwin; they shared a grandfather—Erasmus Darwin—but had different grandmothers. Galton it was who coined the term "eugenics."[1] The contrasting view to eugenics came to be called euthenics, which sought to improve our lot through environmental reform.

Galton's statistical research on the subject of human heredity persuaded him such qualities as intelligence were inherited, not learned. If good characteristics could be bred for, then that was what our species should be doing. Galton was a sort of Christopher Monckton (see p. 00) *predux*, albeit with genuine scientific credentials. It was not part of his psychological makeup to doubt his own conclusions. His ideal of humanity was, unsurprisingly, the Anglo-Saxon model—people like Francis Galton, in fact. When eugenics became fashionable on the European continent, the preference was for Nordics, supposedly the pure descendants of the almighty Aryan race.

Galton was almost obsessively a racist. In his biography, *Extreme Measures* (2004), Martin Brookes examines this aspect in some detail. As noted above, Victorian England was by and large a racist society; even so, Galton was extreme. And yet his achievements shouldn't be underestimated. Among much else, he identified the weather pattern called the anticyclone, devised the symbology we still use on meteorological maps, did pioneering work on the use of fingerprints, derived the statistical concept of the correlation coefficient, devised and used the psychological investigatory techniques of word association and the psychological questionnaire, and got closer than cousin Charles did to working out the basic principles of heredity/genetics. He invented the randomized, placebo-controlled double-blind clinical trial. He received the Royal Geographical Society's Gold Medal in 1852 for his pioneering explorations in southern Africa and in 1854 a Silver Medal from the French Geographical Society. From the Royal Society he received a Gold Medal in 1886, the Darwin Medal in 1902, and the Copley Medal in 1910. He was knighted in 1909 for services to science. They don't make scientific careers much more glittering than that.

The trouble was that he also did things like inventing a lidded hat for hot weather (you could pull a cord that hung from the brim so the top of the hat would hinge upward and allow cooling breezes to circulate around inside) and making himself a special pair of spectacles so he could read the newspaper underwater. And, while he pioneered twin studies, he cherry picked his own results, so his research in this area was valueless.

In an 1865 essay called "Hereditary Character and Talent" Galton spelled out his reasons for vigorously supporting eugenics:

> If a twentieth part of the cost and pains were spent in measures for the improvement of the human race that is spent on the improvement of the breed

> of horses and cattle, what a galaxy of genius might we not create! We might introduce prophets and high priests of our civilization into a world as surely as we can propagate idiots by mating crétins. Men and women of the present day are, to those we might hope to bring into existence, what the pariah dogs of the streets of an Eastern town are to our own highly bred varieties.[2]

In this essay he examined the careers of distinguished men and noticed how often there were family connections; the Darwins (his own clan) and the Huxleys were just two examples of distinguished lineages. Galton seized upon this as proof of his ideology that merit, intellect, sagacity, and the like were hereditary characteristics; it never occurred to him that his own data could be interpreted equally well—if not better—to imply that these qualities are likely to arise from education and the family environment, that networking and nepotism play a major role in opening doors to achievement, and that it is far easier for those born of privilege to make it in the world than for those born in poverty. For Galton to have considered any of this would, of course, have been to challenge his own position as a natural member of the elite.

He amplified his thesis in *Hereditary Genius: An Inquiry into Its Laws and Consequences* (1869), surveying a large sample of those he selected for their eminence . . . or did he? His selection was so idiosyncratic—Leonardo da Vinci didn't make the cut—that it seems either he regarded aristocratic birth as being a sufficient quality for a man to be considered eminent, in which case his argument was invalid because circular, or he was including only those who would support his case and ignoring the others, in which case his argument again fell. Despite its obvious biases, the book was very well received by the intelligentsia amid a general feeling that Galton had proved his thesis.

Some disagreed. In *Applied Sociology* (1906) US sociologist Lester Ward took Galton to task, revisiting those cases Galton had cited as exemplary demonstrations of the inheritance of genius, and showing that in every instance the individual involved had had the advantages of education, opportunity, and, in one way or another, privileged upbringing.

Not long after the end of the century there were, in consequence of Galton's endeavors, what we can with hindsight recognize as auguries of dire events to come. In 1904 the anthropologist and racist Alfred Ploetz, author of *Grundlinien einer Rassenhygiene* ("Racial Hygiene Basics"; 1895) and a fawning correspondent of Galton's, started up the first German eugenics journal, *Archiv für Rassen-und-Gesellschaftsbiologie* ("Archive for Racial and Social Biology"); and in the following year he founded the Deutsche Gesellschaft für Rassenhygiene ("German Association for Race Hygiene"). Much later, in 1933, Ploetz would be one of the first to welcome Hitler's ascent to power, on the basis that now at last the question of racial hygiene would be brought to the forefront of German politics. One of Ploetz's most influential admirers was Heinrich Himmler.

In 1904 the first research fellowship in eugenics was established, at University College, London. Galton funded this himself, and in return UC allotted one of its properties in Gower Street as space for the UK's Eugenics Records Office; in 1906 this became the Francis Galton Laboratory of National Eugenics. Despite the name, the story of this institution is dominated not by Galton but by the statistician Karl Pearson, whom UC appointed as the first Galton Professor of Eugenics in 1911.

Another player in the game was the Eugenics Society, which still exists in the form of the Galton Institute. Founded in 1907 as the Eugenics Education Society, this had Charles Darwin's son Leonard as its president from 1911 to 1925 and spawned branches across the UK. In 1910 the two institutions had a very public spat over a piece of research done by Pearson's team at the Galton Laboratory which concluded that children of alcoholic parents seemed to suffer no consequential physiological or mental impairment. This flew straight in the face of the ideological preconceptions of the Eugenics Education Society. The problem for the latter was that, say what they liked, the Galton Laboratory was actually *doing science*—however misguided—and to its credit was publishing results even when these went against the grain. The Eugenics Education Society, by contrast, was just an advocacy group.

In the US the cause of eugenics was for the most part inseparable from white racism and the notion of American Exceptionalism. Eugenics was promoted through a bewildering barrage of pseudoscholarly societies, conferences, and affiliated journals. Trying to pin them all down is like trying to keep track of all the different fissioning and metastasizing UFO societies.

The American Breeders' Association (ABA) was founded in 1903 by Willet M. Hays with its original stated aims being to

- determine the laws of inheritance in animals and plants;
- learn the application of these laws to increase the intrinsic commercial and artistic values of living things; and
- help bring about this desired improvement through associated effort.

Not long after formation, however, it was joined by the committed eugenicist Charles Benedict Davenport—a zealot who more than once referred to eugenics as a "religion." So convinced was he of the paramountcy of heredity that he even declared prostitution an inherited characteristic. Davenport was director of the ABA's Eugenics Section, and soon made it a power base. In short order he created a scad of committees whose names, listed, read today like something from a Monty Python skit; there were committees on the Heritability of Feeblemindedness, on Sterilization and Other Means of Eliminating Defective Germ Plasm, on the Heritability of Criminality, and many more. Committee members included luminaries like David Starr Jordan (who served as chairman of eugenics), Alexander Graham Bell, Luther Burbank, Henry Fairfield Osborn, and Vernon L. Kellogg. Davenport was a prime mover behind most of the US eugenics organizations of the era. His *Heredity in Relation to Eugenics* (1911), cited by an estimated one-third of US high school biology textbooks in the interwar years, included such forebodings as (p. 219):

> [I]t appears certain that, unless conditions change of themselves or are radically changed, the population of the United States will, on account of the great influx of blood from South-eastern Europe, rapidly become darker in pigmentation, smaller in stature, more mercurial, more attached to music and art, more given to crimes of larceny, kidnapping, assault, murder, rape and sex-immorality and less given to burglary, drunkenness and vagrancy than were the original English settlers.

By 1915, Davenport's Eugenics Section had become such a force within the ABA that the latter changed its name to the American Genetic Association. At the same time, the ABA's journal, *American Breeders' Magazine*, became the *Journal of Heredity*. Both the American Genetic Association and the *Journal of Heredity* (now published by Oxford University Press) still exist, although their eugenics days are long behind them.

Further magazines and newsletters promoting eugenics included *Eugenical News*, published primarily by the Eugenics Research Association, with the Eugenics Record Office and others; it was actively racist, regarding black people as inferior to whites and interbreeding as leading to racial suicide. While the remedy recommended by its writers was usually repatriation (with the obfuscation typical of racists everywhere over the matter of how you repatriate people who were actually born in the country you're planning to expel them from), the language used was often tantamount to an incitement to violence.

The Eugenics Record Office was founded in 1910 at Cold Spring Harbor Laboratory in Cold Spring Harbor, New York, by Davenport with funds donated by Mrs. Mary Harriman, widow of railroad magnate Edward Henry Harriman; later funders were the Rockefeller family and the Carnegie Institution. Throughout its existence the ERO's director was Harry Hamilton Laughlin, a foremost advocate of compulsory sterilization, which he promoted through the ERO. The "model law" he drafted for this campaign proposed as subjects for sterilization alcoholics, epileptics, the blind and the deaf, the feebleminded and the insane, criminals, people suffering deformities, indigents, and anyone else Laughlin thought undesirable. Though Laughlin and the ERO were among the rankest of pseudoscientists, many politicians took them seriously; eighteen states passed laws based on the "model" Laughlin had drawn up. Abroad, the Law for the Prevention of Hereditarily Diseased Offspring, passed by the German Reichstag in 1933, was likewise based on Laughlin's proposal.

Where the American Genetic Association might retain some claims to racial objectivity, the US Galton Society, founded in 1918, was overtly racist from the start. Alongside a roster of usual suspects, like Davenport and Osborn, were people like Lothrop Stoddard, soon to be author of *The Rising Tide of Color against White World-Supremacy* (1920) and a man whose racism was considered extraordinary even in that racist era, and the equally racist and virulently antisemitic Madison Grant, famed for his *The Passing of the Great Race, or The Racial Basis of European History* (1916).

Both Stoddard's and Grant's books were published by Charles Scribner's Sons, a firm that seems to have been attempting to corner the racist market. The Spring 1917 Scribner catalogue, describing Grant's recently published *The Passing of the Great Race*, featured an approving review by Theodore Roosevelt: "It is the work of an American scholar and gentleman; and all Americans should be sincerely grateful to you for writing it."[3]

A few decades later, *The Passing of the Great Race* gained another politically powerful fan: Adolf Hitler told the author during a visit by Grant to Germany that he regarded the book as his "Bible." In fact, one could argue that it was the fact that Grant's proposals were so enthusiastically adopted by the Nazis that put the almost-final nail into the coffin of US eugenics, even before the full, horrific extent of the tragedy became known.

In the 1920s, US eugenicists naturally seized on the newly emergent fad of IQ testing as a further weapon in their arsenal. Of course, neither the eugenicists nor many of the testers were much perturbed by the fact that no one knew what intelligence actually *was*. Nor did anyone know how the IQs being measured might relate to intelligence. The journalist Walter Lippmann was moved to write a series of articles that appeared in *New Republic* in October and November 1922 debunking the claims of the testers. When one enthusiast, Lewis H. Terman, published with a straight face his discovery that only three out of twenty children tested in an orphanage were "fully normal," Lippmann was incandescent in his derision:

> "The orphanage in question," [Terman] then remarks, "is a reasonably good one and affords an environment which is about as stimulating to normal mental development as average home life among the middle classes." Think of it. Mr. Terman first discovers what a "normal mental development" is by testing children who have grown up in an adult environment of parents, aunts and uncles. He then applies this footrule to children who are growing up in the abnormal environment of an institution and finds that they are not normal. He then puts the blame for abnormality on the germplasm of the orphans.[4]

Some of the arguments put forward by the eugenicists in support of the importance of heredity *über alles* do beggar belief. In his paper presented to the 1915 National Race Betterment Conference, "Natural Selection in Man,"[5] Paul Popenoe, while rejoicing in high rates of infant mortality among the poor as "a 'weeding out' of the unfit" (so why bother improving conditions for the poor? went the argument), presented the unusual hypothesis that those who died of tuberculosis did so not primarily because of the activities of *Mycobacterium tuberculosis* but through being hereditarily disposed to have low resistance to the illness. Of course, people vulnerable to tuberculosis were a drag upon the progress of the rest of the species, so . . .

In another instance, the eugenicist Carl C. Brigham, in his *A Study of American Intelligence* (1923), faced the conundrum that black people in the northern US states systematically out-performed those in the South. The reason might seem obvious to us, but not to Brigham: No, the disparity arose surely because migration to the northern states was an intelligent thing for a black person to do, and so, not unnaturally, the descendants of such émigrés were brighter than their stay-at-home counterparts.

One of the triumphs scored by Madison Grant and his ilk was the Johnson–Reed Act, passed in 1924 (and not repealed until 1952). The act's aim was to restrict the immigration of people of "undesirable" stock. As well as advocating racial segregation—primarily to reduce miscegenation—Grant prescribed sterilization of everyone he deemed inferior. The first state sterilization law was passed by Indiana in 1907; over the next couple of decades, thirty-one other state legislatures followed suit. And it was surprisingly easy to be an undesirable: The Iowa sterilization law of 1913 embraced "criminals, rapists, idiots, feeble-minded, imbeciles, lunatics, drunkards, drug fiends, epileptics, syphilitics, moral and sexual perverts, and diseased and degenerate persons."[6]

The US was, although the first, far from the only country to be deceived by eugenic pseudoscience into introducing a campaign of forced sterilization. There was a particularly poisonous program in Sweden that lasted from the 1930s into the 1950s and then was until recently somewhat written out of the country's history books. In the UK, Winston Churchill tried to include in the 1913 Mental Deficiency Act a measure permitting forced sterilization, but it was defeated by a public campaign. Further countries with sterilization programs included Australia, Canada, Czechoslovakia, Estonia, Finland, Iceland, Japan, Norway, Panama, Peru, Switzerland . . . and of course Germany. The motives weren't always racist or antisemitic. The main concern of the Norwegian racial eugenicist Jon Alfred Mjoen was interbreeding between Nordics and Lapps. There are examples of more recent sterilization campaigns, but these haven't been for eugenic reasons.

For the first decade or two, few of the US states in which forced sterilization was official policy actually carried out many sterilizations. There was public uncertainty over such a measure, with many believing it could be construed as a cruel and unusual punishment. Things came to a head, though, with the Carrie Buck case, brought before the Supreme Court in 1927.[7] Buck was an unmarried teenage mother who'd been placed by the State of Virginia into a home for epileptics and the feebleminded because she'd borne a bastard. Now the state wanted to sterilize both Buck and her child so they'd not produce further offspring, who'd likely be similarly promiscuous—promiscuity being heritable. The main reason the case got as far as the Supreme Court was that the state had managed to recruit Charles Davenport of the Eugenics Record Office as an expert scientific witness. The Court found in favor of the state, with Judge Oliver Wendell Holmes writing the majority statement. This was the sign for the states to become less timorous about forcible sterilization. It's estimated that by 1945 over forty-five thousand Americans, about half being inmates of state mental institutions, had been sterilized. Most enthusiastic was California, which contributed about half of this total.

And the Supreme Court's ruling was noticed abroad, too, notably in Germany. On Hitler's accession to power in 1933, one of his government's first acts was a sterilization law based on California's; physicians were obliged to report to a Hereditary Health Court anyone they encountered who was "deficient." In 1934 the American Public Health Association lauded the Nazi law as a template for sensible, socially beneficial, science-based health policy; the *New England Journal of Medicine* and the *New York Times*—a strong supporter of the US sterilization laws—likewise enthused. By 1940, nearly four hundred thousand Germans had been sterilized. Far from being horrified, US eugenicists criticized the US for failing to match Germany's sterling example.

The US connections to the eugenics-driven mass-sterilization programs in Nazi Germany, and eventually to the Holocaust, were for decades after World War II generally downplayed, like some ghastly parental misdeed that should be kept from the children. As we saw, Hitler gained considerable inspiration from Madison Grant's 1916 book *The Passing of the Great Race*. In Grant's view, if the only way to stop the unfit from breeding was to kill them, so be it. Grant was a liberal alongside some. Alexis Carrel, winner of the 1912 Nobel Physiology or Medicine Prize and on the staff of the Rockefeller Institute for Medical Research, wrote in *Man the Unknown* (1935) that society should simply slaughter criminals and the insane rather than pay for prisons and asylums. A few years later, between 1940 and 1941, the Nazis would slaughter some seventy thousand of the mentally ill, mainly in Poland, and mainly to save money.

Financial support for German eugenics endeavors flowed in from the US, too. In 1925 the Rockefeller Foundation, to promote eugenics research, gave $2.5 million to the Munich Psychiatric Institute, plus further money to the Kaiser Wilhelm Institute for Anthropology, Human Genetics, and Eugenics.

Between 1914 and 1928 the number of US colleges and universities offering courses in eugenics increased from 44 to 376. What's especially worrying is the ease with which the eugenicists infiltrated the US public educational system, influencing educators to impart their pseudoscientific and often highly racist views to youthful, malleable minds. In *Inheriting Shame* (1999) Steven Selden spends some time (pp. 63–83) analyzing the pervasiveness of eugenical ideas in US high school biology textbooks between 1914 and 1948. His analysis makes chilling reading. A recurrent theme in these books involved the supposed type cases of the Jukes and Kallikak families.

Richard L. Dugdale's *The Jukes: A Study in Crime, Pauperism, Disease and Heredity* (1877) described research done by Dugdale, a volunteer inspector for the New York Prison Association, into the history of a Hudson Valley family called the Jukeses (not their real name), which seemed to have produced an unduly high number of criminals through the generations. When he calculated the financial burden the Jukeses had placed on society he found it was $1.3 million (about $21 million in today's terms). Dugdale himself was open-minded as to whether the Jukeses history of miscreance arose through heredity or environment, coming down if anything on the side of the latter: Each new generation of Jukes kids got a lousy upbringing.

In 1911 his original notes were discovered and sent to the Eugenics Record Office. There, Arthur H. Estabrook took on the task of updating the study. Estabrook's *The Jukes in 1915* (1915) contained his estimate of the further horrific price the taxpayer was going to have to pay for the Jukes family's continued existence. At the 1921 Second International Congress of Eugenics, held in the American Museum of Natural History, New York, the Jukeses were given a full display as prime targets for eugenic removal. Indeed, until the effective demise of the US eugenics movement after the horrors of Nazi Germany had been revealed, the Jukeses were still being used as an example of why eugenics was desirable.

Early in this century researchers were able, through getting hold of Estabrook's papers, to find out who the Jukeses had actually been. While the family had, sure enough, contained plenty of bad hats, quite a few of its scions had been upstanding members of society. The reason some had a criminal record was simply stark poverty.

A similar buzz surrounded *The Kallikak Family: A Study in the Heredity of Feeble-Mindedness* (1912) by the psychologist Henry H. Goddard. Although more recent analysis of Goddard's research shows absurd methodological failures, in its day this was another significant prop to the eugenics movement.

After the end of World War II, the popularity of eugenical ideas, including enforced sterilization—which at times had enjoyed a 66 percent public approval rating—took an abrupt nosedive, and fortunately they have remained in a fringe position ever since. They still resurface from time to time, though. One amusing instance of a late efflorescence was described by David Plotz in a series of articles for *Slate* and a subsequent book, *The Genius Factory* (2005). It concerns the dreams of Robert Klark Graham, an entrepreneurial optometrist and keen eugenicist, to set up a sperm bank stocked by Nobel laureates. In 1980 he announced to the world his Repository for Germinal Choice.

With technical assistant Stephen Broder, he collected specimens and dealt with applicants for the premium sperm. It all seems to have been very much a cottage industry; for example, Plotz describes (p. 38) an incident in which the two men, having collected material from physicist William Shockley and another laureate in San Francisco, had reached the airport ready to fly home when they realized (a) it was a bad idea to pass semen specimens through an x-ray security screener, and (b) airline and security personnel were likely to look askance at hand baggage consisting of an insulated flask containing ampoules of a mysterious substance and leaking wisps of vaporizing liquid nitrogen.

Graham's venture might have been the most prominent example of what it's tempting to call neo-eugenics, but it's by no means the only one. In 2006 John Glad published his pro-eugenics *Future Human Evolution: Eugenics in the Twenty-First Century*. He makes this point (p. 34): "The question is whether parents have a moral right to bring children into the world who will be disadvantaged by their heredity." It's a refreshingly thoughtful observation in a society that is all too keen to stress the rights of parents while often oblivious to those of their offspring. The problem with it, though is that Glad is displaying the knowledge of genetics that one might expect from a retired languages professor, which is what he is.

14

SOCIAL DARWINISM

The growth of a large business is merely a survival of the fittest, the working out of a law of nature and a law of God. . . . The American Beauty Rose can be produced in the splendor and fragrance which brings cheer to its beholder only by sacrificing the early buds which grow around it. This is not an evil tendency in business. It is merely the working out of a law of nature and a law of God. . . . The day of combination is here to stay. Individualism has gone, never to return.

—John D. Rockefeller Jr.[1]

A primary reason the importance of the Social Darwinism fad has been so underestimated is that, during the decades of its greatest influence, it was not known by that name—or, really, by any name at all. Its adherents merely assumed it was an obvious corollary of Darwin's evolution by natural selection, even though Darwin himself, in *The Descent of Man* (1882), had specifically said the opposite. According to the *Oxford English Dictionary*, the term first appeared in 1877 in the *Transactions of the Royal Historical Society*, in an essay called "The History of Landholding in Ireland" by Joseph Fisher. Although it appeared occasionally thereafter, it didn't really enter popular use until the appearance in 1944 of the book *Social Darwinism in American Thought* by Richard Hofstadter. It's important to realize that the proponents and antagonists discussed here rarely if ever used the term. We will, however.

Social Darwinism is the set of belief systems that takes as its mainspring the concept of the survival of the fittest—a phrase that's often assumed to have been coined by Darwin but which in fact was probably the invention of the UK philosopher Herbert Spencer, the recognized father of Social Darwinism.[2] Spencer had speculated about evolution even before *Origin of Species* appeared in 1859. In his essay "A Theory of Population, Deduced from the General Law of Animal Fertility" (1852) he put forward the idea that hardship worked toward improving the human species, and that it was because of past hardships that we are so far advanced beyond our prehistoric ancestors. The concept sounds closer to Darwinian natural selection than perhaps it is, because Spencer in fact espoused Lamarckism (see page 148). He is often quoted as saying, in his essay "The Development Hypothesis," that "[t]hose who cavalierly reject the Theory of Evolution as not being adequately supported by facts, seem to forget that their own theory is supported by no facts at all." It's a very good point. But the version quoted above appeared in 1891, when the essay was reprinted in Spencer's *Essays Scientific, Political & Speculative*. In its original appearance, in the *Leader* for March 20, 1852, the sentence read: "Those who cavalierly reject the Theory of Lamarck as not being adequately supported by facts . . ."

What Spencer thought he was doing was grafting ideas from evolution onto socioeconomic prescriptions. In his view, social groups—from full-scale societies on down to their smallest elements—could be considered as if they were living creatures. Not only was it evident to him these entities would act in accordance with evolution's laws, with the fittest thriving and the weakest dying off, it was his view that this was a good thing. It seems odd that a man reportedly so brilliant could be so ethically deluded.

Karl Marx was another philosopher who saw the competitiveness of laissez-faire capitalism as at least an analogue of natural selection—although, even as he largely endorsed Darwinism, he excoriated the political scheme supposedly developed from it.

In her *Darwin's* Origin of Species*: A Biography* (2006, p. 64) Janet Browne is succinct in her discussion of Darwin's views on Spencer's philosophy: "Darwin had never taken any of his writings seriously." And T. H. Huxley was no great fan, either. In the US, Lester Ward pointed out that, despite its being a matter of dogma among the Social Darwinists that only through competition can a better form of the organism emerge, you have to look no farther than the domesticated cow or the cultivated fruit tree, for both of whom every possible effort has been made to remove competition from the picture, to realize the falsehood of the claim.

The naturalist Prince Peter Kropotkin added another way in which the analogy between Darwinian evolution and the evolution of societies breaks down. By its very essence, Social Darwinism assumes the survival of the fittest applies to members (or groups of members) of a single species, who are therefore in a constant state of competition against each other. In nature, however, the prime competition is *between different species*, as they jostle to best exploit the environment; within a species, cooperation is often widespread.

Yet another aspect where the biological/societal analogy breaks down is that in Darwinian terms the "fittest" are those best suited for the environment (or for a change in the environment), who are far from necessarily the strongest, most ruthless, and most predatory.

Overall, the great lacuna in the Social Darwinist pseudoscience was that its proponents ignored—or were ignorant of—those parts of Darwin's thinking that recognized the importance of cooperation (and, by extension, of such characteristics as altruism[3]) in evolution. It seems an astounding omission from a supposedly science-based philosophy. Human civilization depends entirely on the structures known as societies, which structures themselves depend absolutely on cooperation between their members. Without cooperation we'd still be at best in the Stone Age. To ignore this—as not only did the Social Darwinists but also, later, Ayn Rand with her adolescent philosophy Objectivism—is surely to be blinded by one's ideology.

Spencer's philosophy took little root in his native UK. In the US it was a different story. The appeal of his ideas here was that they told the wealthiest and most powerful members of society exactly what they wanted to hear: that what they had been doing, no matter how self-serving, exploitative, and destructive of the souls of their fellow human beings it might be, was, according to science, virtuous in that it was to the benefit of the species, and so they should carry on doing it. Not only was might right, it was ethically correct that this should be so. Again, we see a parallel with Objectivism.

The degree to which some Americans revered Spencer's ideas could be embarrassing. Frederick Augustus Porter Barnard wrote: "As it seems to me, we have in Herbert Spencer not only the profoundest thinker of our time, but the *most capacious and powerful intellect of all time*."[4] Perhaps Spencer's most important advocate and elaborator in the US was the political/social scientist and Yale professor William Graham Sumner. Sumner was no recognizer of gray areas, and also a man prone to offsetting false alternatives. A comment William James made about Spencer could as well have been about Sumner: "[O]ne finds no twilight region in his mind. . . . All parts of it are filled with the same noonday glare, like a dry desert where every grain of sand shows singly and there are no mysteries or shadows."[5] Sumner was also, in his unshakable belief that social evolution could come about only with geomorphological slowness, a great promoter of the philosophy that George Orwell would much later lampoon in a quite different context as "jam tomorrow": "Let every man be sober, industrious, prudent, and wise, and bring up his children to be so likewise, and poverty will be abolished in a few generations."[6]

While today it's social conservatives and the devout who tend to treat Social Darwinism as yet another evil to lay at the feet of the demon Darwin, we should note that Social Darwinism in the US was a conservative movement, not just in its ideas of "the survival of the fittest" but in the notion that social change should be, like physical evolution, something that took place very gradually and over a long period of time. Further, it was not an atheistic movement: Although some atheists and agnostics were attracted to it, far more numerous were Christian preachers and theologians, one of the foremost being Henry Ward Beecher.

The seeds of Social Darwinism's demise lay in the philosophy itself. Essentially, its adherents were preaching that all was right with a world in which a very few immensely powerful, immensely rich, and immensely ruthless people prospered at the expense of the vast mass that was the rest of us. Some members of vast masses can be astonishingly gullible when it comes to such propositions—witness the recent spectacle of whipped-up mobs of Tea Partiers furiously demonstrating in favor of policies specifically designed to perpetuate their deprivation—but sooner or later there comes a general cottoning-on moment, when the hoodwinked, even though they may never admit it, see through the hoodwinkers. That moment came in the US for Social Darwinism in the years around 1900, and the reversal was dramatic enough that, not so many years later, one of the main popularly perceived reasons for entering the war against Germany was that the Germans under Kaiser Wilhelm were an international menace because of, and their militarism a result of, their obsessive devotion to Social Darwinism.

In *A Darwinian Left* (1999) bioethicist Peter Singer, while promoting the notion that Darwinian ideas can profitably be applied to social structures—in other words, advocating what we might call a left-of-center Social Darwinistic approach—pinpoints this disabling flaw in the reasoning of Spencer and his followers, both contemporary and modern.

What the Social Darwinist movement had done in the US, though, was poison the name of Darwinism—at least among those who couldn't be bothered to evaluate Darwinism itself rather than the Spencerian corruption. William Jennings Bryan (see pages 184–85) was one such. There were those in the Arab world, too, who fell for the fallacy that Spencer's social philosophy was a necessary corollary of Darwinism—that the two were really the same thing. As Adel A. Ziadat remarks in *Western Science in the Arab World* (1986, p. 24), "Arab thinkers concentrated their discussion on Social Darwinism [even though they] nevertheless referred to many other aspects of the theory."

We could make a very strong case that we've seen since the 1980s in the US and elsewhere a resurgence of Social Darwinist ideals, even if not in name, as the gap between rich and powerful and poor and powerless has grown near-exponentially. To adopt the symbolism of the Social Darwinists for a moment, and to build on from Rockefeller's comment with which we opened this discussion, we could observe, too, that the end product of social survival of the fittest is the corporation, vast and growing ever vaster as it devours and assimilates its smaller competitors on its way toward that Holy Grail of corporatehood, the complete monopoly—all the while becoming less and less good at doing what it was originally brought into existence to do. It's not the wily, fast-moving, flexible mammals that these monolithic corporations remind us of so much as the mammals' precursors on the animal kingdom's throne: the dinosaurs.

15

IT'S THE ECOLOGY, STUPID

> *Liberals don't care about the environment. The core of environmentalism is a hatred for mankind. They want mass infanticide, zero population growth, reduced standards of living, and vegetarianism. Most crucially, they want Americans to stop with their infernal deodorant use.*
>
> —Ann Coulter, "Global Warming: The French Connection"[1]

Ann Coulter's views on environmentalism and environmentalists may seem extreme, but they lie on the same spectrum as those of many other commentators and many members of the public. Conservation measures and protection of the environment are seen not as sensible insurance policies to help guarantee such trivia as that the water we drink and the air we breathe are unlikely to kill us and that our kids and grandkids will, with luck, inherit from us a world worth living in—or at a least a world that *can* be lived in—but rather as a scheme mounted by amorphous groups of "liberals" or "socialists" or even "islamofacists" to inhibit our freedoms. It's hard to dream up a plausible motive for this supposed scheme—but, hey, what are conspiracy theorists *for*? Clearly those grim-faced treehuggers are hoping to ship us all off to concentration camps where we will be forced to plant organic vegetables.

In late May 2010, as BP's Deepwater Horizon blowout continued to pump up to twenty thousand barrels of oil a day into the Gulf of Mexico, Rep. Don Young (R-AK) gave us an ecology lesson:

> This is not an environmental disaster, and I will say that again and again because it is a natural phenomena [*sic*]. Oil has seeped into this ocean for

> centuries, will continue to do it. During World War II there was over 10 million barrels of oil spilt from ships, and no natural catastrophe. . . . We will lose some birds, we will lose some fixed sealife, but overall it will recover.[2]

Young's denial was in harmony with a chorus from Rush Limbaugh on upward to the effect that the disaster—or nondisaster—was all the fault of wicked environmentalists: Had BP been allowed to drill on land in an area of natural beauty or of scientific value, or close offshore, or in your backyard (not Limbaugh's), the company wouldn't have been "forced" to sink its well in such deep waters, with all the risks entailed. Limbaugh actually suggested President Obama had arranged the accident in order to advance an environmentalist agenda. Though most of his blowhard talk-radio colleagues balked at this, the idea was widely bruited that Greenpeace might have done it. The apotheosis of this clamor came when Rep. Joe Barton (R-TX) made a public apology to BP that they'd been required to contribute to the cost of the cleanup.

A May 2009 Pew Research Center survey showed a depressing tendency: 83 percent of Americans think we need stricter laws and regulations to protect the environment, but people don't want to pay for this. Just 51 percent agreed protecting the environment should be a priority if it involved slower growth and/or job losses, and 49 percent agreed we should be willing to pay a higher price to protect the environment. Ah, magical thinking again. If you wish for it to happen it'll somehow just all get better on its own.[3]

The single most important environmental challenge facing us today is climate change. All others are dwarfed by it. And yet, because the earth is a system in which everything is interconnected, all others also affect and are affected by climate change. We cannot consider it in isolation. Our enormous population directly impacts climate change, not just because of the carbon footprint each individual human being leaves in terms of energy use and the like but also because of deforestation to create extra agricultural land, the methane pumped into the atmosphere by our seemingly ever-spreading factory farms, and so forth. And it works the other way round, too: Climate change destructively affects the environment. Most importantly, perhaps, climate change and pollution are acting in deadly tandem to kill our oceans, and thereby much of our food supply.

While the news media give some coverage (too much of it denialist) to the issue of climate change, that of the catastrophic deterioration of our oceans seems largely hidden, as if it were a story with implications too dreadful for journalists and politicians alike to face. Even after the Deepwater Horizon oil spill, there was relatively little notice paid to the possibility that the species upon which the region's fishing communities depend *might never return*. In general, we, the public, are aware fish stocks are declining and reefs are imperiled, but few of us know by quite how *much*. A study done at Dalhousie University, Nova Scotia, in 2003[4] found that, by comparison with the 1950s, thanks to both overfishing and the destruction of ecosystems, just 10 percent of seafloor fish like cod, haddock, flounder, and pollock remain, with a similar figure for the bigger fish like tuna, shark, and swordfish.

We know about the dangers to us of rising levels of mercury in the oceans, leading to potentially hazardous measures of this toxin in many of the fish we eat, and we've seen the tragic pictures of ocean wildlife being killed by our plastic bags and other litter—and of course our oil spills—but many of us are blithely ignorant of the existence of dead zones in the oceans: large areas, often outside river mouths, where no life at all above the microbial level can survive. The estuary dead zones are caused by nitrogen compounds from fertilizers, sewage, and agricultural waste being washed into the sea. These compounds hugely stimulate the growth and hence the oxygen consumption of marine plants and bacteria; eventually, deprived of oxygen, the water can no longer support life. Even before the Deepwater Horizon spill, large tracts of the Gulf of Mexico were dead because of this effect.

And then there's the contribution of climate change. Rising levels of CO_2 in the atmosphere are leading to greater amounts of CO_2 being dissolved in ocean water[5] to form carbonic acid. The rate of increase in the acidity of the oceans, unless arrested, is widely estimated to be such that, by 2050, it will be enough to dissolve the skeletons and shells of many marine organisms—assuming they could anyway survive in those acidic conditions.[6]

Even more immediately, there's another consequence of the warming of ocean waters. One reason the world's fish stocks are plunging is that warmer waters encourage various disease organisms. The earth's ocean life is suffering a sustained series of plagues.

Because of the interconnectedness of environmental issues, and because of the paramount importance of climate change, it seems to make sense in this book to discuss environmental denial largely in the context of AGW denial. In this brief chapter, therefore, we'll look at just a couple of other aspects.

The problem of global warming is much concerned with the infrared—longer-wavelength—light we receive from the sun. We also receive ultraviolet, or UV, light, which is at too *short* a wavelength for us to see. Short wavelength implies high energy, and the high energy of UV photons gives them special properties: It is the UV in sunlight that causes human skin to tan . . . but also to burn. More seriously, UV light has been correlated with higher rates of skin cancer. Luckily, when it interacts with the oxygen in the atmosphere, UV breaks down the standard O_2 molecule to its two constituent atoms; these atoms then have a tendency to hook up with other O_2 molecules to create O_3 (ozone) molecules. The O_3 molecule has the useful property that it absorbs UV light. Thus, the incoming UV light creates the very buffer, the ozone layer, that protects us from receiving overdoses of UV down here at ground level. Without the ozone layer, eventually the surface of the earth would be sterilized by these and other high-energy particles from the sun and other cosmic sources.

In the 1970s, scientists in the Antarctic who'd been studying the ozone cycle there discovered something alarming. During the Antarctic winter there's no new ozone formed over the region because there's no incoming UV; the ozone layer over the polar region dissipates during winter and then starts rebuilding again in spring. What the scientists noticed was that, with each new summer, the rate of replenishment of the ozone layer was decreasing. From this they deduced something was affecting the levels of atmospheric ozone worldwide, and this something proved to be the group of chemicals called CFCs (because they contain chlorine, fluorine, and carbon), then in widespread use in fridges, cooling systems, aerosol spray cans, cleaning liquids, etc. These chemicals were getting into the upper atmosphere, where they were being broken down by incoming UV radiation into their constituent elements. Two of these, the chlorine and the fluorine, just happen to be efficient catalysts in the conversion of O_3 ozone molecules to ordinary O_2 oxygen ones. Hence the depletion of the ozone layer.

Fortunately, those Antarctic observers warned us in time. In 1987 the Montreal Protocol on Substances that Deplete the Ozone Layer was negotiated; this called on the countries of the world to phase out the use of CFCs in orderly but fairly rapid fashion. Every country in the UN has signed the

Montreal Protocol,[7] and the ozone depletion has now been halted. Because of the long lifetimes of chlorine and fluorine in the upper atmosphere, it's estimated the ozone layer won't fully recover until about 2050; inhabitants of some southern parts of Australia may continue to experience elevated skin-cancer rates until then.

Three of the scientists involved in working out the mechanism whereby the CFCs were causing ozone-layer breakdown—Paul Crutzen, Mario Molina, and Frank Rowland—shared the 1995 Nobel Chemistry Prize for their work, and who could dissent? They'd contributed to heading off a global catastrophe at the pass. All in all, this was a triumph for science.

What's all too easily forgotten is that there were denialists at work then, too. In exactly the same way that oil companies today try to protect their profits by pretending AGW is a hoax, so the chemical industry in the 1970s attempted to rubbish the results of the scientific research into the thinning of the ozone layer. The main corporate criminal was DuPont, which had a large share of the global market in CFCs. DuPont founded a denialist front group, the Alliance for Responsible CFC Policy, which claimed the science was not yet solid enough to justify taking any action—think how the economy might suffer! As late as 1987, the year the Montreal Protocol was drawn up, DuPont was telling Congress of its supposed corporate belief that "there is no immediate crisis that demands unilateral regulation."[8]

Even after the Montreal Protocol had come into force, DuPont kept up the pretense. Company chairman Richard E. Heckert, on March 4, 1988, reportedly wrote to Congress that "DuPont stands by its 1975 commitment to stop production of fully halogenated chlorofluorocarbons if their use poses a threat to health. . . . At the moment, scientific evidence does not point to the need for dramatic CFC emission reductions. There is no available measure of the contribution of CFCs to any observed ozone change."[9]

Within two weeks, NASA's Ozone Trends Panel, on which a DuPont scientist had served, delivered its report into the latest conclusions about the effects of the CFCs on the ozone layer, and DuPont had to back down. Within a further week the company had instituted its decade-long phasing out of their manufacture. Despite all the phony belligerence hitherto, DuPont was thus one of the first in the chemical industry to take action in response to the report.

How much of the science involved did the public absorb? The evidence suggests not much. According to James Hoggan and Richard Littlemore in

Climate Cover-Up (2009, p. 154), "When the David Suzuki Foundation commissioned a national study in 2006 on Canadian understanding [of] and attitudes toward climate change, more people blamed global warming on the ozone hole than on any other factor." And two comments on a post at *American Thinker* in December 2010 paint a similarly gloomy picture of public education on this—or any other—scientific front:

> *montegoblack:* How much did we spend to ban R13 refrigerant from air conditioners and replace with R34 to fix the ozone hole? We now know ozone comes from the Sun and when the Earth is tipped in the winter the ozone can not reach the northern hemisphere. No ozone from the Sun, ergo ozone hole.

> *Wes M.:* You know, it is actually interesting that one comment mentioned the ozone hole that we were all warned about having catastrophic consequences by the Greenies: I'm not an Ivy league educated scientist . . . but if the greenhouse gases that are supposedly so hazardous are gathering in the atmosphere and warming the planet, why wouldn't they exit through that giant hole they swore up and down existed? You can't have both. Either there is a hole, or the gases are trapped. Time for them to make up their mind.[10]

There's a prevalent meme that overly enthusiastic environmentalists, influenced by Rachel Carson's *Silent Spring* (1962), pressured the US Government and hence the governments of the world into instituting a global ban on DDT, in consequence of which malaria levels have risen to match those of the bad old days, or even higher, and millions of kids have died.

Carson's concern was that the indiscriminate use of pesticides would cause environmental devastation through building up toxins in the air, water, and soil as well as directly impacting wildlife; toxins in the air, water, and soil end up being in us, too, which is in many cases not a good thing. Further, blanket use of pesticides soon becomes counterproductive, since the pests fairly quickly develop resistance to the pesticide, so that you're left looking for something else you can use. The way to use pesticides is, therefore, with intelligence rather than just the brute-force-and-ignorance approach so much in vogue when Carson was writing. In the particular case of DDT, Carson thought it might be a carcinogen, a worry that was not at all unreasonable at the time.

The Stockholm Convention of 2001 bars the use of DDT for agriculture precisely because intensive use, through encouraging the development of DDT-resistant *Anopheles* mosquitoes, accelerates the spread of malaria. There has never been any such thing as the worldwide ban on DDT that the antienvironmentalists claim.

Those antienvironmentalists are fond of pointing to the experience of Sri Lanka when trying to create their narrative of overenthusiastic environmentalism. Sri Lanka, we're told, had, thanks to DDT, reduced the annual rate of 2.8 million cases of malaria in 1948 to a mere seventeen in 1963. Then the spraying stopped. By 1969 the rate was back up to 2.5 million cases and it has never since subsided. This is blamed on environmental activists' success in getting *the US* to institute a ban on DDT as an agricultural pesticide in 1972, several years *later*.

The way DDT is used in countering malaria is to spray in a targeted fashion for a few years, then monitor for a few more years. Assuming the malaria-carrying mosquitoes have indeed been eliminated, one can reckon the disease is for the present no longer a problem in this particular area; any individual outbreaks that might happen thereafter can be dealt with on a case-by-case basis, as with any other illness. This was the plan instituted through much of the tropics during the 1950s. Its funders, led by the US, failed to take account of the fact that, at the same time as the targeted antimalarial efforts would be going on, so too would be the blanket application of DDT as a pesticide to crops. This latter practice completely subverted the antimalarial endeavor. The program therefore had to be redesigned and extended in midstream; unfortunately the funds dried up and, in many cases, political turmoil complicated matters further. In the final reckoning, then, the program to eradicate malaria was one of humankind's failures; modern efforts concentrate primarily on limiting and controlling the disease, and DDT plays a part in those efforts.

Since the narrative created by the antienvironmentalists relies on plot points that include a worldwide ban which has never in fact existed, causing disastrous consequences years before it's supposed to have come into effect, it's hard to understand why the meme is so thriving.

16

SO, WHAT WAS THE WEATHER LIKE IN 2010?

We can't point to a single piece of weather—a hurricane or a flood or a drought—and claim it to have been caused by global warming; that isn't the way weather and climate work. What we can look at, however, are *patterns* of extreme weather, including not just those we might expect from the term "global warming" (many climate scientists wish they could go back in time and introduce the term "global weirding" instead) but also cold snaps, as the restructuring of atmospheric systems draws polar air into areas where hitherto it has not strayed, and torrential rain and snowstorms, as a warmer atmosphere takes up more moisture over the oceans only to precipitate it elsewhere. Tornadoes occur when winds hit masses of moist air, so we might expect the higher levels of atmospheric moisture to result in increased frequency and ferocity of tornado attacks. During 2010, many clamorous voices tried to tell us climate change was a myth. The idea that climate change was *already happening* was shouted down.

It seems a useful idea to survey the weather events of 2010—the year nearest at hand, as it were—to see if any extreme weather patterns might seem detectable. We should certainly remember that, just as a few chest pains aren't a definitive indication you have a heart problem, so a single year is not a long enough period to claim definitive proof that global warming is underway. But, if your chest pains persist, you'd obviously be wise to see a doctor rather than persuade yourself there might be a whole string of alternative possible causes.

So, what kind of weather events were we witnessing? Here's a rapid review, with some other events for context.[1]

January

In southeastern Brazil in early January, flooding and mudslides killed at least seventy-six people and caused up to $145 million in damage. Heavy rainfall in Kenya's Rift Valley killed thirty-eight people, did extensive damage to crops and livestock, and triggered concerns of a cholera outbreak; this followed record droughts in the region during 2009. Israel and Egypt suffered their worst flooding in over a decade. Flooding and mudslides in Peru and Bolivia killed at least thirty.

Parts of Europe and North America suffered a prolonged cold snap thanks to high pressure over Greenland deflecting the jet stream farther south than usual. Such situations are a product of the Arctic Oscillation, whereby atmospheric pressure systems around the pole occasionally switch, and in so doing affect northern hemisphere weather. The *Daily Mail*, a UK newspaper that has led a campaign against climate science, interpreted the cold weather instead as the beginning of a new ice age, and cited Mojib Latif of the Intergovernmental Panel on Climate Change (IPCC) in support of the claim; less prominently featured by the *Mail* was Latif's angry protest that his remark had been cherry picked and he believed no such thing.

February

In February several provinces in China were suffering their worst drought in over half a century, and over 7.5 million people in southern China were without drinking water other than what they could buy. Agricultural losses were estimated at $556 million. A heat wave in Brazil killed thirty-two in Santos alone. On Java, at least forty-six died in a landslide triggered by heavy rain. On Madeira Island, flash floods and landslides killed at least forty-two and inundated Funchal—the island's worst natural disaster since the nineteenth century.

On the fourth, the Utah House Natural Resources Committee passed a resolution to the effect that "climate alarmists' carbon dioxide-related global warming hypothesis is unable to account for the current downturn in global temperatures" and urging that the EPA not regulate greenhouse gases "until a full and independent investigation of the climate data conspiracy and global warming science can be substantiated." Rep. Mike Noel (R) elaborated the thinking behind this move: "Sometimes when we don't have all the answers, we need to have the courage to do nothing."[2]

On the ninth, the BBC reported: "Springtime in the UK is starting on average 11 days earlier than 30 years ago, causing natural food chains to become disrupted, a study suggests."[3]

March

The month began with the burial by rain-triggered landslides of three Ugandan villages, killing some three hundred. In Kazakhstan heavy rains and melting snow caused floods that killed upward of forty. Australia experienced astonishing storms; at Perth Airport 15.6mm of rain fell in just seven minutes, equal to the average for a full month. The Harmattan—an annual wind that carries dust from the Sahara—for once blew into Nigeria, blotting the sun from the sky; it's expected this will become the new norm, as desertification brings the Sahara ever closer to Nigeria. Cyclone Ului equaled the record set in 2005 by Hurricane Wilma for the fastest intensification of a storm system, within just twenty-four hours going from a mere tropical storm to a category 5 cyclone, its maximum sustained wind speeds increasing from 100kph to 260kph. Two major nor'easters created havoc along the northern Atlantic seaboard of the US, causing about a dozen deaths. New Jersey's largest utility company said the storm was the worst in its history, with 420,000 power outages.

On the fourth, the National Science Foundation announced:

> [R]esearch results show that the permafrost under the East Siberian Arctic Shelf, long thought to be an impermeable barrier sealing in methane, is perforated and is starting to leak large amounts of methane into the atmosphere. Release of even a fraction of the methane stored in the shelf could trigger abrupt climate warming.[4]

What's startling is that our knowledge of this danger is so recent it wasn't even mentioned in the IPCC's 2007 report.

April

Many Caribbean islands were in the grip of a record drought, and drought was afflicting parts of China, too, where conditions were weakening trees' ability to resist the attacks of the pine caterpillar; by official estimates, in the

Guangxi Zhuang region alone, some 36,000 hectares of forest were plagued by the pest and over 2,300 hectares of forest had been destroyed. The northeastern US was hit by a heat wave that brought record temperatures to several areas. Toward the end of the month, a surprise snowstorm struck parts of the region. Delhi suffered its highest April temperature in almost sixty years—a full seven and a half degrees above the norm! Dozens died in rain-triggered landslides in northeastern Peru. About 250 died in mudslides in Rio de Janeiro after the heaviest rainfall the city had experienced in nearly half a century.

May

On the second, lightning from severe storms in Bangladesh killed at least seventeen and injured about fifty others. A few days later, at least seventy were killed by a storm in southwestern China that included the region's first-ever tornado (by the end of the month the region's death toll from storms would increase by forty-five) and a few days after this at least forty-three died in India in storms. Late in the month, parts of the US and Canada suffered a heat wave that shattered temperature records. Northern India and Pakistan were having a very similar if even more extreme experience, with over 260 heat-related deaths in India and at least another 18 in Pakistan. Meanwhile, heavy rains caused floods and mudslides in Kenya that killed nearly a hundred, with extensive devastation of agricultural land. Major flooding in northern Afghanistan and southern Tajikistan killed at least 124. May saw flooding also in the Czech Republic, Slovakia, Hungary, Serbia, and Poland, with at least twenty deaths in the latter; a further round of rainfall in the region two weeks later caused more flooding and more deaths. Sri Lanka suffered its worst flooding in half a century, with twenty deaths reported. In the eastern part of the Democratic Republic of Congo twenty-seven died in a rain-triggered mudslide. Back in the US, a storm system in the Lower Mississippi Valley killed twenty-nine, broke all sorts of rainfall records, created widespread flooding, and generated dozens of tornadoes.

June

Fifty-three provinces in northern and northeastern Thailand were declared disaster areas as the region suffered its worst drought for twenty years; it was reckoned the rice harvest the following August would be halved. The UK announced its driest January–June since 1953 and second driest since 1929.

In southern China, landslides and floods killed at least 430; overall, the flooding affected 69 million, with damages estimated at $12.1 billion. In Arkansas, heavy rains caused the Little Missouri to burst its banks in the Ouachita National Forest, killing twenty. Phenomenal rainfall and subsequent landslides and floods killed at least fifty-eight in Bangladesh, at least sixty-three in Burma, at least forty-six in India's Maharashtra state, at least seventy-two in Brazil, at least thirty in Ghana, and at least twenty-four in northern Romania. France, in its worst flooding since 1827, suffered at least twenty-five fatalities.

The US heartland saw a number of deaths in an unusually active tornado season, the most dramatic event being in North Dakota, where a storm system spawned twenty tornadoes in a single four-hour period. Late in the month, Hurricane Alex, a relatively rare June hurricane in the Atlantic basin, became the strongest such since 1966's Hurricane Alma.

A Cornell University study indicated that warming was threatening supplies of maple syrup.[5]

July

Bolivia suffered severe drought, with consequent devastation of agriculture, a major wildfire, and a major dust storm. A state of emergency was declared in California as three huge fires raged.

Russia was in the grip of a major heat wave, enduring its hottest summer on record. Moscow recorded its highest temperature ever: 39°C, beating the previous record, 36.8°C, set ninety years before. Some sixteen hundred drownings were reported across the country during July as people sought to escape the heat. Drought—the worst since 1972—and wildfires destroyed nearly 20 percent of Russia's crops in July alone. At least 40 people died in the 948 wildfires that devoured about 26,000 hectares; those near Moscow were joined by peat fires, resulting in the city's smog level being 5 to 8 times the norm.

Finland likewise experienced its hottest recorded temperature, 37.2°C, beating the ninety-six-year-old existing record, 35°C. Eastern North America, too, suffered a heat wave, with hundreds of local temperature records being broken in the US and Canada. In Ottawa, Montfort Hospital announced that the fifth had seen a new record for arrivals. Northern China was another region suffering a heat wave; Beijing recorded two new records on the fifth: its highest July temperature, 40.6°C, since 1951, when measurements began, and its highest consumption of tap water since 1910, which was when the city *got* tap water. The high temperatures and drought brought another problem: swarms of locusts. In Inner Mongolia locusts destroyed nearly 4 million hectares of grassland.

Meanwhile, Antarctic air was reaching up through South America, with hundreds of people perishing in consequence.

In India, the monsoon was late; once it came, flooding caused an estimated seventy-six fatalities. In China, at least 147 died in torrential rainfall and flooding. Flooding was also to blame for the death of three people in Vietnam, along with seven in Saudi Arabia and at least eight in both Japan and Cameroon; in Cameroon, the wet conditions also promoted a cholera outbreak that killed scores. A hailstorm damaged every single house in Vivian, South Dakota, so massive were the stones. Bronx County, New York, witnessed its second tornado in recorded history; seven people were injured.

Pakistan saw the first of the torrential rainfall and flooding that would dominate its August.

August

In July and August the monsoon rains caused extreme flooding in Pakistan, the worst in eighty years, killing as many as two thousand and affecting an estimated 14 million. The agricultural heartlands of Punjab and Sindh were hard hit. Two million fled Upper Sindh province, as the Indus threatened to burst its banks. In India flooding and mudslides killed at least 132. Elsewhere, at least a dozen were killed by flooding and mudslides in the northern Turkish town of Gundogdu, as were at least thirty-four across Nicaragua.

China suffered its worst flooding in a decade. The floods killed at least fifteen hundred and likely well over two thousand (estimates differ), and displaced millions; one massive flood-induced landslide in Gansu province covered an area 500m wide and 5km long, killing upward of a thousand. On August 9, at least eleven died in Central Europe in consequence of major flooding there, and in the Sahel, of all places, at least two hundred died because of torrential rainfall and a consequent cholera outbreak. The flooding, coming after the droughts of 2009, resulted in agricultural devastation; over half the population of Niger was reported starving.

The biggest Arctic iceberg since 1962—250 square kilometers—calved off the Petermann Glacier, northwest Greenland.

Indonesia suffered extreme weather conditions, with the *Jakarta Globe* for August 19 bearing the headline: "'Super-Extreme' Weather Is the Worst on Record."[6] Less obvious was that sea temperatures were far above normal for August: 28°C to 29°C as opposed to 24°C to 26°C. Because of warm waters in the Andaman Sea up to 60 percent of the corals there suffered bleaching, according to the Wildlife Conservation Society. The loss of corals harms the marine ecosystem and in due course leads to reduced fish catches.

In Moscow it was reported in August that about seven hundred were dying daily in the city, twice the usual number; the difference was believed to be a consequence of the prolonged heat wave affecting the city, with temperatures sometimes topping 40°C—Moscow's average August temperatures are usually more like 21°C. Many of the deaths were drownings when people drank vodka and then went swimming to try to get cool. All told, deaths from the Russian heat wave were estimated to top fifteen thousand—seven thousand in Moscow alone. The Russian grain harvest was "devastated," the falloff reportedly being as much as 38 percent. President Dmitry Medvedev, who had up until this point declared himself an AGW "skeptic," rather abruptly changed tune.

A heat wave and drought afflicted British Columbia, with rainfall across the province being 50 to 75 percent lower than normal. Numerous temperature records were broken both here and in neighboring Alaska.

In the *Columbia Journalism Review* Curtis Brainard patted journalists on the back for knowing better than mere climate scientists what was going on:

> More and more, reporters have been asking whether or not climate change could be responsible for this summer's extreme weather. Thankfully, most have resisted the temptation to pin the events directly to global warming, placing them in proper climatic context instead.[7]

Raining on Brainard's parade, "The climate is changing," commented Jay Lawrimore, head of climate analysis at the National Climatic Data Center in Asheville, NC.[8] "Extreme events are occurring with greater frequency, and in many cases with greater intensity."

September

In late September torrential rains triggered a landslide that killed up to thirty north of Medellin, Colombia. In Chad, flooding killed at least twenty-four and displaced almost seventy thousand, with a flood-related cholera outbreak reported to have killed another forty-three; 31,500 hectares of cropland were destroyed. At least sixty died from flooding and landslides in the Indian state of Uttarakhand, where rainfall was about 22 percent above normal levels.

A quite different weather-related disaster was happening in late September in California: a heat wave with temperatures above 38°C, which led to a number of (attributed) deaths; the state experienced its warmest day since 1877, when records began. Because people's overuse of air conditioners strained power equipment, thousands of Los Angelinos were for a while deprived of electricity. Through much of the summer California had been ravaged by wildfires. Janet Upton, a deputy director with Cal Fire, pointed out that "of the 20 largest wildfires in California since 1932, 11 occurred in the past decade. . . . And eight occurred in the last four years."[9] The consensus, she added, was that climate change was responsible for this upswing.

Colorado suffered two major wildfires; the first, near Denver, was the most destructive in the state's history.

On the sixteenth, a sudden storm in New York City produced winds up to 200kph and two tornadoes, one each in Brooklyn and Queens. This brought the 2010 tally of tornadoes to hit the city up to three; according to NOAA, "Prior to 2010, only seven tornadoes had been recorded in the city since 1950."[10]

On the twentieth it was reported that "[t]he carbon dioxide emissions from burning fossil fuels have melted the Arctic sea ice to its lowest volume since before the rise of human civilisation, dangerously upsetting the energy balance of the entire planet."[11]

October

On the sixth, the region surrounding Bellemont, northern Arizona, suffered no fewer than eight tornadoes. The annual average for the entire state is four. The strongest storm ever recorded in the Midwest—widely described as a "land hurricane"—rampaged across America's heartland on October 26 and 27. On the twenty-sixth, 24 tornado reports were logged alongside 282 reports of damaging high winds. Extreme drought gripped parts of nine US states from the Southeast to the lower Midwest.

Northern and western Brazil suffered their worst droughts for four decades. Several Amazon tributaries almost completely dried up.[12] Floods and landslides killed at least sixty-four in central Vietnam, and another thirty-four thousand had to be evacuated. In China, some 450,000 had to be evacuated because of flooding. In West Papua over a hundred died when a river burst its banks. Bangladesh had its driest monsoon season since 1994, but then October 7 through 9 saw downpours that killed seventeen and displaced about half a million. At least ten died in Burma from the same storm system. Thailand saw at least fifty-nine die from flooding, with 3 million displaced. Flooding in Benin caused at least forty-three deaths, with hundreds affected by a consequent cholera outbreak. Flooding killed at least twelve around the earthquake-afflicted Port-au-Prince, Haiti's capital.

Senate and House candidates across the US who accepted the science of climate change were facing turbulence among the ranks. One such was Democratic incumbent Baron Hill of southern Indiana, who attracted hostility because of his support for the House's cap-and-trade bill.

> A rain of boos showered Mr. Hill, including a hearty growl from Norman Dennison, a 50-year-old electrician and founder of the Corydon Tea Party.
>
> "It's a flat-out lie," Mr. Dennison said in an interview after the debate, adding that he had based his view on the preaching of Rush Limbaugh and the teaching of Scripture.[13]

Late in the month it was reported that the richer nations were covertly reneging on promises made at the 2009 Copenhagen climate summit to give developing nations $30 billion to help adapt to climate change.[14]

November

The BBC reported that the Met Office, the UK national weather service, thought 2010 was likely to be the hottest year on record, then spent one-third of its report airing the opinions of "climate contrarian" John Christy.[15]

Russia, still recovering from the summer heat wave, was experiencing temperatures ten degrees above the average for this time of year. Pause for a moment: ten degrees is the sort of difference you find in apocalyptic science-fiction novels. All over Russia, animals were failing to hibernate because of the warm winter.

On the third, some forty people died in Costa Rica because of floods and landslides. At least thirteen died because of torrential rain in the Philippines during the early part of the month, and nearly half a million had to be evacuated. On the fourteenth, a rainstorm killed three in Belgium and northern France; the Belgian authorities said flooding there was the worst in half a century.[16]

Hurricane Tomas caused flooding and mudslides in Haiti that killed at least twenty. Cyclone Jal killed at least twenty-two in Sri Lanka and at least eleven in India. Perth, Australia, suffered its hottest November on record.

An Oxfam report released at the end of November[17] estimated that, in the first nine months of 2010, twenty-one thousand people had died worldwide as a result of weather events, against a total of ten thousand for the whole of 2009.

December

In the early part of the month Israel's Mount Carmel region experienced the worst forest fire in the nation's history. Over forty people died[18] and about four million trees were destroyed. Drought hit Somalia's central region due to the failure of the annual Deyr rains. Flooding afflicted the Balkans, with over twelve thousand people having to be evacuated in Albania and some fourteen thousand hectares of agricultural land lost. Three died in a rain-triggered landslide in the Bosnian city of Tuzla. La Niña took its toll in South America. Flooding and landslides in Venezuela killed at least thirty-four and compelled a hundred thousand people to evacuate their homes. In the Colombian town of Bello, near Medellin, a landslide killed at least sixty-three in Colombia's heaviest rains on record. In mid-December record rains also fell in California, Washington, Arizona, Nevada, and Utah; California's Laguna Beach and others suffered mudslides and landslides, and on a single day Los Angeles received as much rainfall as it normally experiences in six months.

Australia was counting the cost of its worst plague of locusts in decades; this, combined with drought and floods, meant the country was looking at a bill of A$3 billion in terms of lost wheat alone. In fact, we may all have to become accustomed to more frequent plagues of locusts: "[A]n alarm bell for yet another serious consequence of climate change," commented Ge Quansheng of the Institute of Geographical Sciences and Natural Resources Research, Beijing.[19]

Locusts weren't Australia's only problem. "Australia's most populous state [New South Wales] declared a natural disaster and flood warnings were issued for two other regions as rains swept the nation's east coast, damaging crops and driving up global wheat prices," announced *Bloomberg News*.[20] Soon the flood crisis was spreading through the states of Queensland, South Australia, and Victoria. Later in the month, Western Australia had its worst flooding in half a century; flooding in Queensland—enhanced by Cyclone Tasha—covered an area larger than France and Germany combined and required extensive evacuations, with floodwaters reaching up to 9m. As much as 10cm of snow fell in parts of New South Wales and Victoria, raising the tantalizing prospect of a White Christmas . . . in the height of the Australian summer.[21]

The Queensland flooding continued into the new year, with Brisbane having to be largely evacuated. Pastor Daniel Nalliah offered a reason: Australia's foreign minister, Kevin Rudd, had been insufficiently supportive of Israel during a visit there. "It is very interesting that Kevin Rudd is from Queensland. Is God trying to get our attention?"[22]

Severe winter weather gripped much of Europe, killing at least 114 people in Poland. Earlier, on the eighth, Scotland suffered its heaviest snowfall since 1963 and the army had to be called in to help cope. On the eighteenth, some 12cm of snow fell on London's Heathrow Airport in a single hour while County Tyrone recorded the lowest temperature ever for Northern Ireland: –18°C. The US, too, was experiencing exceptional levels of precipitation; on the twelfth, the roof of the Minneapolis Metrodome collapsed under the weight of snow. By Christmas it was Russia's turn: Freezing rain disrupted Moscow's airports and cut off electricity for over four hundred thousand people in the immediate region.

On New Year's Eve, unusually warm air in southern and central regions of the US fueled a flurry of rare winter tornadoes that killed three people in Arkansas and four in Missouri.

Meanwhile, in late 2010, our glorious leaders jetted into Cancún, Mexico, for a climate conference at which they spent several enormously expensive days deciding to do almost nothing at all about any of this. The mainstream media achieved a self-fulfilling prophecy by declaring in advance that the conference would be a nonevent and then, in large numbers, failing to turn up for it.[23]

In January 2011 it was announced that, according to US government figures, 2010 had equaled 2005 as the hottest year on record. The World Meteorological Organization went further, declaring 2010 the hottest year since record keeping began in 1850. It was also—and this too is symptomatic of global warming—the wettest year on record.

Bearing in mind that the weather events listed above are just a fairly random selection, is this, we must ask ourselves, a picture of a planet that's in the best of climatic health?

But it wasn't just weather catastrophes that characterized 2010. There was gloomy news from a wide variety of other theaters. In July appeared "Global Phytoplankton Decline over the Past Century"[24] by Daniel G. Boyce et al., which details a decline in oceanic phytoplankton of some 40 percent since 1950. Phytoplankton are the fundamental building block of marine food chains; if they're in trouble, so are all other marine organisms—and us too. But phytoplankton play another important role: They absorb CO_2 and release oxygen. It used to be said that phytoplankton were responsible for about half the world's supply of fresh oxygen, an estimate clearly now becoming problematical.

Perhaps the biggest climate-related disaster of 2010 was, however, the capture of the US House of Representatives by a Republican Party rich in AGW deniers.[25] Even before the November midterm elections, the GOP had been saber rattling about the primary antiwarming measure the Obama administration had been able to push through absent any serious Senate determination to confront the problem: the classification by the Environmental Protection Agency of CO_2 as an atmospheric pollutant, so the agency could begin to crack down on excessive emissions. Reported the *Los Angeles Times*:

> The attack, according to senior Republicans, will seek to portray the EPA as abusing its authority and damaging the economy with needless government regulations.

> In addition, GOP leaders say, they will focus on what they see as distortions of scientific evidence regarding climate change. . . .
>
> Several key Republican Congressmen—most notably Rep. Darrell Issa (R-CA), who could take over the chairmanship of the House Oversight and Government Reform Committee—have said they plan to investigate climate scientists they contend manipulated data to prove the case that human activity is contributing to global warming. . . .
>
> House Republicans like Issa and James F. Sensenbrenner Jr. (R-Wis.) of the Select Committee on Energy Independence and Global Warming have criticized the EPA for basing its endangerment finding on what they consider flawed research. Republicans assert that the science on climate change is not yet "settled," despite the vast global scientific consensus about its human causes.[26]

Climate scientists divide roughly 98 percent to 2 percent over the reality of human-caused global warming, yet somehow politicians who lack scientific training *know better*. As Ronald Brownstein commented in the *National Journal*, "The GOP is stampeding toward an absolutist rejection of climate science that appears unmatched among major political parties around the globe, even conservative ones."[27] One of the first deeds announced by the incoming Republican leadership was to abolish the House Select Committee on Energy Independence and Global Warming, which had been created by the outgoing Speaker, Nancy Pelosi, in 2007.

A very odd aspect of this is that, at state level, many Republican legislators are in the forefront of actions to ameliorate global warming. There are of course exceptions, like New Jersey's Governor Chris Christie, who in 2010 diverted to other purposes the entirety of the state's income from the Regional Greenhouse Gas Initiative, a cap-and-trade system whose profits are supposed to be directed toward improving energy efficiency, cutting CO_2 emissions, investing in renewables, etc.[28] By contrast, Governor Arnold Schwarzenegger of California was a champion in the war against AGW, battling hard to give the state better auto-emissions standards and supporting other useful moves. He wasn't alone. If Texas were a country, it would be sixth on the list of those producing the most energy from wind power. Nevada is building geothermal energy plants. Of the thirty-one states to have enacted laws requiring increases in the use of renewables, eleven are generally identified as Republican. Houston, capital of the oil industry, is trying to make itself a green city. And so on.[29]

Yet according to Representative John Shimkus (R-IL) we really have no need to worry about the perils of global warming. After all, as he told a 2009 hearing of the House Energy and Commerce Committee's Energy and Environment Subcommittee:

> The Earth will end only when God declares it's time to be over. Man will not destroy this Earth. This Earth will not be destroyed by a Flood. I do believe that God's word is infallible, unchanging, perfect.[30]

At the same hearing he explained some science:

> If we decrease the use of carbon dioxide are we not taking away plant food from the atmosphere? We could be doing just the opposite of what the people who want to save the world are saying. . . .
>
> Today we have 388 parts per million in the atmosphere. I believe in the days of the dinosaurs, where we had the most flora and fauna, we were probably at 4000 parts per million. There is a theological debate that this is a carbon-starved planet, not too much carbon.[31]

It's interesting Shimkus used the term "*theological* debate" in connection with climate science. Theology has absolutely no relevance to considerations of the ideal level of CO_2 in the earth's atmosphere.

And the response of the news media to the weather of 2010?

> Analysis of DailyClimate.org's archive of global media coverage shows that journalists published 23,156 climate-related stories in English last year—a 30 percent drop from '09's tally.[32]

17

GLOBAL WEIRDING

It is an excruciating experience to watch the planet fall apart piece by piece in the face of persistent and pathological denial.

—Ross Gelbspan, *Boiling Point* (2004), p. ix

. . . and none of this snake oil science stuff that is based on this global warming, Gore-gate stuff that came down where there was revelation that the scientists, some of these scientists were playing political games.

—Sarah Palin, Southern Republican Leadership Conference, April 8, 2010

It took 50 million years for life on earth to recover after the Permian extinction, for which extinction global warming seems the prime culprit.

In *Collapse* (2005) Jared Diamond notes a number of instances of even localized climate changes having dramatic consequences. Drought, he says, was probably the final straw for the Mayan civilization. Overpopulation had led to competition for ever-scarcer resources, so the collapse would likely have come in due course anyway, but a major drought was the last straw. One of Diamond's comments (p. 177) resonates alarmingly with our current situation: "[W]e have to wonder why the kings and nobles failed to recognize and solve these seemingly obvious problems undermining their society. Their attention was evidently focused on their short-term concerns of enriching themselves."

That the atmosphere should display a greenhouse effect was first posited by French physicist Joseph Fourier in 1824. Calculation showed him that, at its orbital distance from the sun, the earth should be far cooler than it is. The reason for the discrepancy, he speculated, might be that the atmosphere acted as a kind of blanket. By the time John Tyndall started working on the subject in the 1850s it was fairly widely accepted that a greenhouse effect operated in the atmosphere; he devised an experiment to prove it. He hypothesized that the major greenhouse gas—to use the modern term—was water vapor. In the 1890s Svante Arrhenius correctly deduced that carbon dioxide (CO_2) played a major role, and made a surprisingly good calculation of the relationship between atmospheric CO_2 levels and the extent to which they raised (or lowered) global temperatures. We can now even study the greenhouse effect in the atmospheres of other planets.

So our knowledge of the greenhouse effect has a long history, and the basic science can be said to be very well settled. This didn't stop prominent AGW denier Peter Hitchens from claiming in 2001, "The greenhouse effect probably doesn't exist. There is as yet no evidence for it."[1]

Of the gases in the earth's atmosphere, oxygen and nitrogen don't absorb infrared, and so don't contribute to the greenhouse effect. The important atmospheric gases that do absorb infrared light are water vapor (H_2O), carbon dioxide (CO_2), and methane (CH_4). Because there are relatively small amounts of these gases in the atmosphere, the influence of the greenhouse effect is quite sensitive to changes in them. And, of course, there are feedback effects in operation: Increase CO_2 levels and average temperatures rise; because of the rise in temperatures there's more evaporation from the ocean surfaces into the warmer air, so there's more water vapor in the atmosphere; because of the heightened levels of water vapor, average global temperatures can be expected to rise . . .[2]

Yet a further consideration is that CO_2 dissolves more readily in cold seawater than in warm. Thus, if you warm the oceans, they release more CO_2 into the atmosphere—another piece of unfortunate positive feedback! Oxygen, too, is less soluble in warm water than in cold; the warming of the oceans is depleting the water of oxygen down to depths as great as 500m, creating dead zones where no fish can survive.

Deniers are fond of pointing to large, CO_2-induced climate changes in the remote past and observing that, since humans weren't yet around at the time, these could hardly have been of anthropogenic origin—so how can we rationally conclude the climate changes currently underway are caused by us? This is tantamount to saying that, since you can point to plenty of examples of people dying naturally, the maniac on the loose with an AK-47 is surely harmless.

Sometimes it's easy to think a couple of degrees either way can't make too much of a difference. But a rise of six Celsius degrees, as some models anticipate, would be comparable to the temperature rise at the end of the last glaciation. Global mean temperatures were then about five to six degrees cooler than they are now. We've seen how the face of the earth can be transformed by just those few degrees.

Still roaming the stratigraphic column, many AGW deniers point to ancient instances when atmospheric CO_2 levels were higher than they are today without inducing climate catastrophe. True, such instances exist. However, the further back in time you go—and here we're talking about millions of years—the cooler the sun was, and that affects the equation. The warmer the sun, the greater the impact any increase of greenhouse gases will have. Also, because of the warmer sun, the longer it will take for the planet to recover from a climate catastrophe.

The grimmest possibility is that the planet will suffer a runaway greenhouse effect—what James Hansen calls the Venus syndrome. The planet Venus is about the same size as our own, but about one-third closer to the sun. It has an extraordinarily dense atmosphere, made up primarily of CO_2 with small amounts of other gases, notably nitrogen. Because of this dense atmosphere, Venus is in the grip of runaway greenhouse conditions from which it will almost certainly never emerge. The atmospheric pressure at the surface is about ninety times that on earth—roughly the same as you'd experience here if you were under about a kilometer of water. The surface temperature can be over 450°C, which is considerably hotter than your kitchen oven gets; the heat is so intense that surface rocks glow red from their own heat. The thick clouds in the upper Venusian atmosphere consist of sulfur dioxide and, in droplet form, sulfuric acid. And yet, at the outset, Venus was probably not unlike the earth. It almost certainly had oceans of water; it may even have sustained life, or at least the early building blocks thereof. But when the oceans evaporated, filling the atmosphere with water vapor, a greenhouse effect was initiated . . . and the rest is history.

Obviously, because of earth's greater distance from the sun, we wouldn't expect quite so extreme an outcome to a terrestrial runaway greenhouse effect. But we could expect something similar. And certainly we couldn't expect any life to survive on earth above the bacterial level, if at all. But how likely is it our planet could suffer such an effect? James Hansen, in his *Storms of My Grandchildren* (2009, p. 236), is gloomy:

> I've come to conclude that if we burn all reserves of oil, gas, and coal, there is a substantial chance we will initiate the runaway greenhouse. If we also burn the tar sands and tar shale, I believe the Venus syndrome is a dead certainty.

As a measure of how much worse our current situation may be than we think, the IPCC's Fourth Assessment Report (2007) estimated that the Arctic could see essentially ice-free Septembers by "the latter part of the 21st century." That was four years ago. More recent estimates are putting the first ice-free September perhaps as early as 2013 or 2014. Though deniers attack the IPCC as alarmist, the truth is it's not nearly alarmist enough. For a picture of what's *really* going on in the Arctic, try the National Snow and Ice Data Center's page "Sea Ice" at http://nsidc.org/sotc/sea_ice.html.

Another unequivocal sign of global warming is to be found in temperature measurements in the permafrost. Overall, since the early 1980s the surface of the Alaskan permafrost has warmed by nearly two degrees, and in some places by twice that. Such results bring us well into the kind of temperature ranges that, when forecast for *later in the century*, the modern breed of AGW denier tends to describe as "alarmist."

One further AGW-related problem facing the people of the Arctic is that, as habitats creep poleward, the Inuit are now seeing bird species for which no native word exists.

It is currently estimated that, by about 2025–30, Glacier National Park in Montana will no longer contain any glaciers. The glaciers in the Himalayas may not last much longer. And, years before then, Mount Kilimanjaro's famous snow-covered peak will be snowless.

As we saw in the last chapter, climate change is bringing us an increasing number of ever worse natural disasters, from floods to storms to droughts to heat waves and more. In the near future, there will also be disasters brought about by the human response to the natural catastrophes—wars: wars over increasingly scarce water or food, wars over agricultural land, wars because of attempted migrations by climate refugees, civil wars as scores of states collapse. The possibilities for conflict are endless. In *Climate Wars: The Fight for Survival as the World Overheats* (2010) the journalist and military historian Gwynne Dyer lays out a number of plausible scenarios for potential future conflict. Clearly the purpose of his exercise is not to predict particular conflicts but to demonstrate the *kinds* of strife we can expect to suffer—all these in addition to the calamities inflicted upon us by the climate itself.

In November 2000 Andrew Dlugolecki of the Commercial Union, General Accident, and Norwich Union (CGNU) insurance group told the international climate change summit at The Hague that, unless checked, by 2065 the cost of damage inflicted by climate change will exceed the wealth of the world. Of course, that's a piece of mathematical modeling: such a thing will never come to pass because—which was the brunt of Dlugolecki's remarks—civilization will have collapsed long before we reach that point.[3]

According to the Energy Information Administration's *International Energy Annual 2006*,[4] each US citizen was in 2006 responsible for 19.78 metric tons of CO_2, each Australian for 20.58, each Brit for 9.66, each German for 10.40, each Russian for 12.00, and each Canadian for 18.81. As for those countries of whose CO_2 emissions US politicians and pundits most complain? For China the relevant figure was 4.58 and for India 1.16. Both countries have huge populations; per capita—and it's per capita that matters—the developed nations' CO_2 emissions are far greater. Venture capitalist John Doerr has pointed out that the annual expenditure by the US government on researching renewable-energy technologies, about $1 billion, is roughly the same as that earned by ExxonMobil *per day*.[5]

The environmental devastation being perpetrated on US soil by special interests, especially those associated with the fossil fuel and livestock industries, is so gross that it's no exaggeration to say, were this to be done to us by a foreign power, we'd declare war on it. Yet those who argue against this destruction of the national heritage and assault on the common good are habitually lambasted as anti-patriotic.

One further measure of the US's reluctance to face up to the facts on global warming is that, while George Monbiot's book *Heat: How to Stop the Planet from Burning* was first published in the UK in 2006 by a major publishing house, Allen Lane, with a paperback a year later from Penguin, and was regarded as an important piece of publishing, with wall-to-wall reviews, in the US its first publication had to wait until 2007 and then was from a small nonprofit publisher, South End Press of Cambridge, Massachusetts.

Yet there have been periodic efforts by US administrations to face up to the problem. The first National Academy of Sciences investigation of the possibility of global warming took place longer ago than most people imagine, during the presidency of Jimmy Carter. The panel said it appeared that, if CO_2 emissions continued to rise at then-present rates, the consequences could be disastrous. The panel also stressed: "We may not be given a warning until the CO_2 loading is such that an appreciable climate change is inevitable."[6] Three decades later we still have politicians telling us we should hang on a bit longer until "the science is settled."

Though Carter's initiative went nowhere—and Reagan removed the solar panels Carter had symbolically placed on the White House roof—there have been other attempts to wake US politicians up to the threat of climate change. The executive summary of *Global Climate Change Impacts in the United States* (2009), the report of a two-decade study by the US Global Change Research Program, formed in 1989 by President George H. W. Bush, says:

> The global warming observed over the past 50 years is due primarily to human-induced emissions of heat-trapping gases. These emissions come mainly from the burning of fossil fuels (coal, oil, and gas), with important contributions from the clearing of forests, agricultural practices, and other activities. (p. 9)

The political impact of the report seems to have been approximately nada.

In May 2010, the National Research Council—the research arm of the National Academy of Sciences—released at the request of Congress a suite of three studies concerning the science for and imminence of climate change, what could be done to ameliorate it, and what could be done to minimize the damage. The NRC's language was fairly uncompromising: "A strong, credible body of scientific evidence shows that climate change is occurring, is caused largely by human activities, and poses significant risks for a broad range of human and natural systems."[7] The reaction by the mainstream media to the studies, which were given a major rollout by the NAS, was tepid. Questioned soon after the studies' release by the *Columbia Journalism Review*'s Brett Norman on this, various editors and journalists gave various excuses, but the standard one seemed to be—even as the *science* correspondents were stressing the importance and uniqueness of the studies—that "we've heard it all before." And it seems many of our lawmakers, whom we pay to read reports like these, in this instance simply didn't bother, because mere months later an already grotesquely weakened Senate bill promoting cap-and-trade measures to reduce CO_2 emissions was allowed to wither and die.

For an October 2010 Pew Research Center survey[8] respondents were asked, "Do scientists agree the earth is getting warmer because of human activity?" Just 44 percent said yes.

In 2004 Naomi Oreskes surveyed 928 randomly selected scientific papers that contained the words "global climate change" and discovered not one concluded global warming wasn't happening or that our use of fossil fuels wasn't primarily responsible.[9] More recently, a 2010 report published in *Proceedings of the National Academy of Sciences*[10] estimated that 97 to 98 percent of the world's actively publishing climatologists were in accord about climate change and supported the "primary tenets" of the current IPCC report; that is almost certainly a higher percentage than that of cosmologists who accept the Big Bang theory. The possibility of the scientific consensus on climate change being wrong is somewhat lower than the likelihood that millions of Americans are being abducted by UFOs for sexual experimentation. But the investigators went further, evaluating the expertise of the dissenters through their publications and citations histories. They discovered the discrepancy in expertise between the supporters and the "skeptics" was huge. Basically, the more science a scientist does that's relevant to the discussion, the greater the certainty he or she has about imminent climate change; the "skeptics" tend not to have published much of relevance, if at all.

The reaction of the AGW deniers was as one might expect. Blogger Joanne Nova was typical: "The survey is a loose proxy for government grants. . . . R.I.P. The Scientific Method. Hello totalitarian government."[11]

All through human history until about 1850, the level of CO_2 in the atmosphere held steady at about 280 ppm. Since then it has increased to about 380 ppm, seemingly entirely because of human activities, and this increase—already 36 percent—shows no signs of stopping. Far from it, the increase is accelerating: many predictions suggest that by the end of this decade the level of CO_2 in the atmosphere will have *doubled* from that 280 ppm base rate. The 2010 level was already higher than any in the past 20 million years.

One of the earliest to warn of the dangers of CO_2 emissions fueling global warming was John H. Mercer of the Institute of Polar Studies, Ohio State University. The abstract of his 1978 paper "West Antarctic Ice Sheet and CO_2 Greenhouse Effect: A Threat of Disaster"[12] states: "The computed temperature rise at lat 80°S could start rapid deglaciation of West Antarctica, leading to a 5m rise in sea level." He adds, "One of the warning signs that a dangerous warming trend is under way in Antarctica will be the breakup of ice shelves on both coasts of the Antarctic Peninsula, starting with the northernmost and extending gradually southward."

Such estimates as a 5m rise in sea level have been widely denounced by AGW deniers and in the media as "alarmist," with the more palatable figure of 60cm (2 ft.) in the twenty-first century being preferred. A consensus guesstimate among climate scientists appears to be more like *at least* 2m—enough to inundate cities like London, New York, Bangkok, Venice, Cairo, Shanghai . . . not to mention the Maldives and most of Bangladesh. Since the Antarctic contains some 90 percent of the world's ice, the substantial melting of this is obviously going to have a huge influence. It's not just that the world's oceans will have far more water in them; the additional quantities will be fresh rather than salt, leading in the shorter term to upsets in the present regime of ocean currents. Also, the disappearance of so much ice will noticeably affect the earth's reflectivity: the less light/heat reflected back into space, the more planetary warming.

So where do we stand now? Far earlier than Mercer predicted there have been dramatic changes. The northwestern Antarctic Peninsula is among the fastest-warming places on earth, and eight of the peninsula's ice shelves have already collapsed, either fully or partially. Ninety percent of the 244 glaciers along the peninsula's western shore are in retreat. Temperatures recorded at a base in the region show an increase in average winter temperatures over the past sixty years of some six Celsius degrees and an increase in average year-round temperatures of about 2.75 degrees. The former is the killer: With more ice melting in the summer, warmer winters mean less of it refreezes in each year's winter. Satellite data show that today sea ice covers the Southern Ocean off the peninsula's west for *three fewer months annually* than in 1979.[13]

In terms of sea level, there's an important point that tends to get overlooked. Over the past few millennia, so far as the development of human civilization has been concerned, the significant aspect of sea level has been not its height but the fact that it has been *stable*. The stability of sea levels has encouraged the building of coastal cities, for example. As sea levels continue to rise in the future, the obvious (although astonishingly arduous) course might seem to be to relocate those cities further inland. But how would you choose a suitable new site for a city if there were every possibility you might have to move it *again* in a few decades' time? This is no idle question. The extent to which our civilization depends on our ports is clearly enormous. Where do you build a port if the coastline keeps moving? Also, your city has sewers; as the water level rises these will eventually flood and back up. Cholera, anyone?

And this isn't a temporary problem. Ice sheets can melt relatively rapidly. Restoring them to their previous levels (if possible at all), as winter snowfall annually accretes just a bit more than springtime melting, can take millennia.

One natural phenomenon that acts to counter global warming is the presence of aerosols (droplets) in the upper atmosphere. These reflect sunlight back into space, and much has been made by deniers of the way erupting volcanoes, because of their sulfur dioxide emissions, create huge quantities of aerosols in the stratosphere; could we be saved by a few hefty volcanic eruptions? After the 1991 eruption of Mount Pinatubo, scientists studied the cooling effect of the aerosols it created. The effect was quite large, but within about two years the droplets had largely fallen out of the air and things had returned to normal. By comparison, a CO_2 surge in the atmosphere won't dissipate for centuries or longer. So, yes, the natural operations of volcanoes could help us, but only if they were to erupt so frequently and in such numbers that life would become intolerable for other reasons.

By and large, bacteria and protists prefer a warm environment. So do insects. Thanks to the warming that's occurred already, infectious diseases that used to be found only in tropical or subtropical areas are now expanding toward the poles; already regions like New England and Michigan—and even Toronto!—are suffering occasional outbreaks of diseases like malaria and dengue fever. In the Australian state of Victoria, "[t]here has been a significant rise in the number of reported cases of Ross River fever and Barmah Forest virus as unseasonably wet and warm conditions in the first two weeks of January provided perfect breeding conditions for mosquitoes."[14] We can expect such trends to continue, with other "tropical" diseases—yellow fever,

West Nile fever, etc.—following along closely behind. All over the world, malaria is returning to regions where it had been thought extinct.

A further disease we can expect to expand through North America over the next few years is Lyme disease, spread by ticks. Warmer winters in areas like New England have failed to kill off the requisite number to keep the tick population at acceptable levels.

Another insect enjoying the warming of the planet is the bark beetle. One reason California wildfires have become so much more intense is that they've had so much more tinder to work with. The population of bark beetles has spiraled upward, leading to a hugely increased number of trees being killed. In April 2008 the Canadian Broadcasting Corporation (CBC) reported estimates that, by 2013, 80 percent of British Columbia's pine forests will have been destroyed by this pest, a figure even more horrifying in that as recently as the early 1990s British Columbians were barely aware of the insect's existence. And, as the CBC goes on to report, there's a feedback effect at work here: "The beetle's damage to the forests has even had an impact on the release of greenhouse gases into the atmosphere, according to Natural Resources Canada. The death of trees normally involved in capturing carbon has instead released carbon into the atmosphere."[15]

Global warming is causing US deforestation, too. It's worth putting some figures to this. Between 1997 and 2001 the US lost about 1 million hectares of forest annually, on average, due to insects and disease; the equivalent annual figure for 2002 through 2007 was nearly 3.25 million hectares. As for wildfires, the annual loss in the US was about 1.5 million hectares on average through the 1990s; for the five years leading up to 2008 the figure was 3.25 million hectares.[16] These are not gradual rises.

According to the World Health Organization:

> Measurement of health effects from climate change can only be very approximate. Nevertheless, a WHO quantitative assessment, taking into account only a subset of the possible health impacts, concluded that the effects of the climate change that has occurred since the mid-1970s may have caused over 150,000 deaths in 2000.[17]

Launching *The Anatomy of a Silent Crisis* (2009), a report prepared by the Global Humanitarian Forum, former UN Secretary General Kofi Annan cited similarly grim figures: "The findings of this report indicate that every year climate change leaves over 300,000 people dead, 325 million people seriously affected, and economic losses of US$125 billion. . . . In the next twenty years those affected will likely more than double—making it the greatest emerging humanitarian challenge of our time."[18]

It's obviously not just by wildfires that the world's forests are threatened. The clearing of forest lands for agriculture is another source of anthropogenic increase in atmospheric CO_2. Precise figures aren't available, but the effect is not negligible. A widely predicted consequence of global warming is the death of the Amazon rainforest. This region's role in stabilizing the composition of the atmosphere is well known; less discussed is what would happen if it fails, which reports indicate it's close to doing. Rotting trees release CO_2. The death of the Amazon rainforest would, according to a 2006 calculation, add 10 percent to current manmade emissions for a period of some three-quarters of a century.[19]

While there are plenty of reasons to support forest conservation—from the promotion of biodiversity and clean air on down—it's not certain we could rely on maintaining and increasing the earth's plant life as a way of soaking up excess atmospheric CO_2. There's a limit to the amount of CO_2 we can expect the earth's vegetation to absorb, and we're already fairly close to it. One seven-year experiment, done by a team headed by Duke University's William Schlesinger, involved treating pine trees to a diet enriched in carbon dioxide; the conclusion was that, in essence, while in theory the trees could absorb more, in practice they couldn't swallow it fast enough. So, while halting deforestation is helpful, it doesn't offer any permanent solution.

This somewhat stymies one of the more ludicrous claims of the denialists—that increased atmospheric CO_2 levels are to be welcomed because CO_2 is "plant food." This idea seems first to have been actively promoted in the 1991 video *The Greening of Planet Earth*, which was funded by the Western Fuels Association, an advocacy group for coal suppliers, through its Greening Earth Society front organization. The video painted a glowing picture of a future earth in which, thanks to the use of fossil fuels, there would be burgeoning crops and abundant vegetation. The argument is roughly the same as the notion that, if two beers make you happy, twenty beers will make you happier, not sick.

In 2005 a discussion forum organized by the UK's Royal Society, "Food Crops in a Changing Climate," brought together, from all over the world, scientists in the fields of meteorology, climate science, and agriculture. It concluded very emphatically that the promises of improved agricultural yields from higher CO_2 levels were, for various reasons, fallacious. Perhaps most important, a few days of scorching temperatures can so devastate crops as to wipe out any moderate gains there might have been during the year. And we all know how well crops flourish in droughts.[20]

In 2010, a further—and one would hope fatal—blow was dealt to the "plant food" notion by research done at the University of California, Davis, by a team led by Arnold Bloom. It was already known that heightened CO_2 levels, through inhibiting photorespiration, promoted photosynthesis in the short term; yet the improved yields we might have expected simply don't happen. The plants' nutritional content falls. It seems that, as ambient CO_2 levels rise, the ability of plants to process nitrates from the soil declines.[21]

In November 2009 a zipped folder containing just over 60 Mb of data was placed in several online locations by a hacker whose identity may never be known; whoever it was, it seems evident this was no enthusiastic amateur but someone acting with an educated purpose: to do as much harm as possible to the public reputation of climate science. Sir David King, the UK's Chief Scientific Advisor for seven years until 2007, claimed the hack "bore all the hallmarks of a coordinated intelligence operation—especially given their release just before the Copenhagen climate conference in December."[22]

The folder contained about one thousand e-mails and about three thousand other files hacked from the UK's Climatic Research Unit's backup server. Soon the various AGW-denialist factions—notably bloggers Steve Mosher, Steve McIntyre, Ross McKitrick, Patrick Condon, and Anthony Watts—were hard at work to spin the material into evidence of a grand conspiracy.

It has to be admitted up front that Phil Jones and his colleagues at the University of East Anglia's Climatic Research Unit (CRU) and around the world were guilty of occasional waspishness and impatience at the purblindness of the denialist movement and its facilitators in the media. Professionals in any sphere, exchanging views in what they assume to be private, say dreadful, bitchy things. Whodathunkit?

Less trivially, Jones had been digging his heels in against a Freedom of Information Act (FoIA) request to reveal these e-mails. It seems he was advised that to release the e-mails might contravene the UK's Official Secrets Act, or at the very least that he should delay until the matter was cleared by those further up the bureaucratic food chain. Another

FoIA-related factor was that, under the rules, it took about eighteen hours' work to process each FoIA request, and the CRU had received about sixty such requests from around the world in July 2009 alone. With a staff of just thirteen, all busy on research, it was little wonder a backlog had built up—especially since some of the requests were for data already publicly available![23]

Naturally, the above aspects were spun out of all recognition by the denialist movement, including prominent antiscience politicians in the US, like Joe Barton (see pages 292–93) and James Inhofe (see pages 290–92). It takes but one moment's reflection to realize these two supposed crimes of the CRU scientists say nothing whatsoever about climate science. Nature doesn't care if sometimes those investigating her are fractious or ill advised.

But the spinners seized far more dishonestly on other aspects. In a much earlier letter to *Nature*[24] Jones and colleagues had drawn attention to the fact that the relationship between tree-ring growth and ambient temperatures, a useful tool for deducing past air temperatures, had more recently begun to break down, for reasons unknown.[25] Essentially, the air temperatures calculated from more recent tree-ring studies had begun to disagree with those measured by thermometers. At the time nothing seemed controversial in this study. In one of the hacked CRU e-mails, Phil Jones mentioned a mathematical means he used to marry the older, tree-ring based calculations to the more recent direct measurements in such a way as to ensure they correlated properly—in other words, to ensure there was nothing potentially misleading in the presentation of the unified data. He said the math was needed to "hide the [illusory] decline," and described the technique as a trick—i.e., as a nifty bit of math. His uses of terms like "trick" and "hide the decline"—unlikely to be misconstrued in professional communication—were cast as clear evidence of a global conspiracy.

Many sections of the mainstream media reported the spin as if it were gospel, especially those rightwing newspapers that had anyway been looking for some kind of support for their AGW denialism. Here's James Delingpole in the UK's *Daily Telegraph*:

> If you own any shares in alternative energy companies I should start dumping them NOW. The conspiracy behind the Anthropogenic Global Warming myth (aka AGW; aka ManBearPig) has been suddenly, brutally and quite deliciously exposed. . . (Hat tip: *Watts Up With That*).[26]

The blog *Watts Up With That?* (see page 298) is a notorious hotbed of irrational AGW denialism. No serious journalist would rely upon it as a source of information.

The consequence of irresponsible journalists whipping up public indignation against the CRU scientists was that the latter became a target of a sustained campaign of vilification. The scientists received relentless hate mail and numerous death threats.[27] Jones later admitted he'd been more than once driven close to suicide. It's all very well for journalists and politicians to claim these horrors are the responsibility of the crazies directly involved, as if this somehow exonerates their own actions—as in Bill O'Reilly claiming the murder of abortion doctor George Tiller in 2009 had nothing to do with the campaign O'Reilly had been waging for months on Fox News against "Tiller the Baby Killer." Yet the journalists and politicians who whip up this vitriolic antiscience atmosphere know full well there are crazies out there who may regard the pronouncements of their heroes as license to take action.

One aspect that seems particularly to have incensed some of the public is the occasional example of a scientist expressing contempt toward said public—to the effect that we're by and large stupid, ignorant, hysterical sheep. Further disdain was reserved by the scientists for the mainstream media. How appalling that such attitudes should be held, let alone expressed! Although, in light of what happened . . .

Since the opening salvo of the Climategate farrago there have been no fewer than four official vindications of the CRU scientists and their scientific work: The Parliamentary Science and Technology Committee exonerated Phil Jones (he has "no case to answer" and his "reputation . . . remains intact"), the International Panel set up by the University of East Anglia under Professor Ron Oxburgh exonerated the CRU researchers ("their work has been carried out with integrity"), the Muir Russell Review did likewise, and in September 2010 the Conservative-led UK government stated flatly that "[t]he Government agrees with, and welcomes, the overall assessment of the Science and Technology Committee that the information contained in the illegally-disclosed emails does not provide any evidence to discredit the scientific evidence of anthropogenic climate change."[28]

It wasn't just the CRU scientists who suffered. The violent fringe of the AGW denial movement declared open season on the CRU's counterparts on the other side of the Atlantic, and again the death threats proliferated. On July 5, 2010, Leo Hickman reported in the *Guardian*: "Another [US] climate scientist, who wished to remain anonymous, said he had had a dead animal dumped on his doorstep and now travels with bodyguards."[29]

The principal victim of the US witch-hunt was Michael Mann, the climatologist who'd led the team that derived the famous Hockey Stick graph. A comment Mann made was widely bruited by the denialists as evidence of dishonesty: "As we all know, this isn't about truth at all, it's about plausibly deniable accusations." What was evident in context was that the phrase "plausibly deniable accusations" described the attacks being carried out by the deniers. It's hard to find a more blatant example of the fine art of quote mining.

Much of Mann's work had been done at the University of Virginia; but his current employers, the Pennsylvania State University, received the brunt of the rumormongering:

> Beginning on and about November 22, 2009, The Pennsylvania State University began to receive numerous communications (emails, phone calls and letters) accusing Dr. Michael E. Mann of having engaged in acts that included manipulating data, destroying records and colluding to hamper the progress of scientific discourse around the issue of anthropogenic global warming from approximately 1998.[30]

The university duly investigated, and Mann was cleared on all counts. His travails were not, however, over (see pages 75–76).

When MSNBC.com's Newsvine section ran an online poll on July 7, 2010, just after the third UK exoneration of the CRU scientists, the extent of the damage done by the denialist dishonesty was manifest:

> *Yes*, the panel was fair in reproaching their behavior while upholding key data. 36.9%/8,227 votes
> *No*, I still believe those scientists fabricated data to support their beliefs on man-made warming. 63.1%/14,085 votes[31]

The comments by the "No" voters are predictably idiotic, a tangled knot of scientifically illiterate conspiracy theories, but perhaps still not so idiotic as the rebuttal of the Oxburgh report by the Australian denialist blogger Andrew Bolt (see page 303), which seems to be that, er, Lord Oxburgh rides a bicycle.[32]

Professor Scott Mandia did some online research of the major news media to find out how many sources made headline news out of the Climategate accusations and how many made headlines out of the exonerations.[33] The results were depressing:

> Dr. Phil Jones accused of wrongdoing resulted in 263 headlines in the first 42 days after the story broke. . . . When Dr. Jones was exonerated there were only 24 headlines in the first 19 days after exoneration.[34]

Mandia then ran comparisons between coverage of the Climategate accusations and that of the disclosure that Koch Industries (see pages 274ff.) had been funding a massive campaign of misinformation about climate change; the results were equally gloomy.

As the *New York Times* caustically observed, albeit many months too late, Climategate was "a manufactured controversy."[35]

A little noticed disclosure in the explosion of Wikileaks revelations in November/December 2010 was that climate denialists had made an attempt to hack the computers of the US State Department's Bureau of Oceans, Environment, and Science (OES). Several officers of the OES were sent PDF attachments that concealed malware designed to spy on their computers. US law defines such an attempt as an act of terrorism. Nothing was done.

> As a matter of fact, carbon dioxide is portrayed as harmful!
>
> But there isn't even one study that can be produced that shows carbon dioxide is a harmful gas. There isn't one such study because carbon dioxide is not a *harmful* gas, it is a *harmless* gas. Carbon dioxide is natural. It is not *harmful*. It is part of Earth's life cycle.
>
> And yet we're being told that we have to reduce this natural substance and *reduce* the American standard of living to create an arbitrary reduction in something that is naturally occurring in the earth.
>
> —Michele Bachmann (R-MN), speaking before the US House of Representatives, April 22, 2009[36]

The Hockey Stick graph first appeared in a 1998 paper by Michael E. Mann, Raymond S. Bradley and Malcolm K. Hughes;[37] it plots derived northern hemisphere temperatures going back to 1440, and got the "Hockey Stick" appellation because the otherwise essentially horizontal line's right-hand end, representing the most recent temperatures, crooks dramatically upward. In a

1999 paper[38] the same authors projected the history further back, to 1000 CE; and it's this form of the graph that was taken up by the IPCC for its 2001 Assessment Report.[39] There has been some controversy in scientific circles over the methodology used by Mann and his colleagues in deriving the data, especially toward the earlier period covered by the work; various scientific investigations and reanalyses have shown, however, that overall the graph is accurate and in particular that there's nothing wrong with the interpretation of data concerning the graph's right-hand end: Temperatures *are* rising rapidly.

Outside scientific circles, the known controversy over the earlier sections of the temperature history has allowed professional denialists to muddy the issue about the latter part. In particular, one reanalysis of the data, by former mining prospector Steve McIntyre (see page 303) with "editorial assistance" from economist Ross McKitrick, has claimed that Mann and his colleagues seriously distorted the information and really we have nothing to worry about.

Another person unqualified in the field of climate science is statistician Edward Wegman, who in 2006 was called in by two Republican congressmen, Joe Barton (see pages 292–93) and Ed Whitfield, to head a team tasked to produce a Committee on Energy and Commerce report into the Hockey Stick; also part of the committee's remit was to look into some aspects of the Climategate farrago. A few weeks before the Wegman report was delivered to Congress, a National Research Council (NRC) panel that had studied the issue produced its own report, *Surface Temperature Reconstructions for the Last 2,000 Years*,[40] which concluded there was nothing much wrong with Mann's conclusions (and certainly no dishonesty involved). The Wegman report, by contrast, hailed the rival McIntyre/McKitrick analysis as "vibrant and compelling"! It added that Mann's analysis didn't support the contentions that the 1990s were the hottest decade of the millennium and 1998 its hottest year.

This was all very damning, yet hardly had the ink dried on the banner headlines that Barton and Whitfield had ensured the report would receive than climate scientists were raising objections. The NRC panel's chairman, Gerald North, was among those who criticized the Wegman conclusions. And, as investigators dug, it emerged there'd been some rigging of Wegman's results. For example, it seems the main conduit of scientific information to the panel members was one of Barton's staffers. While the Wegman panel was undertaking its review it had been in contact with

McIntyre. And the panel's remit was largely defined by a presentation McIntyre and McKitrick had made in Washington a month beforehand, sponsored by the Cooler Heads Coalition and the George C. Marshall Institute.[41]

Further, Ray Bradley, a climatology professor at the University of Massachusetts, Amherst, and one of Mann's coauthors on the relevant papers, complained that large parts of Wegman's report had been plagiarized from a climatology textbook he, Bradley, had written. *USA Today* called in three plagiarism experts to evaluate the claim, and they found plenty of evidence to support it, including many examples of word-for-word "lifting."[42] At the time of writing, Wegman is under glacially slow investigation by his employer, George Mason University, over the plagiarism.[43] The university's Law and Economics Center and Mercatus Center receive funding from ExxonMobil.

It's nigh impossible to escape the conclusion that the Wegman report—the whole exercise—was nothing more than "a facade for the climate anti-science PR campaign," as computer scientist John Mashey described it in the subtitle to his monumental, and damning, investigation *Strange Scholarship in the Wegman Report*,[44] the first to reveal, *en passant*, the extent of the plagiarism.

In short, two politicians had, at enormous taxpayer expense, rigged a political stunt. On whose behalf, aside from their own? It seems to have been blogger Tim Lambert who created the moniker "Joe Barton (R-Exxon)."

Since incapable of offering a genuine scientific counter to the contents of the various IPCC reports, the denialist community decided the next best thing was to mount a smear campaign against the IPCC's chair, Dr. Rajendra Pachauri. Interestingly, Pachauri's appointment to the chair of the IPCC in April 2002 was the brainchild of the Bush administration, whose advisers presumably looked at his career working with electricity utilities and his position as head of the Tata Energy Research Institute[45] (TERI), and assumed he'd be their man. They assumed wrong.

On December 20, 2009, the UK *Sunday Telegraph* published a long article, "Questions over Business Deals of UN Climate Change Guru Dr Rajendra Pachauri" by Christopher Booker and Richard North, which claimed the chair of the IPCC was making "millions" by exploiting his position:

> What has also almost entirely escaped attention, however, is how Dr Pachauri has established an astonishing worldwide portfolio of business interests with bodies which have been investing billions of dollars in organisations dependent on the IPCC's policy recommendations. . . . One subject the talkative Dr Pachauri remains silent on, however, is how much money he is paid for all these important posts, which must run into millions of dollars. Not one of the bodies for which he works publishes his salary or fees, and this notably includes the UN, which refuses to reveal how much we all pay him as one of its most senior officials.[46]

There was lots more, and all of it bullshit. The most obvious lie is that "this notably includes the UN, which refuses to reveal how much we all pay him as one of its most senior officials." As is widely known, the work Pachauri does for the UN he does for free. For the rest, an audit of Pachauri's personal finances showed that between April 3, 2008, and December 31, 2009, he received from TERI only his salary, a touch over $70,000 per annum. Any money earned through TERI-related work for other organizations went not to him but to TERI. During the period scrutinized he achieved a Lifetime Achievement Award from the Environment Partnership Summit; the monetary part of this, some $4,500, he passed to TERI.[47]

The audit did reveal he had some other earnings, unrelated to his TERI work: stipends from work done for the Indian government, book royalties, lecture fees, etc. For the period in question—twenty months—these payments totaled about $3,500. This does not "run into millions of dollars," as Booker and North would have you believe.

And as the *Sunday Telegraph* would have you believe. In the same way that the *Sunday Times* refused to retract a piece by Jonathan Leake for months after it had been proven false (see page 301), the *Telegraph* refused to retract until Pachauri called in his lawyers.

While the smear has been removed from the *Telegraph* site, it's easy enough to find copies of it all over the web.[48] Among those sites still eager to promote the lie are the Heartland Institute, the Science and Public Policy Institute, *Watts Up with That?*, Prison Planet, and Roger Pielke at Climate Change Fraud. Further, you can still find on the *Telegraph* site such equally malignant specimens as Booker's "The 'Anomalies' of Dr Rajendra Pachauri's Charity Accounts."[49]

Not long after the Booker/North smear, the *Telegraph* was denigrating Pachauri again, this time through its daily organ. A January 30, 2010, article by Robert Mendick and Amrit Dhillon, "Revealed: The Racy Novel Written by the World's Most Powerful Climate Scientist," reported that Pachauri's debut novel *Return to Almora* (2010) contains—gasp!—racy passages. Amazingly, the piece was published as part of the *Telegraph*'s "Climate Change" coverage.[50]

On October 14, 2010, the BBC reported that "Dr. Rajendra Pachauri confirmed his intention to stay in the chairman's post until the next assessment is published in 2014."[51] He's a brave man.

Various writers have pointed to the seven stages of AGW denialism:

- There's no such thing as global warming.
- Global warming's happening, but it's not our fault.
- Okay, so a *little* of it's our fault, but not so much as to make any difference.
- Our contribution *is* big enough to make a difference, but global warming's actually a good thing.
- Global warming's not such a good thing after all, but there's nothing we can do about it.
- We could have done something about it, but it's too late now.
- Barack Obama was born in Kenya.

The core arguments used by AGW deniers boil down to a relatively small number of recurring memes.

A favorite is that warming is a result of variations in the sun's output. This has been shown false. The sun *is* heating up, but it's an effect noticeable over millions of years, not decades. If it were true the sun were going through a temporary hot phase, all parts of the atmosphere would be heating up. In fact, the troposphere is warming and the stratosphere cooling, exactly as predicted for a greenhouse effect.

Another old saw is that computer models are unreliable, so why believe anything the modelers say? In his *Storms of My Grandchildren* (2009) James Hansen tends—with emphatic caveats—to agree with the criticism, but points out (p. 44): "Fortunately Earth's history allows precise evaluation of climate sensitivity without using climate models."

Trotted out almost as often as Christopher Monckton are the thousands of coal miners who will "lose their jobs" if the world starts moving toward a saner energy regime. Is there a political solution? Retraining the miners to build windmills or install solar panels would seem a good start.

Then there's the confusion between weather and climate. "In a single day in February, New York got its fourth-deepest snowfall since 1869. Baltimore got more snow in February than in any other month in recorded history. I wish there were global warming," remarked Ann Coulter in her think piece "Global Warming: The French Connection,"[52] seemingly ignorant that a predicted consequence of global warming is that we can expect heavier rain- and snowstorms.

An argument used in claims that the world was warmer around the time of Christ is that Hannibal was able to bring his elephants and army across the Alps in winter. In fact, Hannibal adopted this stratagem precisely because the Romans thought it couldn't be done—and lost half his army in the process.

The "skeptics" also point to the Medieval Warm Period, which lasted from about 1000 to about 1300. After it came the Little Ice Age, whose peak lasted until about 1700 and which had not entirely tailed off by the early twentieth century; it's not unreasonable to posit that what brought it to an end was the Industrial Revolution and the ever-increasing levels of CO_2 pumped into the atmosphere. Some deniers suggest the warming we've witnessed in the past few decades is just a natural recovery from Little Ice Age conditions to the norm. But the term "Little Ice Age" is misleading. It's been estimated that global average cooling during the Little Ice Age was about 0.5 Celsius degrees, probably less. Any warming back up was over long ago. As for the vaunted Medieval Warm Period? Average temperatures seem to have been *lower* then than what we've known since about the 1960s.

Some people detest Al Gore so much they deny AGW simply because he warns about it. He's often accused of making millions from his campaigning on the climate issue; in fact, he has donated all royalties from both book and DVD versions of *An Inconvenient Truth* (2006), plus his Nobel Peace Prize award—and even his annual salary from venture capital firm Kleiner, Perkins, Caulfield, and Byers—to environmental causes. Well, in that case, goes the chorus, *An Inconvenient Truth* is riddled with errors. The few trivial errors in the movie don't dent his argument. So denialists have had to invent some, and parrot them all over the Internet. Here's a comment by "Luddhunter" pulled from Tim Lambert's *Deltoid* blog:

> I guess I am just too dumb to understand why I am confused when Oreskes' vaunted IPCC predicts 23 inches of ocean rise over 50+ years, and Pope Gore's prediction in his propaganda flick is 20 feet, over perhaps less than 50 years.[53]

As Lambert points out, the IPCC estimate of 23 inches (58cm) in the short term omits (because no one knows the timing) the melting of the polar icecaps, which would add about 20 feet (6m) to sea levels. Gore gives this as the figure consequent upon melting of the icecaps, and says nothing about it possibly happening within fifty years—as "Luddhunter" would know if he'd watched the movie.

The notion, exemplified by "Luddhunter," that finding a single tangential error or lacuna demolishes an entire edifice is common throughout science denialism. As Francisco J. Ayala points out in a different context, in *Darwin's Gift to Science and Religion* (2007, p. 142), "[W]e do not know all the details about the configuration of the universe and the origin of the galaxies, but this is not a reason to doubt that the galaxies exist or to throw out all we have learned about their characteristics."

Another recurring meme is wide-eyed astonishment that a quick back-of-the-envelope calculation serves to show how all the "alarmist" stuff produced by climate scientists is purest hokum. It seems not to register that climate science is complex enough that the relevant calculations require more than the back of an envelope. Even a *very big* envelope.

And finally there are all the scientists who dissent from the consensus, like those on Senator James Inhofe's celebrated list. Even if these scientists have remotely relevant expertise, do genuinely dissent, and are alive to say one way or the other (see page 292), they're not in fact *engaging in argument* with the consensus. Anyone can sign a petition saying the world is flat; arguing the case is a bit trickier. The public, however, sees the lists and assumes there's substantive argument over climate change within the scientific community. The organizers of such lists, the politicians who vaunt them, and indeed the participants themselves, are mounting a grossly cynical deceit. It's an equal deception that, by and large, the mainstream media don't call them on it.

Useful summaries of the standard "skeptic" arguments, and of the reasons why they're nonsense, have been posted online by, among others, John Cook[54] and Coby Beck.[55]

There is conflicting data on that. A lot of data I've seen shows that perhaps we don't even have global warming. In fact I've read some articles that it's actually decreasing, that we have climates getting colder. . . . We do need to promote a clean world. I think everybody wants clean air and clean water. But I think, for the most part, here in the United States, we are leading the way in that area.[56]

—Vicky Hartzler, successful GOP congressional candidate in Missouri, 2010

[T]he world is not primarily wrestling at this moment with a climate crisis. . . . The world is in the grip of a crisis of political inaction, a crisis of political leadership—or a complete lack thereof.

—James Hoggan, with Richard Littlemore, *Climate Cover-Up* (2009, pp. 225–26)

A factor that bedevils conversation about what's being done to counter climate change is greenwashing: Politicians, organizations, and corporations know there's a crisis yet are reluctant to face the necessary political or economic sacrifices; they therefore make all the right noises, announce a few measures . . . and do nothing. The Bush administration often named its environment-harming measures in Orwellian fashion: The Clear Skies Initiative of 2003 loosened controls on industrial polluters. The Obama administration has made a certain amount of progress but seems to lack the guts to tackle the problem head on and so has resorted to subtler forms of greenwashing.

In the UK, successive Labour and (to date, briefly) Coalition governments have offered a worked example of greenwashing in their claims that, between 1992 and 2004, the nation's CO_2 emissions dropped by a small but creditable 5 percent. This figure is actually correct, so what's the complaint? The decrease in the UK's CO_2 emissions is a product of the country's manufacturing decline: UK companies are shifting their manufacturing abroad—i.e., exporting their CO_2 emissions. According to Professor Corinne Le Quéré of the University of East Anglia,[57] if the UK's CO_2 emissions are calculated in terms of the goods actually consumed in the country we see not a drop but a 12 percent *increase* over the years in question.

We're accustomed to the parade of international conferences at which the major nations of the world bicker over carbon credits and cap-and-trade commitments and essentially get nowhere. And so everyone poses smiling for the photographers and plans are laid for the *next* conference, at which most certainly "progress will be made."

The first Earth Summit, organized by the UN, was held in Rio de Janeiro in June 1992, with the aim of thrashing out the details of the UN Framework Convention on Climate Change. The aim of the Framework Convention was to work toward "stabilization of greenhouse gas concentrations in the atmosphere at a level that would prevent dangerous anthropogenic interference with the climate system." One of its champions was President George H. W. Bush; when he presented the agreement to the US Senate some weeks later, it was passed with unanimous approval. Since then, we've regressed.

Even at the outset, it became evident that the commitment of the developing nations—scheduled to take the initial brunt of curtailing CO_2 emissions—was more talk than deeds. The 1997 meeting at Kyoto was intended to tighten up the agreement such that the "participating countries"—165 had ratified the Framework Convention—actually *participated.* A point made by the developing countries was that, if one thinks of the maximum safe atmospheric levels of CO_2 and the other greenhouse gases as being a resource, the rich countries had *already used up their share* of that resource in building their current industrialized economies. It was only fair the remainder of the resource be disproportionately allocated to the poorer countries so they, too, could build their economies. This argument was (at least seemingly) taken to heart by members of the EU, whatever their left–right positions on the political spectrum, but seems to have gone almost ignored in the US.

The Clinton administration signed the Kyoto Protocol in 1998, but, realizing it would never get the measure through the US Senate as then constituted, it delayed ratification. The problem was that some of the richest people on earth had woken up to the fact that countering climate catastrophe was going to involve sacrifice not just by poor people in distant parts of the world. In July 1997 Senator Robert Byrd (D-WV) and Senator Chuck Hagel (R-NE) had introduced an extraordinarily disingenuous and destructive resolution to the Senate, the so-called Byrd–Hagel Resolution. This said, in effect, that the US should refuse to honor any agreement that required it to reduce greenhouse-gas emissions unless similar obligations were imposed on the developing countries. In essence, the Senate, which passed this resolution 95–0, was saying to the people of the developing countries that they must stay poor in order that the richer elements of US society could continue to enjoy their unsustainable lifestyles.

In March 2001, the Bush administration reneged on the US commitment to the Kyoto Protocol (and on its own campaign promises), blaming the protocol's provisions to exempt developing nations from the first round of emissions cuts. It was at about this time, too, that George W. Bush began to produce the specious argument that the science on climate change was "not settled." In November 2002, strategist Frank Luntz issued a now-infamous advisory memo to the Republican Party called "The Environment: A Cleaner, Safer, Healthier America,"[58] in which he stressed that the "not settled" lie was the one to use.

Instead of the Kyoto Protocol, in 2003 the administration presented its very own pretend plan for action on climate change. "Drafted with the help and approval, among others, of the Edison Electric Institute, the American Petroleum Institute, and the Southern Company, as well as the National Mining Association, the Portland Cement Association, the American Iron and Steel Institute, the American Chemistry Council, the Aluminum Association, the Association of American Railroads, and the American Forest and Paper Association,"[59] this essentially presented the kind of voluntary self-regulatory scheme that served us so well on Wall Street in the years leading up to 2007. Also, calculations of emissions reduction were based on a concept called carbon intensity: Rather than any reduction per capita, reductions would be calculated against economic output—the greater the economic output, the greater the permitted level of emissions. Since it's in the nature of economies to grow rather than shrink, effectively what the plan promised was an overall *increase* of CO_2 emissions by, according to some estimates, about 14 percent.

A further parody of Kyoto came from the Bush administration with the establishment of the Asia Pacific Partnership on Clean Development and Climate, which involves some countries that are serious about the issue and some, like Australia (under the AGW-denying John Howard government of the day) and the US, which are not. The purpose of the partnership was to stop India and China from signing any binding emissions treaty so the US could then trot out the excuse that, without those countries having signed a binding treaty, it would be senseless for the US to do so.[60]

Since the late 1990s, each succeeding international convention to discuss global warming seems to start with lower aspirations than the last and finish having achieved not even those. A few months before the Copenhagen Climate Conference in December 2009 Kofi Annan said: "The alternative [to Copenhagen's success] is mass starvation, mass migration, and mass sickness."[61] The attendees opted for the starvation, migration, and sickness.

In December 2010, at the climate summit in Cancún, there occurred one of those moments of bitter irony that seem to toll like knells. Japan, evidently weary of other countries pretending to be taking action on climate change while all the time doing nothing, opened the negotiations by flat-out rejecting any extension of the Kyoto Protocol after its expiration in 2012. Days later, Russia followed suit. By the end of the summit, a weak agreement had been reached between the participating countries, and yet again there were "progress has been made" apologetics. So, in honesty, we're back to business as usual—the sure route to disaster.

It's all a bit too late for some low-lying communities around the world. In November 2000 the first evacuees from the Duke of York Islands, Papua New Guinea, were taken to nearby New Britain; eventually the islands' entire population of some forty thousand will have to be found homes for elsewhere, as their archipelago is being submerged by the rising sea. The tiny nation of Tuvalu, population about 12,500, is similarly threatened. Again, our glorious leaders may not care too much about small, remote island nations. And it seems they're prepared to simply ignore the calculation that the first one meter rise in sea level—which is beginning to look like an absolute minimum for the foreseeable future—will take out 6 percent of the Netherlands and perhaps 20 percent of Bangladesh.[62] The Nile delta will become uninhabitable. And as for the US?

> A one-meter sea level rise would wreak particular havoc on the Gulf Coast and eastern seaboard of the United States.
>
> "No one will be free from this," said Overpeck,[63] whose maps show that every U.S. East Coast city from Boston to Miami would be swamped. A one-meter sea rise in New Orleans, Overpeck said, would mean "no more Mardi Gras."[64]

One politician who—sporadically at least—has spoken out strongly about the need to act on climate change is Osama bin Laden. In a 2002 "Letter to the American People" he commented: "You have destroyed nature with your industrial waste and gases more than any other nation in history. Despite this, you refuse to sign the Kyoto agreement so that you can secure the profit of your greedy companies and industries." And in a 2010 audio recording he stated an evident truth: "Speaking about climate change is not a matter of intellectual luxury—the phenomenon is an actual fact. All of the industrialized countries, especially the big ones, bear responsibility for the global warming crisis."[65]

Evident the truth might be, but immediately rightwing US commentators, logic be damned, seized on the support of bin Laden as yet further "proof" that AGW was just a liberal conspiracy. The conservative *RedState* site blustered, "Al Qaeda leader Osama bin Laden expresses views that are nearly indistinguishable from those of radical environmentalists, European anarchists, or many Congressional Democrats," with commenter "ericstenner" asking rhetorically, "So what is the difference between bin Laden and Al Gore?"[66] (Science blogger Ed Brayton responded bluntly: "You need someone to tell you the difference between Al Gore and the world's most famous terrorist? Then you're an idiot."[67]) The *Drudge Report* and Fox News joined the inane howling.[68] Oh yes: It's our freedoms they hate us for, not our CO_2 emissions.

> *This week, the House of Representatives is moving ahead on historic legislation that will transform the way we produce and use energy in America. We all know why this is so important. The nation that leads in the creation of a clean energy economy will be the nation that leads the 21st century's global economy.*
>
> Barack Obama, press conference, June 23, 2009

About 80 percent of the CO_2 added to the atmosphere each year is the product of burning fossil fuels. Global CO_2 emission from fossil fuels approximately quadrupled between 1950 and the present, from about two billion metric tons of carbon to about eight billion. The biggest rate of annual increase was between 1950 and 1973; there was a slowing between 1973 and 2003, but since then emission rates have been accelerating again, largely because of the increased use of coal (especially by China) as other fossil fuels have become more expensive. Of all the fossil fuels, coal is the dirtiest. Per unit of energy extracted, burning coal adds about twice as much CO_2 to the atmosphere as burning natural gas, and about 30 percent more than burning oil. Many authorities expect our coal burning to double over the next twenty years, a circumstance that's obviously untenable unless some highly effective means of controlling CO_2 emissions can be found.

There are plenty of possibilities to be explored for finding a cleaner way of burning coal, but until a concerted effort is made at a governmental and/or international level, progress may be fatally slow. The one that's talked about most is carbon capture and storage (CCS), but an MIT report called "The Future of Coal"[69] (2007) brings gloomy news on this front. Essentially, there's no chance of the first CCS plants being on stream until about 2030—and that's assuming there's any concerted action toward making that happen. Other estimates of when CCS might be practicable are just as pessimistic, if not more so: "The energy company Shell, though enthusiastic about the technology, doesn't foresee CCS being in widespread use until 2050," reported Fred Pearce in *New Scientist* in March 2008.[70]

Where, then, did the claim come from during the November 2008 election that the US could meet its energy needs while restricting its CO_2 emissions using this marvelous substance called "clean coal"? It came out of the imagination of an industry front group called American Coalition for Clean Coal Electricity (ACCCE), headed by one Stephen L. Miller—who had earlier headed the front group Americans for Balanced Energy Choices and before that the front group Center for Energy and Economic Development. ACCCE paid for massive TV ad buys, and seems even—one hopes only temporarily—to have suckered Barack Obama and Joe Biden into believing the fibs. Public opposition to coal-burning power plants halved.

According to James Hansen, James Lovelock, and many others—whether climate scientists, environmentalists, economists, or otherwise—the only reasonable short-term solution is to expand our use of nuclear fission as a source of energy.

Far too much gets written about the dangers of nuclear power, feeding the understandable fears of a public that still remembers the consequences of the Chernobyl meltdown of 1986 (the UN estimate of the death toll from this accident is that it may eventually be as high as 4,000, but other estimates are higher) and the Three Mile Island meltdown of 1979 (estimated consequential death toll: one, maybe).

When the Fukushima Daiichi nuclear plant was struck by a Richter 9 earthquake and consequent tsunami on March 11, 2011, the world was soon abuzz over the potential dangers of nuclear power; the real wonder was, however, that despite being subjected to a "one-thousand-year event" and stresses far beyond those for which it had been designed, the plant survived sufficiently for a rescue and damage-limitation venture to be mounted. (It should be mentioned, too, that the siting of the plant in a tectonically active area close to the shore was profoundly foolish.) At the time of my writing (August 2011), the Japanese government is still maintaining an evacuated safety zone around the plant. In no wise should we attempt to portray the Fukushima crisis as anything other than a major one; at the same time, it's worth observing that, to date, the number of known casualties is zero. We've no real way of estimating what a final toll might be, but early indications seem to be that it'll be low rather than high. The message to be taken from the March 2011 accident is really, then, that nuclear power plants are safer than we might think.

A death toll of about 4,000 since the nuclear industry began seems pretty small beer by comparison with the total deaths caused since 1979 by our coal use—from mining accidents to widespread respiratory and other diseases. And it's estimated that another industrial disaster just two years before Chernobyl, not involving nuclear material at all, will eventually cause five times as many deaths. This was the tragedy at Bhopal, India, which involved a chemical plant. We hear few calls saying we should stop using chemicals.

Nuclear power's risks also seem trivial when set alongside the annual casualties incurred by coal mining. More than half the world's mining deaths occur in China; in recent years the figure for that country alone has been running into the thousands, according to official figures that may well be underestimates. (Some reckonings put China's annual death toll from mining accidents as high as 10,000.) Also capable of taking a huge toll are sludge spills. The sludge or slurry is liquid coal waste left over after mined coal has been washed to remove impurities; these impurities typically include mercury, arsenic, beryllium, cadmium, thallium, nickel, and selenium. The usual method of dealing with the sludge is to dam it into reservoirs while someone tries to think of a way of dealing with it. When the dams burst the results can be horrific. In one famous incident in 1966 a sludge spill at Aberfan, Wales, killed 144 people, including 116 children who'd been in classes at the local school.[71]

And still we haven't considered those countless premature deaths due to respiratory diseases and the like . . .

The technology of fission power has moved ahead vastly since the time of Chernobyl: to pretend that decades-old safety standards are the same as those today is as if to use the Comet airplane disasters of the early 1950s in assessing the dangers of air travel. Even if the new-style reactors *weren't* hugely safer than those of yore, they would *still* offer a better option than continuing to burn coal. All in all, the idea of nuclear power as a dangerous option seems increasingly remote from the reality. If Green campaigners find the concept of nuclear energy ideologically impure, they should perhaps re-examine their ideology.

We hear a lot about the economic costs of tackling the shift to cleaner forms of energy. Three US government agencies have provided estimates of what these costs might mean for the average household. According to the Congressional Budget Office, the annual figure for 2020 per household might be about $160, or 44¢ per day. The Department of Energy's estimate came in much lower, at $83 a year, or a daily 23¢. The EPA gave a range of estimates: an annual $80–$111, representing a daily 22–30¢.[72] And what the three agencies didn't provide estimates for were the likely *savings* US households might make because of the switch to greener energy, most notably in the costs of healthcare. Add in improved national security and the creation of what the University of Massachusetts's Political Economy Research Institute estimated in a 2009 study[73] to be 1.7 million manufacturing, service, and other jobs associated with the new technology—the green-collar workforce —and it seems like a bargain.

What sort of capacity could solar power have? In theory, filling the Sahara with solar panels could serve the entirety of Europe with electricity, while also providing an economic boom to some of the poorest countries in the world. The Gobi could be put to similar use for China, and so on. In other words, world electricity needs could be supplied by using those areas that at the moment are by definition useless to us and where, too, there's a maximum of sunshine and interruptions are rare. The only trouble with the theory is the matter of getting the electricity from the desert to your TV set without a massive loss of energy along the way. New types of cables now under development may go a long way toward solving this problem. Of course, setting up these immense installations would require a measure of international resolve that's so far lacking.

The US is covered with houses that are connected to a regional grid of some kind and have roofs that could usefully be adorned with solar panels. It seems odd that the power companies have been so slow to recognize in those currently empty roofs the power plants of the future. They could do deals with homeowners to put solar panels there, to the benefit of power company, homeowner, and local workforce alike. In fact, a few similar arrangements have been sorted out with commercial operations—a number of Walmart stores are now covered in solar panels financed by Morgan Stanley, while the public schools in San Jose have roof panels financed by Bank of America. The SolarTech Energy Corporation is looking to find ways of extending such schemes to smaller organizations and individual householders through "creating an online marketplace to match consumers with financing and solar suppliers. Eventually the group envisions individual homeowners being able to switch to solar without any front-end cost."[74] But if banks like Morgan Stanley and Bank of America and specialist entrepreneurs like SolarTech reckon there's money to be made in them thar roofs, how come the power companies themselves are being so slow?

The world's largest seller of solar-power systems is, surprisingly, an oil company: BP, through its subsidiary, BP Solar. As long ago as 1997 BP recognized the company was eventually going to have to shift from fossil fuels to renewable energy sources. A similar decision was made in 2001 by Royal Dutch/Shell, which increased its investment in renewables by a factor of two (albeit from a fairly low base figure). To put it mildly, the progressive measures taken by these two giants of the oil industry aren't matched by ExxonMobil. Interviewed by the magazine *Chief Executive* in October 2002, ExxonMobil president Lee Raymond produced a classic of mind-numbing, xenophobia-fanning disingenuity:

> The mainstream of some so-called environmentalists or politically correct Europeans isn't the mainstream of all scientists or the White House. The world has been a lot warmer than it is now and it didn't have anything to do with carbon dioxide.[75]

Presumably Raymond was talking about the earth immediately after its formation, when the crust was molten and asteroid impacts were commonplace. Otherwise, of course, those warmer periods of the planet's history were intimately associated with atmospheric CO_2 levels.

What made his statement all the more ludicrous was that, just a few months earlier, in May 2002, at the annual shareholders' meeting, the activist group Campaign ExxonMobil had forced onto the agenda a motion calling on the company to cease its disinformation campaign about AGW and to start investigating its options for developing renewables. Shareholder motions challenging a company's leadership are quite frequently permitted onto the agenda of such meetings with the principal purpose of roundly defeating them so everyone can talk about shareholder democracy without actually having to make any changes. A 5 percent vote is regarded more seriously. This one got 21 percent support. The effect seems to have been to make the company tone down its public disinformation campaign; its covert campaign, operating through astroturf organizations and sponsoring bad science, seems to continue unabated.

In addition to solar and nuclear energy, there's wind power. It's estimated that wind farms could possibly supply some 15 percent of our global energy needs. Unfortunately, recent years have seen an upswing in the number of nimbies (not-in-my-backyarders). The EPA has offered a way around this. Because of lax regulations concerning pollution, there are currently some seven million hectares of land in the continental US that are so toxic as to be unusable for any purpose—you can't grow crops on them and you can't build houses on them. But, assuming the workers involved were equipped with hazmat suits, they could be ideal places for the installation of wind farms. And the land would be *cheap*!

Powering cars with other than fossil fuels is not a new idea. As long ago as 1912 Rudolf Diesel predicted that vegetable oils would one day be as widely used for cars as gasoline, and even earlier, at the 1900 Exposition Universelle (World's Fair) in Paris, he demonstrated a car engine that used peanut oil. Today, owners of diesel cars can buy equipment that will permit them to run on used vegetable oil—oil that, at the moment, many restaurants and diners are only too willing to give gratis to anyone who'll take the stuff away. It doesn't take more than a moment's thought to realize this recycling is economically and morally preferable to deriving biofuels like ethanol from specially grown food crops that could instead be used to feed hungry people.

There are other worries about "biofuel farming." The environmental impact of ethanol use depends considerably upon the source of the ethanol. In an idealized situation, biofuel use would be carbon neutral; but that's unattainable in the real world. The factors affecting the desirability of your biofuel include the crop you use and where you grow it. Thus, for example, the use by Brazilian vehicles of ethanol made from Brazilian sugarcane gives about a 90 percent reduction in greenhouse emissions, whereas use in US autos of ethanol derived from US corn renders only about a 13 percent reduction.[76] The ideal source for biofuel, all other things being equal, would be a plant that grows rapidly in conditions other plants would merely sneer at, and which has no use as a foodstuff (or anything else)—a weed, in other words. Step forward, kudzu.

In his 2004 State of the Union address, George W. Bush outlined the country's new energy policy. The fuel of the future was to be hydrogen, which is plentifully available—you can extract it from seawater—and has the advantage of being clean, its combustion product being, once more, water. It's in the "once more" that the fallacy lies. If we take the whole process involved—extraction from seawater, compression (so your car doesn't need a tank the size of a house), transportation, distribution, and finally combustion—it's evident that what we in fact have is a very energy-expensive method of converting water into water. Perhaps Bush was confusing the use of hydrogen as a fuel with the process of hydrogen fusion. Who knows?

Some of the energy technologies already on offer are pleasingly offbeat. Denmark, Germany, and the Netherlands are pioneering the concept of "eco-friendly garbage"—power stations that use garbage as their fuel, thereby cutting down on the environmental devastation of landfills and also, in part because the incinerators bear an array of appropriate filters, on emissions into the atmosphere of toxins and greenhouse gases. Electricity bills, too, benefit from the new technology. Curiously, among the most vocal opponents to the US developing a similar program are some of the environmental groups, which feel the focus for dealing with waste should be to minimize it through recycling. Since the European principle is to recycle what can be recycled and generate "green" energy from as much as possible of the rest, it's hard to follow the logic.[77]

One of the more inventive approaches to the CO_2 problem has been a project involving scientists and engineers from three UK universities to develop automated technology to absorb greenhouse gases using porous filters and then, powered by solar energy, to recycle them into plastics or even car fuel. "The facts are that CO_2 is rather diluted in the atmosphere and its chemical reactivity is very low. By combining clever material design with heterogeneous catalysis, electrocatalysis and biocatalysis, we aim at developing an effective carbon neutral technology."[78]

An unusual means of slowing down warming is being essayed by Russian physicist Sergey Zimov, who's reintroducing selected wild animals into a vast tract of the remote Siberian tundra in hopes their presence will re-create the conditions pertaining there at the end of the last glacial period. In particular, during winter they should trample flat the snow that would otherwise insulate the frozen ground and promote the release of methane.[79]

A little discussed means of reducing one's own carbon footprint is through a change in diet. In "Diet, Energy, and Global Warming"[80] Gidon Eshel and Pamela A. Martin, of the University of Chicago, conclude that, since a plant-based diet requires the expenditure of roughly one-quarter the energy required for one rich in red meat, shifting from a carnivorous to a vegetarian lifestyle is almost equivalent, in terms of carbon savings, to swapping one's gas-guzzling Chevrolet® Suburban® SUV for a Toyota® Prius®.

And then there are schemes like cap-and-trade, which seems to be about as far as most politicians can think. Cap-and-trade is better than nothing, but only just. George Monbiot observed in *Heat* (2007, p. 212), "Buying and selling carbon offsets is like pushing the food around on your plate to create the impression that you have eaten it."

James Hansen prefers a fee-and-dividend scheme like the one already successfully introduced in British Columbia.[81] The idea is that the tax on gasoline and the like is raised very considerably—an extra dollar or two on the gallon, perhaps. However, at the end of each month (or quarter, or year) the revenue raised is equally shared among the citizens. Driving a hybrid sparingly would mean you'd get a big percentage of your fuel costs returned to you; your neighbor, with his gas-guzzling SUV, wouldn't do so well. This offers a tremendous inducement to make fuel economies. Obviously, some fine-tuning would be required to make sure the system was fair—a farmer has no choice but to use far more gasoline than an inner-city dweller—but the principle seems attractive, at least at the national level; at the international level it's less useful, in that it doesn't address the need to generate money to help developing countries make the transition to clean energy sources.

Another idea that has been promoted as an alternative to cap-and-trade is contraction and convergence. The "contraction" part involves establishment of a schedule for international reduction of CO_2 emissions. For each year, the total permissible amount of CO_2 emissions is divided by the current population of the earth so that every individual has an equal CO_2 entitlement, as it were. Multiply the entitlement by the population of any given nation, and you have that nation's ration for the year. The "convergence" component of the expression is that, while the richest nations would have to get their act together pronto, the poorest nations would be entitled to emit modestly more CO_2 than they currently do. The richer would have to tighten their belts while the poorer would have room to improve their lot.

18

MARKETING CLIMATE DENIALISM

But the real fear driving climate alarmists wild is that a more rational approach to the fundamentalist religion of global warming may be in the ascendancy—whether in the parliamentary offices of the world's largest trading bloc or in the living rooms of Blacktown.

—Miranda Devine,
"Beware the Church of Climate Alarm" (2008)[1]

Deniers of anthropogenic global warming have managed to conceal from large swaths of the public an incontrovertible fact: that, within the scientific community, the argument that "global warming isn't happening and if it is it'll be good for us" *has already been lost*. In science, when a hypothesis makes predictions that reality confounds, it's the hypothesis that's ditched, not reality. The prediction of the deniers—climate change won't happen—has indeed been confounded.

It's astonishing quite how unanimous the view is not just among climate scientists but also among those professional and governmental organizations whose interests are most closely involved. As Frances Beinecke summarized in *Clean Energy Common Sense* (2010, p. 7), "It isn't often that the Pentagon, the U.S. Department of Commerce, the United Nations, the National Academy of Sciences, the European Union, the Environmental Protection Agency, the National Intelligence Council, and the Shell Oil Company agree on anything." To that list she could have added the CIA, which launched its Center on Climate Change and National Security in September 2009 to analyze the threats posed to the US by worldwide instabilities consequent upon global warming.

The illusion that there might be serious controversy about global warming within the scientific world is maintained thanks to the attention given by the media to the very loud voices of a few contrarians. Undeniably there *are* a few scientists who quite genuinely doubt the reality of AGW, or at least that it's anything for us to worry about—just as there are some who question the Big Bang theory. The trouble is that we can't tell the genuine dissenters from the shills. We do know who some of the latter are, though. There's a wonderfully useful interactive map at Greenpeace's ExxonSecrets site (http://www.exxonsecrets.org).[2]

One of the earlier front organizations for the fossil fuels industry was the Information Council on the Environment (ICE), which was set up in 1991 with $500,000 from the Western Fuels Association, the National Coal Association, and the Edison Electric Institute. Guided by the public relations firm Bracy Williams and the polling firm Cambridge Reports, ICE mounted a campaign to cast doubt on climate science, to (in its own terms) "reposition global warming as theory (not fact)" by comparing those worried about global warming to flat-earthers, for example. Like APCO a year or two later (see below), ICE discovered that the public was unimpressed when industry spokespeople questioned science but could be mightily impressed by scientists who did the same, and, further, that the public was fairly unsophisticated about the *choice* of scientists—it was, in effect, the white coat that was important rather than the actual expertise. Thus, ICE's science advisory panel contained rightwing environmental scientist Patrick J. Michaels, geographer Robert C. Balling Jr., and physicist Sherwood B. Idso (see pages 288ff.), with physicist S. Fred Singer (see pages 286ff.) also playing a major role.

The ICE endeavor didn't last long. Some of its strategy memoranda were leaked to the press, and the dishonesty of the campaign's premises laid bare for all to see. Michaels promptly distanced himself, claiming to be shocked by the turpitude.

No sooner had ICE disappeared than the next relevant industry effort began, this time organized not by the fossil fuels industry but by one seemingly unconnected with climate, the tobacco industry. By the late 1980s/early 1990s it was becoming obvious that smoking—even secondhand smoking—was a killer. As we saw in an earlier chapter (see page 94), the manufacturer Philip Morris called in public relations company APCO. APCO advised that

Morris should start *merchandizing doubt* as a way to deceive the US public. Recognizing, if these efforts focused only on the tobacco issue, that the public would smell a rat, APCO determined to erode the credibility of US government scientific agencies on other scientific issues. The three further areas of science on which it focused were nuclear waste disposal, biotechnology, and global warming. Since then, like Topsy, the network of fake think tanks and astroturf organizations has just growed, with the fossil fuels industry becoming ever more deeply engaged.

Here's a partial list of organizations that have been funded—in some cases entirely—by ExxonMobil: Accuracy in Academia, Accuracy in Media, Advancement of Sound Science Center, Advancement of Sound Science Coalition, Air Quality Standards Coalition, Alexis de Tocqueville Institution, Alliance for Climate Strategies, American Coal Foundation, American Council on Science and Health, American Enterprise Institute for Public Policy Research, American Enterprise Institute–Brookings Joint Center for Regulatory Studies, American Friends of the Institute for Economic Affairs, American Petroleum Institute, Annapolis Center for Science-Based Public Policy, Arizona State University Office of Climatology, Aspen Institute, Association of Concerned Taxpayers, Atlas Economic Research Foundation, Capital Research Center and Greenwatch, Cato Institute, Center for Environmental Education Research, Center for the Study of Carbon Dioxide and Global Change, Chemical Education Foundation, Citizens for the Environment and CFE Action Fund, Clean Water Industry Coalition, Committee for a Constructive Tomorrow, Consumer Alert, Cooler Heads Coalition, Council for Solid Waste Solutions (there's an obvious joke to be made here), Earthwatch Institute, Environmental Conservation Organization, Foundation for Research on Economics and the Environment, Fraser Institute, George C. Marshall Institute, Global Climate Coalition, Greening Earth Society, Harvard Center for Risk Analysis, Heartland Institute, Heritage Foundation, Hudson Institute, Independent Commission on Environmental Education, Institute for Biospheric Research, Institute for Energy Research, Institute for Regulatory Science, Institute for the Study of Earth and Man, James Madison Institute, Lexington Institute, Locke Institute, Mackinac Center, National Council for Environmental Balance, National Environmental Policy Institute, National Wetlands Coalition, National Wilderness Institute, Property and Environment Research Center, Public Interest Watch, Reason Foundation, Science and Environmental Policy Project, Tech Central Science Foundation; and add in journals like *Climate Research Journal* and *World Climate Report* and the website junkscience.com.[3]

That website—junkscience.com—is worth a little more attention. In predictable Orwellian fashion, the site reserves the term "junk science" for genuine science that's disliked by its corporate paymasters, describing as "sound science" all sorts of nonsense that, if believed by the gullible, will help this year's quarterly results. The site is run by Steven Milloy, who started it in 1992, while working for APCO. In 1997, he was appointed executive director of the tobacco industry's TASSC (see page 95); by 1998 TASSC had taken over the funding of www.junkscience.com. Milloy's weekly column, called Junk Science, appears on the website run by Rupert Murdoch's Fox News, and essentially regurgitates the website. Milloy is also, it seems, responsible for two organizations not mentioned in the above list, the Free Enterprise Education Institute and the Free Enterprise Action Institute, both funded by ExxonMobil. Milloy has no scientific qualifications yet has played an enormous part in manipulating the media's—and hence the public's—impression of the science concerning climate change.

The Global Climate Coalition, founded in 1989 by a group of nearly fifty corporations and trade associations, mainly from the auto and fossil fuel industries—was a major player in the AGW-denial stakes, representing in total over six million businesses. Around 1999 it began to fall apart, as giants like British Petroleum and Shell made their exits. Texaco soon followed suit, as did major car companies: Ford, General Motors, DaimlerChrysler. It seemed finally industry was beginning to get the message.

Unfortunately, that message didn't penetrate the Bush administration—hardly a surprise, since the man Bush appointed as associate undersecretary at the Department of Energy, Jeffrey Salmon, came from the ExxonMobil-funded George C. Marshall Institute. In particular, it didn't impress Vice President Richard Cheney.

In 2004 Greenpeace was able, through determined exercise of the Freedom of Information Act, to get its hands on (among much else) an e-mail dated June 3, 2002, from Myron Ebell of the Competitive Enterprise Institute to Philip Cooney, a lawyer who—lacking entirely in scientific training and with a career representing the oil industry in its fight against restrictions on greenhouse-gas emissions—must have seemed to the Bush administration an obvious choice as chief of staff of the White House Council on Environmental Quality. The e-mail, leaked to the public through *Harper's Mag-*

azine, demonstrated that the astroturf organization was cozy enough with the White House to dictate policy and discuss the hiring and firing of a federal agency head.[4] It was a marketer's dream come true.

And these groups can be surprisingly inventive. It was the *Wall Street Journal* that tracked down the true source of one supposedly spontaneous piece of satire. The YouTube video "Al Gore's Army of Penguins" attracted over a hundred thousand viewers in 2006, almost all of whom must have been persuaded by its amateurishness that this animated parody of *An Inconvenient Truth* was of populist origin. Not so. In their article "Where Did That Video Spoofing Gore's Film Come From?"[5] Antonio Regalado, Dionne Searcey, and Jeffrey Ball demonstrated that the video had been created by the Right-leaning PR/lobbying firm DCI Group, perhaps at the behest of ExxonMobil. Not only were the "facts" in the video false, that same spirit of falsehood ran right through its presentation.

In another major piece of smokescreenery, in early December 2009, Competitive Enterprise Institute senior fellow Christopher Horner threatened to sue NASA, claiming the space agency had falsified climate change data. Being obvious tripe, this didn't make many headlines . . . except on Fox News, where Horner appeared on both Sean Hannity's program and *Fox & Friends*, the case also being reported on the Fox News website and by host Bret Baier on *Special Report*. As the leftish media watchdog Media Matters for America reported incredulously,[6] not once in any of this coverage was it mentioned that the institute's funding comes primarily from the oil industry.

Although most of the world's fake think tanks and astroturf organizations are based in the US, other countries are unlucky enough to suffer their own infestations. In Canada there was the group called, ironically, Friends of Science, whose funding was originally laundered through an "educational" account created at the University of Calgary by a political science professor there, Barry Cooper; that practice ceased abruptly when the matter was exposed in the press. One Canadian newspaper, however, mysteriously failed to report the story: the *Calgary Herald*, for which Cooper writes op-eds on subjects like climate change. Surprisingly, the University of Calgary didn't fire him. He is a "fishing buddy" of Stephen Harper, Canada's prime minister.[7]

Other Canadian AGW-denying organizations include the Natural Resources Stewardship Project and the International Climate Science Coalition (ICSC). The "ICSC Mission Statement" begins: "The International Climate Science Coalition (ICSC) is an apolitical group of independent scientists, economists and energy and policy experts. . . ."[8] Yeah, right.

Although Greenpeace has for some while chronicled those institutions and individuals who have been covertly funded by ExxonMobil, collating the results at exxonsecrets.org, it was only in March 2010 that the group blew open the lid on an even greater corruption of the public discourse over climate change, this time by Koch Industries—or, to be more accurate, by billionaire brothers David and Charles Koch. In a special report entitled "Koch Industries: Secretly Funding the Climate Denial Machine,"[9] Greenpeace detailed some horrifying stuff. And the sums involved are beyond staggering: Between 1997 and 2008 the Kochs donated $48,510,856 to groups opposing action on climate change, with $24,888,282 coming between 2005 and 2008 alone. These are the sorts of sums that, used wisely, could provide enormous aid to developing countries in making the switch from dirty to clean energy, thereby saving untold lives.

The Koch brothers justify this on the grounds of their lifelong libertarianism. According to Jane Mayer, in a long *New Yorker* exposé,[10] "The Kochs . . . believe in drastically lower personal and corporate taxes, minimal social services for the needy, and much less oversight of industry—especially environmental regulation." These are easy attitudes to hold should you be lucky enough to inherit great wealth. The company has oil refineries in Alaska, Texas, and Minnesota, and has held leases on the environmentally devastating tar sands of Alberta, Canada. Each day its refineries process over eight hundred thousand barrels of crude oil in the US; it also owns a refinery in Holland. Koch Industries owns the companies that make Stainmaster® carpets, Brawny® paper towels, Dixie® cups, and Lycra®.

The Kochs donate also to some worthwhile charitable causes, although usually (so far as I can see) cultural charities—the Lincoln Center, PBS, the Smithsonian. But no amount of charitable giving could possibly compensate for the damage the Kochs have done to our future and our children's future, both directly (Koch Industries is filthy, being among the top ten air polluters in the US; and in 2000 the EPA fined the company $30 million for its role in *three hundred* oil spills) and indirectly, through sponsorship of AGW denial. In the wake of the Greenpeace report the Kochs wailed plaintively that really, really they weren't that bad, honest; the consensus seems to be that, if anything, they're worse.

Who are the beneficiaries? In alphabetical order, here are the organizations Greenpeace discovered the Koch brothers funded between 2005 and 2008 through various "foundations":[11] American Council for Capital Formation, American Council on Science and Health, American Enterprise Institute, American Legislative Exchange Council, Americans for Prosperity Foundation, Americans for Tax Reform, Atlas Economic Research Foundation, Capital Research Center, Cato Institute, Center for the Study of Carbon Dioxide and Global Change, Federalist Society for Law and Public Policy Studies, Foundation for Research on Economics and the Environment, Fraser Institute, Frontiers of Freedom, George C. Marshall Institute, Goldwater Institute, Heritage Foundation, Independence Institute, Independent Women's Forum, Institute for Energy Research, Institute for Humane Studies, John Locke Foundation, Mackinac Center for Public Policy, Manhattan Institute, Media Research Center, Mercatus Center, National Center for Policy Analysis, National Taxpayers Union Foundation, Pacific Research Institute for Public Policy, Property and Environment Research Center, Reason Foundation, State Policy Network, Tax Foundation, Texas Public Policy Foundation, and Washington Legal Foundation.

All but a few received Koch funding as far back as 1997, which is where the Greenpeace survey begins. Some funded during that earlier era but no longer are the Center for the Study of Market Processes, Citizens for a Sound Economy, the Competitive Enterprise Institute, and the Heartland Institute.

Many of the usual suspects we've already come across are associated with some of these groups; for example, Fred Singer's name (see pages 286ff.) pops up in the context of the Institute for Humane Studies, and the Center for the Study of Carbon Dioxide and Global Change is the Idso family "think tank" (see page 288).

Bear in mind these are just the *AGW-denying* organizations funded by the Koch brothers. In addition, Greenpeace unearthed a whole raft of "think tanks" and astroturf organizations designed to smear the policies of the Obama administration and counter the possibility of the American middle class finally having affordable, efficient healthcare, and so on. And this is before we mention the colossal sums spent by Koch Industries on lobbying in the Senate and House against environmental regulations. Then there are the sums contributed to the campaigns of individual politicians by Koch

Industries' political action committee, which, according to Greenpeace, "bundles contributions from Koch employees and their spouses."[12] Here Greenpeace offered two lists of interest:

> *U.S. House Members receiving more than $20,000 from Koch Industries PAC since 2006 election cycle*: Cuellar, Henry (D-TX), Cantor, Eric (R-VA), Barton, Joe (R-TX), Blackburn, Marsha (R-TN), Boehner, John (R-OH), Kline, John (R-MN), Davis, Geoff (R-KY), Boyd, Allen (D-FL), Boren, Dan (D-OK).
>
> *U.S. Senate Members receiving more than $10,000 from Koch Industries PAC since 2004 election cycle*: DeMint, James W (R-SC), Murkowski, Lisa (R-AK), Lincoln, Blanche (D-AR), Burr, Richard (R-NC), Chambliss, Saxby (R-GA), Coburn, Tom (R-OK), Roberts, Pat (R-KS), Thune, John (R-SD), Inhofe, James (R-OK), Vitter, David (R-LA), Cornyn, John (R-TX), Shelby, Richard C (R-AL).[13]

If you recognize your local legislator here, you might want to check their voting habits.

Quite separately, it has been claimed that the downplaying of the dangers of climate change by the Smithsonian's National Museum of Natural History is a consequence of donations received from David Koch, in particular a gift of $15 million toward the David H. Koch Hall of Human Origins. Here, past climate changes are presented as beneficial, because our ancestors, in surviving them, became very adaptable. This is an interesting hypothesis, but hardly relevant to the current situation—even though many visitors to the exhibit will go away thinking it is. The truth is human civilization started becoming a viable proposition about ten thousand years ago when world climate stabilized; agriculture depends on predictable weather patterns, coastal cities depend on stable sea levels, and so on. Before the start of the modern era of climate stability, the climate changes during which our ancestors became so adaptable occurred over thousands or tens of thousands of years when the planet's human population was likely under a million and seems on occasion to have come close to extinction. World population is today seven billion, but the real killer is that we're looking at radical transformation of the earth's climate over at best a few decades—far too short a time for adaptation. Of course, we have technology to help; but, if civilization largely collapses and the world population plummets to, in some scenarios, just a few million, how is that technology going to be supported?

That not all of US industry was following what one might call the Koch/ExxonMobil/Cheney line became particularly apparent in Spring 2007 when ten leading companies, among them Alcoa, Duke Energy, General Electric, and Caterpillar, publicly declared themselves in favor of a national cap on carbon emissions. This seems a genuine move, not greenwashing. These were not voices in the commercial wilderness; over the next few months scores of other companies joined the rising chorus, including Royal Dutch/Shell and all three big US carmakers. Google announced it would discipline its own carbon emissions by accounting as if there were already a cap in place, and began some huge investment in developing renewable-energy technology. The more enlightened US states and US cities introduced emissions-curbing legislation of their own—on occasion in the teeth of retaliatory blood-and-thunderish condemnation by the Bush administration.

Such antiscientific government behavior was supposed to evaporate with the arrival of the Obama administration but, while the rhetoric was abandoned, the lack of urgency persisted. With the return of the House of Representatives to Republican hands in November 2010, we can expect the situation to deteriorate once again (see pages 232ff.).

Also playing their part in the marketing of AGW denial to the US public have been all kinds of Christian fundamentalist and "family values" organizations. (What *is it* about homophobia that makes people deny climate science?) When, in 2004, the National Association of Evangelicals developed and adopted the document "For the Health of the Nation: An Evangelical Call to Civic Responsibility"[14] as an outline of future intentions, there appeared some hope the US fundamentalist movement might have come to its senses on the issue. The doctrine outlined in the document has been called Creation care: the interpretation of scriptures to imply that God expects us to look after his Creation. Among prominent conservative evangelicals to support the measure were Mike Huckabee, governor of Arkansas, and Senator Sam Brownback of Kansas.

This may have represented a false dawn. Soon an evangelical organization called the Interfaith Stewardship Alliance had sprung into existence to counter any sense of environmental responsibility among the devout. Although the sources of funding for this organization are presently unclear, several of the main figures associated with it are usual suspects from ExxonMobil-funded outfits: David R. Legates, Roy W. Spencer, Ross McKitrick . . .[15] Taking as their mission statement the so-called Cornwall Declaration on Environmental Stewardship[16] (which, spawned in 2000, in fact predates "For the Health of the Nation"), this group maintains such positions as: "We deny that Earth and its ecosystems are the fragile and unstable products of chance, and particularly that Earth's climate system is vulnerable to dangerous alteration because of minuscule changes in atmospheric chemistry."[17] Clearly these are faith-based statements, since they have no foundation in reality.

By the end of 2010 what was now the Cornwall Alliance for the Stewardship of Creation created a twelve-part "educational" video series, *Resisting the Green Dragon*, which portrayed environmentalists as adherents of a "false religion" antithetical to Christianity. Among those helping to spread this message in the videos were Tony Perkins, president of the Family Research Council, and Tom Minnery, senior vice president of Focus on the Family. According to the Cornwall Alliance's founder, Calvin Beisner, in a printed accompaniment to the video set, "Some of what goes under the name of 'creation care,' even in evangelical circles, is infected by the false worldview and theology of secular and pagan religious environmentalism."[18]

In *Boiling Point* (2004, pp. 79–80) Ross Gelbspan recounts a conversation he had in 1999 with "a top editor at a major TV network." Why was it, Gelbspan asked, that the network devoted much of its time to coverage of weather disasters yet never made the connection between the increasing frequency and severity of these and global warming? The response came that they'd made this connection exactly once; immediately lobbyists for the Global Climate Coalition (see above) aimed at the network's executives a fusillade of threats to withdraw advertising. The word came down from on high to the journalists that they should stop telling the truth about climate science. So much for "the most trusted name in news."

In 2007, Rupert Murdoch announced climate change was real and dangerous, and his vast media company, News Corp., would make every effort to render all its offices carbon neutral as speedily as possible. One wonders if he ever reads or watches his own product. The number one purveyor of false information in the US is News Corp.'s Fox News, whose star presenters without exception use every opportunity to ridicule AGW. His newspaper the *Australian* leads the charge in its country against the climate consensus; in general it allocates reporting on climate issues to political journalists, who, like noisy drunks in bars, blurt ignorant denial. His London flagship, the *Times/Sunday Times*, while not in the same AGW-denialist league as the *Daily Mail* and the *Daily Telegraph* among UK national newspapers, employs Jonathan Leake (see page 301).

The extent of Fox News's pattern of misinforming its audience came sharply into focus in a World Public Opinion survey published in December 2010:[19]

> Those who watched Fox News almost daily were significantly more likely than those who never watched it to believe that:
>
> - most economists estimate the stimulus caused job losses (12 points more likely)
> - most economists have estimated the healthcare law will worsen the deficit (31 points)
> - the economy is getting worse (26 points)
> - most scientists do not agree that climate change is occurring (30 points)
> - the stimulus legislation did not include any tax cuts (14 points)
> - their own income taxes have gone up (14 points)
> - the auto bailout only occurred under Obama (13 points)
> - when TARP came up for a vote most Republicans opposed it (12 points)
> - it is not clear that Obama was born in the United States (31 points)

The misinformation business is official policy at Fox News, as various internal e-mails, leaked to Media Matters for America, have demonstrated. Particularly notorious is one from December 8, 2010, in which Fox News's Washington managing editor Bill Sammon instructs the troops always to stress the "controversy over the veracity of climate change data"[20]—i.e., the public confusion sown by journalists like those at Fox News.

The paper "Balance as Bias: Global Warming and the US Prestige Press"[21] by Maxwell T. Boykoff and Jules M. Boykoff reports on the authors' analysis of articles about climate change in the *New York Times*, *Washington Post*, *Los Angeles Times*, and *Wall Street Journal* between 1988 and 2002, and finds "the prestige press's adherence to balance actually leads to biased coverage of both anthropogenic contributions to global warming and resultant action." In only 5.88 percent of the articles they surveyed was the focus exclusively on human contributions to warming; in a further 35.29 percent the topic of human contributions dominated, while in 6.18 percent the denialist viewpoint dominated. In just over half the articles, 52.65 percent, the coverage was "balanced"—which is to say the science and antiscience were presented as if equivalent. The bias becomes obvious if you consider that when Naomi Oreskes surveyed papers in the scientific journals over a similar period, 1993–2003, she found *not one* of 928 disputed the consensus (see page 241).

Rightwing newspapers in the UK are as guilty as Fox News of promoting dangerous disinformation about the impending climate crisis, through journalists like Christopher Booker (see page 300) in the *Telegraph* and, at the *Mail*, just about the entire reportorial staff, with the possible exception of the astrologer. The level of scientific ignorance is genuinely staggering. As just a single example, Melanie Phillips wrote in the *Daily Mail* for January 13, 2006, that "if the climate is indeed overheating, that does not mean that man-made emissions are necessarily to blame. Indeed, it is extremely unlikely that they would be since carbon dioxide forms a relatively small proportion of the atmosphere, most of which consists of water vapour."[22]

Although the authors of AGW denialist books and websites like to portray themselves as independent thinkers, the reality is different. In a 2008 paper, political scientists Peter J. Jacques, Riley E. Dunlap, and Mark Freeman analyzed 141 environmentally skeptical books published between 1972 and 2005 and found that over 92 percent of these, mostly published in the US since 1992, were linked to conservative think tanks (CTTs). "We conclude that scepticism is a tactic of an elite-driven counter-movement designed to combat environmentalism, and that the successful use of this tactic has contributed to the weakening of US commitment to environmental protection."[23]

In many instances it wasn't merely a matter of the author being connected to a CTT: The book was *published* by a CTT. And the figures speak for themselves, surely. Out of 141 books promoting the AGW-denialism cause, *just eleven* were seemingly free of a "think tank" association. So much for independence of thought!

19

CLIMATE DENIALISM: DRAMATIS PERSONAE

Who are the main figureheads of the worldwide AGW-denialist movement? Their name is, alas, Legion. Certainly there are far too many, often working with covert or overt funding from the fossil fuels industry, to include any comprehensive listing here. Instead, here is a sort of dramatis personae of some of the major—or at least noisiest—players. I've had to omit many figures who're quite significant in this curious little alternative reality: Tony Abbott, Chris Allen, Robert Balling, Fred Barnes, Joe Bastardi, Glenn Beck, David Bellamy, Chris Christie, John Christy, Ken Cuccinelli, Joe D'Aleo, James Delingpole, Peter Heck, Peter Hitchens, Robert Jastrow, Jonathan I. Katz, Nigel Lawson, Mark R. Levin, Rush Limbaugh, John Mackey, Pat Michaels, Andrew Montford, Marc Morano, William Nierenberg, Richard North, Joanne Nova, Ted Nugent, Sarah Palin, Benny Peiser, Roger A. Pielke Jr., Roger A. Pielke Sr., John Raese, Dana Rohrabacher, Roy Spencer, John Stossel, Graham Stringer, John Tierney, David Whitehouse, Ian Wishart . . . Many can be found in the index.

And nor is there space to deal in detail with all the points put forward by the deniers, or even by all of the *major* deniers . . . and, anyway, it would be a waste of time. What they produce collectively is essentially just noise, the sheer volume of which serves to bamboozle, just as the din of a raucous day in the monkey house can drive the wits out of visitors and attendants alike. Perhaps this is the deniers' intention.

Scientists

The physicist and mathematician **Freeman J. Dyson** is the most prestigious scientist the AGW community considers one of its own; most of his significant work dates from the 1940s and early 1950s. In fact, Dyson accepts the reality of anthropogenic global warming and agrees something needs to be done; in his view, the best treatment would be to plant a trillion genetically engineered fast-growing trees to hugely increase the vegetational removal of CO_2 from the atmosphere. As Isaak Walton might have said, first genetically engineer your tree. Where Dyson differs from climate scientists is in his assessment of the severity of the problems of atmospheric CO_2 and ocean acidification—he believes they're less urgently in need of our efforts than other issues like global poverty—and in what he regards as the unjustified ostracizing by most climate scientists of the few "heretics" in their midst. It would seem Dyson should treat himself to a visit to, say, www.exxonsecrets.org.

Dyson is not, as sometimes claimed, entirely unqualified in climate science. He did some early work on the subject at the Institute for Energy Analysis, Oak Ridge, during the 1970s, investigating the environmental effects of rising levels of atmospheric CO_2. The team concluded that the environmental hazards could be controlled by a mixture of emissions reduction and intelligent management of land use.[1] In the mid-1970s the atmospheric concentration was around 330 ppm. It is today around 390 ppm, and rising.

Richard Lindzen is an atmospheric physicist of some distinction who nevertheless brings to studies of AGW what James Hansen describes as "a theological or philosophical perspective that he doggedly adheres to."[2] His *Wall Street Journal* piece "Climate Science in Denial: Global Warming Alarmists Have Been Discredited, but You Wouldn't Know It from the Rhetoric This Earth Day"[3] was derided as so riddled with errors even the first six words of its subtitle are false. This was the fourth op-ed by Lindzen the *WSJ* had run, each decrying the 98 percent of climate scientists who're concerned about global warming. Cynics pointed out that, if the *WSJ* thought the voice of the dissident 2 percent deserved four op-eds, it was about time the newspaper ran 196 or so op-eds by climate scientists.

In 1997 Lindzen conceded to journalist Ross Gelbspan that his consultancies for Australian and US coal interests and for OPEC companies were worth some $2,500 *per day* to him—this in addition to his academic salary.

Professor of geology at Adelaide University, **Ian Plimer** is the author of the AGW-denialist *Heaven and Earth: Global Warming, the Missing Science* (2009); he has to date published no peer-reviewed papers on climate science. It's his contention that civilizations prosper in warmer times, so why fear global warming, which anyway is a natural process about which we can do nothing?

In particular, he claims, as a geologist, that "[o]ver the past 250 years, humans have added just one part of CO_2 in 10,000 to the atmosphere. One volcanic cough can do this in a day."[4] This is quite simply untrue. Annual CO_2 emissions through burning fossil fuels and changing land use are about *one hundred times greater* than volcanoes can produce even in a volcano-rich year. Plimer has often been corrected on this point, but carries on making it anyway. And the same goes for his repeated assertion that the United States Geological Survey (USGS) data don't take into account submarine volcanic activity; the USGS quite explicitly says they do.

He also repeats some very strange conspiracy-theory notions. "The Greens opt to pressure democratically elected governments to reject a large body of science in favor of authoritarianism and promote policies which create unemployment and economic contraction,"[5] he told an apparently embarrassed audience at the Paydirt Uranium Conference in Adelaide in March 2008. It is, of course, Plimer himself who rejects "a large body of science."

In July 2009 Plimer challenged *Guardian* journalist George Monbiot to a public debate on "Humans Induce Climate Change: Myth or Reality?" and added, "I am happy to fly to London at my expense."[6] After some thought Monbiot accepted the challenge, but on condition Plimer answer beforehand thirteen very specific questions about the content of *Heaven and Earth*. Plimer started wriggling . . . and wriggling . . . The debate never happened, but on December 15, 2009, the two men faced each other on the Australian Broadcasting Corporation's *Lateline*. Here's a sample Plimer answer:

> [W]hat I think is happening is governments just cannot resist the opportunity to tax us more, to set up huge bureaucracies, and this is what Copenhagen's about. It's not about science, it's not about morality, it's not about the Third World—it's about money, and it's about governments putting their hands in our pockets, taking out our money, and having to go through sets of sticky fingers to end up disappearing somewhere else in the world.

No, that isn't an *Onion* parody; it comes from the ABC's own transcript.[7] Similarly bizarre was Plimer's insistence that neither Monbiot nor the moderator, Tony Jones, had read *Heaven and Earth* when obviously both had. A typical exchange:

> *Jones:* [L]ast week we had the World Meteorological Organisation release their annual statement which says the first decade of the 21st Century is likely to be the warmest on record, that 2009 is set to be the fifth warmest year on record. Are they credible?
>
> *Plimer:* A couple of points. That is a projection; we haven't yet finished this year.

This discussion took place on December 15.

One of the eight-strong Advisory Board at the Advancement of Sound Science Coalition (TASSC; see page 95) was **Frederick Seitz**, also chairman of the Science and Environmental Policy Project (SEPP), an organization so rooted in the sciences that it was founded (in the early 1990s) using seed capital and office space provided by the Moonies; it now receives funding from ExxonMobil plus a bunch of rightwing organizations. Seitz served as president of the National Academy of Sciences from 1962 to 1969 and as president of Rockefeller University from 1968 to 1978. During World War II he worked on the atomic bomb project. He worked with Eugene Wigner on the Wigner–Seitz unit cell; in other words, he was a solid-state physicist rather than a climatologist.

Seitz's connection with science denial started around 1978 or 1979, when he became a permanent consultant to tobacco company R. J. Reynolds, his task being to commission research that might help counter the profits-denting scientific consensus about smoking. Over the next decade or so he was responsible for distributing some $45 million to scientific researchers who, astonishingly, could find little evidence that smoking was deleterious to health; just under $900,000 went into his own pocket.[8]

Another phase in Seitz's assault on the integrity of science saw him, in 1984, cofounding the George C. Marshall Institute. Its original purpose was to wage a publicity campaign against scientific criticism of Ronald Reagan's SDI program ("Star Wars"). The view of Seitz and his cofounders—Robert Jastrow and William Nierenberg—was that the Soviet Union's socialism was such a threat that maximum resources should be put into defending the US from it, even if scientific facts didn't support the proposal. Indeed, the staff of the institute traditionally show a fairly cavalier attitude toward facts, many of their data having been found dubious. Where their real skill lies, however, is in propaganda—in persuading politicians, media, and public that there's a powerful body of opinion disagreeing with the scientific consensus.

When the Soviet Union collapsed, it might have seemed the Marshall Institute would pack up its bags—but no. Various discoveries of science—whether about the dangers of smoking or the toxicity of environmental mercury—have an impact upon politics in the sense that they suggest regulation: Rules governing drinking water, for example, are generally agreed to be a good thing. But not by the libertarians associated with the Marshall Institute. To them, governmental regulations of any kind were (and are) anathema. Just like those fringe evangelicals who believe it's okay to lie for Jesus, the Marshall Institute people carried on propagandizing.

In 1994 the institute published Seitz's *Global Warming and Ozone Hole Controversies: A Challenge to Scientific Judgment*, which denied the ozone hole (see pages 216ff.) was a problem. He also became associated with the Oregon Institute of Science and Medicine, a maverick operation founded in 1980 by Arthur B. Robinson; Robinson started his career as a chemist but was drawn by his Christian fundamentalism into science denial. In 1998, when the Oregon Institute and the Marshall Institute got together to publish Robinson's mighty *Research Review of Global Warming Evidence*, the book bore a preface by Seitz. The book's thesis is our old friend: Increased atmospheric CO_2 will create a new Eden. This piece of pseudoscience was produced in a format precisely mimicking the journal *Proceedings of the National Academy of Sciences* and sent out for free to tens of thousands of nonclimatological scientists and graduates who very often, assuming it was from the NAS, took it seriously. The media swallowed the fraud and regurgitated it wholesale for viewers on a nightly basis. Accompanying *Research Review* was a document called the Oregon Petition, written by Seitz. Here's an extract:

> There is no convincing scientific evidence that human release of carbon dioxide, methane, or other greenhouse gases is causing or will, in the foreseeable future, cause catastrophic heating of the Earth's atmosphere and disruption of the Earth's climate. Moreover, there is substantial scientific evidence that increases in atmospheric carbon dioxide produce many beneficial effects upon the natural plant and animal environments of the Earth.[9]

The George W. Bush administration used the petition as part of its excuse for withdrawing from the Kyoto Protocol in 2000.

S. Fred Singer played a significant role in the development of observation satellites and was the National Weather Satellite Service's first director. In 1993 he and Kent Jeffreys produced a paper called "The EPA and the Science of Environmental Tobacco Smoke," which claimed the EPA was rigging the science concerning the dangers of secondhand smoking; the study was funded by the Tobacco Institute and published by a rightwing think tank, the Alexis de Tocqueville Institution. Singer had a fair history in the movement to deny the harmfulness of secondhand smoking (see pages 95 and 270), much of which is detailed, alongside his career in AGW denialism, in Naomi Oreskes's and Erik M. Conway's *Merchants of Doubt* (2010). Naturally Singer was displeased by this book. In a piece for *American Thinker* he sought to belittle Oreskes and Conway, to justify his stance on passive smoking, and to make the curious argument that, since he'd been right all along about smoking (he had?), he was probably also right about global warming—or the lack thereof.

> Oreskes' and Conway's science is as poor as their historical expertise. To cite just one example, their book blames lung cancer from cigarette smoking on the radioactive oxygen-15 isotope. They cannot explain, of course, how O-15 gets into cigarettes, or how it is created. They seem to be unaware that its half-life is only 122 seconds. In other words, they have no clue about the science, and apparently, they assume that the burning of tobacco creates isotopes—a remarkable discovery worthy of alchemists.[10]

I checked. Oreskes and Conway make no such claim. They're referring to a scientifically impenetrable 1979 statement by Fred Seitz that "the oxygen in the air we breathe and which is essential for life plays a role in radiation-induced cancer"[11] and making a guess that he might have been talking about radioactive oxygen in the air: "Oxygen, like most elements, has a radioactive version—oxygen 15—although it is not naturally occurring."[12] The source they footnote for this statement gives the half-life of O^{15}. So:

- O&C don't claim lung cancer from cigarette smoking is caused by O^{15};
- O&C don't say there's O^{15} in cigarettes;
- O&C don't say O^{15} is created in cigarettes;

- O&C, who say O^{15} isn't naturally occurring, very evidently *are* aware of its half-life;
- O&C don't "assume that the burning of tobacco creates isotopes."

There are, at least, no obvious spelling errors in Singer's paragraph.

As early as 1991 Singer coaxed the dying Roger Revelle—among the most significant early figures to warn of elevated atmospheric greenhouse gas levels—and persuaded him to sign on as one of three coauthors of an article that essentially said the science involved was still dodgy. The article, "What To Do About Greenhouse Warming: Look Before You Leap,"[13] appeared just as preparations were being made for the 1992 Earth Summit in Rio de Janeiro. Singer and his cronies publicized Revelle's near-deathbed "recantation"; according to all who knew him, the elderly Revelle hadn't really known what was going on. One of Revelle's PhD students, Justin Lancaster, spoke out intemperately on the subject. Singer, with considerable industry resources behind him, via SEPP, sued, claiming monstrous damages; it was an ugly example of the use of the legal system and superior finances to silence criticism—or, as they say, a "triumph for solid science." Lancaster was forced to retract, but has since withdrawn his retraction.

In *Heat* (2007, pp. 24–27) George Monbiot recounts his early experience with Fred Singer. He traced a set of widely disseminated bogus figures to Singer's SEPP website, where there was loose citation of a 1989 report in *Science*. Monbiot diligently went through 1989's issues of *Science* and found there was no such report. He published this discovery in his *Guardian* column and thought that was the end of the story.

However, one of his readers contacted Singer and asked for an explanation. Singer's first response was to claim Monbiot must be "confused—or simply lying" about the figures having appeared on the SEPP site. The reader persisted: Not only had the figures appeared on the website, they were *still there*. Singer's second response was that they'd been put there in error by a "former SEPP associate"—in fact, his wife—and had now been corrected. A year later, Monbiot checked the site and discovered that, no, they hadn't. I visited the SEPP website in November 2010 and, so far as I can establish, they've now gone. But for how long?

The February 12, 2001, issue of the *Washington Post* carried a letter from Singer claiming it had been twenty years since he'd last received any funding from the oil industry, and that had been for a piece of consultancy work. Unfortunately for his claim, ExxonMobil's own website bore evidence (since removed, but lovingly preserved in PDF form by Ross Gelbspan) that the company had paid $10,000 to SEPP and $65,000 to the Atlas Economic Research Foundation—both owned by Singer—in the year 1998 alone.

Sherwood B. Idso and his two sons **Craig D. Idso** and **Keith E. Idso** have built up a sort of familial cottage industry based on AGW denialism, pushing their ideas through well funded groups like the Center for the Study of Carbon Dioxide and Global Change (CSCDGC)—which Craig founded in 1998 and of which Sherwood has been president since 2001—and the CO_2 Science website, which offers a digest of material favoring denialist views.

Sherwood Idso's 1980 paper "The Climatological Significance of a Doubling of Earth's Atmospheric Carbon Dioxide Concentration"[14] claimed a doubling of atmospheric CO_2 would have little effect. He was by now associated with Arizona State University's Office of Climatology; he and his colleagues would, over the ensuing years, receive over $1 million in grants from the fossil fuel and utility industries.[15] In 1990 he and Robert Balling of Arizona State University's School of Geographical Sciences were sponsored by the Cyprus Minerals Company for a project called "Greenhouse Cooling." Sherwood's 1999 paper "Real-World Constraints on Global Warming"[16] was published by the ExxonMobil-funded Fraser Institute, an indication of where his scientific reputation had migrated. The paper's conclusions—natural negative feedbacks will solve the warming problem for us—fly in the teeth of established climate science.

His books are published either by the Institute for Biospheric Research (president: Sherwood B. Idso) or by Vales Lake Publishing, a small print-on-demand press with an odd list of science-denying items. He took part in the 1991 half-hour video *The Greening of Planet Earth* (see page 245).

Craig Idso served from 2001 to 2002 as director of environmental science at Peabody Energy, the world's largest coal company to remain in private hands. Since then he seems to have devoted the bulk of his energies to the CSCDGC, which has now, according to Greenpeace,[17] become a Koch Industries front group. In 2009 he delivered to the Climate Change Conference organized by the Heartland Institute a paper called "Climate Change Reconsidered"; this proposed, inter alia, that one motivation of the IPCC scientists was their hope of being invited to all-expenses-paid conferences in "exotic locations." More recently, he is one of the listed authors—alongside Fred Singer—of the 880-page Heartland Institute-published *Climate Change*

Reconsidered: The Report of the Nongovernmental International Panel on Climate Change (NIPCC), which you can pick up from Amazon for a nifty $123.00 . . . in paperback.

Number Two Son, Keith E. Idso, is currently the CSCDGC's vice president while also—the mind boggles—serving on the Arizona Advisory Council on Environmental Education.

The résumé of **Sallie Baliunas** reads almost like a parody: An astrophysicist by training, she has been a board member, chair of the Science Advisory Board and senior scientist at the Marshall Institute (ExxonMobil-funded), cohost and environment/science editor at *Tech Central Station* (ExxonMobil-funded), a contributing editor to *World Climate Report* (published by the Western Fuels Association), on the Advisory Board of the UK's Scientific Alliance (secretive, but probably ExxonMobil-funded), an "expert" at the Competitive Enterprise Institute (ExxonMobil-funded), and an authority on global warming and the ozone layer at the National Center for Public Policy Research (ExxonMobil-funded). In 1993 and 1994 she was the Robert Wesson Endowment Fund Fellow at the Hoover Institution (ExxonMobil-funded). There seems to be a theme running through all this . . .

Baliunas's list of publications on climate science, as given on her Marshall Institute bio page,[18] seems impressive—especially for one who has no academic qualifications in the subject. On clicking the links, however, we discover a slightly different story. These supposedly scholarly papers seem not to have appeared in the standard peer-reviewed scientific journals—in other words, they're directed not at the scientific community but at the denialist echo chamber.

The American Astronomical Society awarded Baliunas its Newton Lacy Pierce Prize in 1988, and in 1991 *Discover* magazine singled her out as one of the USA's outstanding female scientists. Since then, the decline.

One of her frequent colleagues is Harvard–Smithsonian astrophysicist **Willie Soon**, who contends climate change is the product of variations in solar output—a contention that has been proven false (see page 254). He is associated with a whole string of fossil fuel–funded organizations, including the Marshall Institute, the American Petroleum Institute, the Fraser Institute, *Tech Central Station*, *World Climate Report*, United for Jobs, the Heartland Institute, the Cooler Heads Coalition, and the Frontiers of Freedom Institute and Foundation Center for Science and Public Policy; the latter was set up in 2002 thanks to a $100,000 grant from ExxonMobil.[19] Even the Center for

Astrophysics and the Smithsonian Astrophysics Observatory, to both of which he and Sallie Baliunas belong, receive funding from ExxonMobil.

He was a coauthor—with Baliunas, David Legates, and Tim Ball—of the non-peer-reviewed paper "Polar Bears of Western Hudson Bay and Climate Change: Are Warming Spring Air Temperatures the 'Ultimate' Survival Control Factor?" (2007), which was largely responsible for the meme that it's not climate change causing polar bear populations to decline.[20] In the acknowledgements section of this paper appeared the sentence: "W. Soon's effort for the completion of this paper was partially supported by grants from the Charles G. Koch Charitable Foundation, American Petroleum Institute, and Exxon-Mobil Corporation." There was, naturally, controversy.

Funding came from the American Petroleum Institute for an earlier Soon–Baliunas paper, "Lessons & Limits of Climate History: Was the 20th Century Climate Unusual?"[21] (2003), an attack on the work of paleoclimatologist Michael Mann, which this time failed to make it into even the non-peer-reviewed section of a journal but was instead published by the Marshall Institute.

When the Soon–Baliunas paper "Proxy Climatic and Environmental Changes of the Past 1000 Years," partly funded by the American Petroleum Institute, appeared in the January 31, 2003, issue of the obscure journal *Climate Research*, there was an outcry. One of the journal's editors, Chris de Freitas, a known AGW denier, had published the piece despite all four of its peer reviewers recommending rejection. Not long after, half the journal's editorial board departed in protest.[22]

The paper had zero effect on the accepted science concerning AGW, but has had a significant political impact: James Inhofe cites it frequently.

Politicians

Chairman of the Senate's Environment and Public Works Committee from 2003 to 2006, Senator **James Inhofe** (R-OK) stands as one of the most spectacular AGW denialists in an age that has seen plenty of them. It is his conviction, real or assumed, that "[w]ith all of the hysteria, all of the fear, all of the phony science, could it be that man-made global warming is the greatest hoax ever perpetrated on the American people? It sure sounds like it."[23] In addition to denying climate science, Inhofe has played, through his committee chairmanship, a major role in ensuring that little or nothing gets done in the US about mercury pollution, which he says is unimportant.

He's associated, according to exxonsecrets.org, with the Competitive Enterprise Institute (funded by ExxonMobil, the American Petroleum Institute, Cigna Corporation, Dow Chemical, EBCO Corp., General Motors, IBM, and others), with the Annapolis Center for Science-Based Public Policy (funded primarily by the National Association of Manufacturers), with the Tech Central Science Foundation (funded largely by ExxonMobil, with further contributions from AT&T, Avue Technologies, Coca-Cola, General Motors, Intel, McDonalds, Merck, Microsoft, Nasdaq, PhRMA, and Qualcomm), and with the Washington Legal Foundation (largely funded by ExxonMobil).[24]

Inhofe maintains that the earth has been undergoing a warming trend since the end of the so-called Little Ice Age. Since this process is purely natural, it's pointless to try to counter it by, say, capping CO_2 emissions.

His single most destructive act has been, arguably, his involvement in the legal campaign to suppress the National Climate Assessment. Commissioned by the George H. W. Bush administration, this was a survey of the likely effects of climate change on the US, region by region. Because of this regional approach, it's deemed the report might have a more immediate impact on US readers than any number of global assessments—that your own neighborhood may become a dustbowl hits harder than hearing millions will starve in Africa—and thus it was subject to attack by US denialists from the outset. The ExxonMobil-funded Competitive Enterprise Institute led the charge in 2000, bringing a lawsuit, with Inhofe and Congresswoman Jo Ann Emerson (R-MO) as coplaintiffs, in an attempt to suppress publication. The incoming Bush Jr. administration dodged the issue by claiming the National Climate Assessment was just an advisory report, not official policy.

This meant the report could still be disseminated. The Competitive Enterprise Institute mounted a second lawsuit, now assisted by the Data Quality Act, which Emerson had smuggled into an appropriations bill and which essentially offers a license to politicians to alter any governmental science report until it says what they want it to say. In consequence of this second lawsuit, the Bush Jr. administration covered the National Climate Assessment with disclaimers and ignored it entirely when formulating its own—laughable—climate policy. That's worth thinking about for a moment: The single most valuable tool in assessing the dangers to this country of climate change has been cast aside because of the activities of a fake think tank and two denialist politicians.

In a long article on Inhofe's antiscience, antienvironment activities, Chris Mooney records that "[s]ince 1999, according to the Center for Responsive Politics, Inhofe has received almost $300,000 in campaign donations from oil and gas interests and nearly $180,000 from electric utilities."[25]

Inhofe maintains a list of "over 650" qualified scientists who dispute the consensus on global warming. Analysts have pointed out a few problems. First, some names appear twice, so "over 650" represents 604.[26] Second, most are unqualified. Third . . . in 2007, meteorologist George Waldenberger wrote to Inhofe's staffers:

> Take me off your list of 400 (Prominent) Scientists that dispute Man-Made Global warming claims. I've never made any claims that debunk the "Consensus."[27]

Guess what? This, however, is now attached to Waldenberger's listing:

> *[Note: There have been questions raised regarding whether Waldenberger belongs in this report. For clarification, please see this January 13, 2008 letter to Waldenberger. (LINK)]*[28]

When I clicked the "*(LINK)*" it didn't work.

In an ABC News/*Washington Post* interview on July 23, 2010, Inhofe came up with a remarkable statement: "I don't think that anyone disagrees with the fact that we actually are in a cold period that started about nine years ago."[29] While it's easy enough to laugh at Inhofe as a caricature of the bought politician, the truth is that, because of the activities of this man, countless people are needlessly going to suffer and die—arguably, many already have. And he has made it clear in pronouncement after pronouncement that he simply does not care.

Inhofe's counterpart in the House of Representatives is the Texan **Joe Barton**, popularly known as "Smokey Joe" because of his environmental record. Until 2006 Barton was chairman of the House Energy and Commerce Committee, and it was from this position that in 2005 he launched a campaign against paleoclimatologist Michael Mann and his colleagues on the Hockey Stick graph, Raymond Bradley and Malcolm Hughes—a campaign that would culminate in the infamous Wegman report (see page 251). Even before things reached that stage, Barton and his staffers had been sending out demands to see the full financial and research records of the three climatologists—a witch-hunt.

Barton's reasons for refusing to accept AGW are the usual litany of errors and misunderstandings. Although denying that our CO_2 emissions might be contributing to the problem, he accepts the planet is warming but reckons this is a good thing—for the "plant food" argument, in other words. As Professor Andrew Dessler, of the department of atmospheric sciences at Texas A&M University, Barton's alma mater, sadly observed, "He listens to the people who tell him what he wants to hear. He has never come to us, and I think that we would be a logical choice if he wants to hear a mainstream climate view."[30]

A 2004 Center for Public Integrity report[31] revealed that by then Barton had, since 1998, received more money in the form of "campaign contributions" from oil and gas companies than any other current US politician, with the solitary exception of President George W. Bush. According to a 2010 estimate, his contributions from the fossil fuel industries now top $1.5 million.[32]

Famously, in the wake of the Deepwater Horizon tragedy of spring 2010, Barton made a public apology to BP for the fact that the company had been expected to contribute to the cleanup expenses; even his Republican colleagues were stunned. His most puzzling scientific speculation is probably this, from early 2009:

> Wind is God's way of balancing heat. Wind is the way you shift heat from areas where it's hotter to areas where it's cooler. That's what wind is. Wouldn't it be ironic if in the interest of global warming we mandated massive switches to [wind] energy, which is a finite resource, which slows the winds down, which causes the temperature to go up?[33]

Christopher Walter Monckton, 3rd Viscount Monckton of Brenchley, has emerged in recent years as the standard-bearer for the AGW denialist Right, touring to debate publicly against scientists. In one such debate, with University of New South Wales computer scientist Tim Lambert (known for the blog *Deltoid*) in February 2010, there was no doubt that, from a scientific point of view, Lambert demolished Monckton. Yet the public perceived it rather differently. In the red corner was a rather hesitant, somewhat rumpled figure who spent a lot of time scowling at his computer screen as he accessed graphics to support fact after verified fact. In the blue corner was someone who spoke fluently without notes and appeared to have huge stores of data at his mental fingertips. It can hardly be doubted that many viewers came away convinced by Monckton's imitation of science.

And they must, too, be impressed by his qualifications. He was, for example, Science Advisor to Margaret Thatcher . . . except that, oops, he wasn't: He was one of her policy advisors, his special area being not science but economics. In fact, she, with a degree in chemistry, is significantly the better qualified in science: Monckton has a BA (Hons) and MA in classics and a diploma in journalism.

And he makes other odd claims. He has repeatedly stated—even in front of the US House of Representatives—that he is a member of the House of Lords, the UK's upper parliamentary chamber. Contacted about this, the House of Lords Information Office said: "Christopher Monckton is not and has never been a Member of the House of Lords."[34] Again, the UK journalist George Monbiot discovered in 2007 that Monckton's Wikipedia page alleged the UK newspaper the *Guardian* had paid Monckton £50,000 as compensation for an article by Monbiot it had published about him. No such payment has ever been made. Monbiot discovered the false statement had been inserted into the Wikipedia article by Monckton himself.[35] The most bizarre claim came in Monckton's autobiographical introduction to "More in Sorrow Than in Anger: An Open Letter from The Viscount Monckton of Brenchley to Senator John McCain about Climate Science and Policy" (p. 2), which was issued under the auspices of the denialist Science and Public Policy Institute (SPPI):

> His contribution to the IPCC's Fourth Assessment Report in 2007—the correction of a table inserted by IPCC bureaucrats that had overstated tenfold the observed contribution of the Greenland and West Antarctic ice sheets to sea-level rise—earned him the status of Nobel Peace Laureate. His Nobel prize pin, made of gold recovered from a physics experiment, was presented to him by the Emeritus Professor of Physics at the University of Rochester, New York, USA.[36]

Plenty of people on both sides of the argument have read this as a claim that Monckton is a Nobel laureate—and that was doubtless the intent. (Although the "Emeritus Professor of Physics at the University of Rochester" is not named in this document, Monckton has elsewhere identified him as David H. Douglass, a solid-state physicist and known climate "skeptic" at the University of Rochester.)

And there's more (p. 2):

> He has lectured at university physics departments on the quantification of climate sensitivity, on which he is widely recognized as an expert, and his limpid analysis of the climate-feedback factor was published on the famous climate blog of Roger Pielke, Sr.

Outside denialist circles, Monckton is *not* recognized as an expert on "the quantification of climate sensitivity" or any other aspect of climate science. As for Pielke's blog, it's famous as a rallying point for AGW denialists.

Barry Bickmore, in his online essay "Lord Monckton: 3rd Viscount of Brenchley, King of Fantasyland" (2010), concluded:

> Many people listen to crackpots like Lord Monckton when it comes to climate science, because (a) they don't trust anything that smacks of environmentalism, and (b) they don't have the background knowledge to tell real science from pseudoscience. However, in this case Lord Monckton has done us the service of making several false claims about non-technical subjects that anyone can easily understand. At least it should be clear to anyone that Monckton should not be treated as a trusted source of information.[37]

In all this, one's inescapably reminded of L. Ron Hubbard's claim (see page 81) to be "one of America's first nuclear physicists."

Returning to that open letter to John McCain, we find that things get no better in its introductory paragraphs. Take this (p. 3):

> If the United States, by the ignorance and carelessness of her *classe politique*, mesmerized by the climate bugaboo, casts away the vigorous and yet benign economic hegemony that she has exercised almost since the Founding Fathers first breathed life into her enduring Constitution, it will not be a gentle, tolerant, all-embracing, radically-democratic nation that takes up the leadership of the world.
>
> It will be a radically-tyrannical dictatorship—perhaps the brutal gerontocracy of Communist China, or the ruthless plutocracy of supposedly ex-Communist Russia, or the crude, mediaeval theocracy of rampant Islam, or even the contemptible, fumbling, sclerotic, atheistic-humanist bureaucracy of the emerging European oligarchy that has stealthily stolen away the once paradigmatic democracy of our Mother of Parliaments from elected hands here to unelected hands elsewhere. For government of the people, by the people and for the people is still a rarity today, and it may yet perish from the earth if America, its exemplar, destroys herself in the specious name of "Saving The Planet."

McCain presumably stopped reading about here, if he ever started. One suspects McCain was not the letter's intended recipient—that Monckton was engaging in a damage-limitation exercise after the presidential candidate AGW denialists were most likely to support (and who was Monckton's own clear favorite, rather than the science-oriented Barack Obama) had publicly declared that AGW was real and that doing something about it was an urgent priority.

Monckton represented the libertarian front group Committee for a Constructive Tomorrow (CFACT)—funded, according to SourceWatch,[38] by the Carthage Foundation, the Sarah Scaife Foundation, Chevron, ExxonMobil, and the DaimlerChrysler Corporation, among others—at the event called Copenhagen Climate Challenge, mounted as an alternative to the UN-sponsored international climate talks in Copenhagen in December 2009. There he famously described a Jewish environmentalist as a member of the Hitler Youth.[39]

Among other items on Monckton's résumé are that he serves as chief policy advisor for the Science and Public Policy Institute (SPPI), whose website claims its aim is "[s]cience based policy for a better world" and which heavily promotes online articles like "Proved: There Is No Climate Crisis."[40] The SPPI was founded in 2007 by Robert Ferguson, previously executive director of the now defunct Center for Science and Public Policy (CSPP), a group under the aegis of the Frontiers of Freedom Institute. Despite this, despite the fact that the SPPI and the CSPP share the same office building, and despite there being no perceptible difference between the e-mail communiqués from the two groups, the SPPI is very firm in its claim there is no connection. Earlier still, Ferguson served successively as chief of staff to three Republican congressmen: Jack Fields, John E. Peterson, and Rick Renzi. The group's chief science advisor is Willie Soon.

Monckton is listed by the Heartland Institute on its roster of "climate change experts"—a group of individuals whose unifying characteristic is AGW denial. The Institute is a rightwing think tank that adopts the position of being antiregulation in every sphere, no matter the human consequences. It is known to have been funded heavily by the tobacco industry (until recently a section of its website was called "The Smoker's Lounge"), but more recently has become secretive about its donors.[41]

In December 2009 Monckton joined the UK Independence Party (UKIP), generally regarded as on the loony fringe of UK politics. In the general election of 2010 he was UKIP's advisor on science policy (he's now its deputy leader). The *Guardian*, as part of a series quizzing the various political parties on their science policy, published a summary of Monckton's responses.[42] Monckton told Brian Cox, concerning climate change, that the party's policy was immediately to cancel all funding for AGW research and convene a royal commission to investigate; only thus could the "rapacious extremists" warning of the threat of climate change be thwarted. As Cox commented:

> Since just about every national and international scientific institution on Earth accepts the evidence of humanity's impact on the climate, such a commission would almost inevitably conclude in science's favour. Ukip doesn't explain what it would do in the event of that outcome, having already crippled scientific research into potentially one of the greatest problems facing humanity this century.

On medical matters, Monckton claimed that linking salt intake to high blood pressure was just a matter of "the medico-scientific community [whipping] up unjustifiable fears" and stated: "The placebo effect is not fully understood, and should not be dismissed in favour of 'scientific' proof." This latter seems to have been a muddled defense of homeopathy.

John Abraham has produced a magnificent rebuttal of Monckton's climate claims (see pages 74–75); also splendid is the "Monckton Myths" section of the *Skeptical Science* blog.[43]

Journalists and Others

Anthony Watts is a TV journalist, a weather presenter who studied electrical engineering and meteorology at Purdue University; there's no record of his having graduated, however, and he's been reticent in discussing this. After a career in local television, in 2004 he moved to radio, joining the Fox News affiliate KPAY (1290 AM) in Chico, California.

He is associated with the Heartland Institute, which published his report *Is the U.S. Surface Temperature Record Reliable?* (2009); in this he argued that the location, orientation, and even the paint of many US weather stations were responsible for those stations registering artificially high temperatures. The National Oceanographic and Atmospheric Administration analyzed his claim and found it baseless.

Watts is best known for his very heavily trafficked blog *Watts Up With That?*, begun in 2006, which provides not just a megaphone for himself but a rallying ground for other AGW deniers, notably Christopher Monckton. The blog played an important role in the Climategate fiasco through its dissemination of the hacked CRU e-mails. Among the claims made in its postings are that Venus (see page 237) is hot not because of the runaway greenhouse effect but because of its atmospheric pressure.[44] In another post, the same author came out with the remarkable statement that "If there were no Sun (or other external energy source) atmospheric temperature would approach absolute zero. As a result there would be almost no atmospheric pressure on any planet"[45]—which is true, insofar as that, at near absolute zero, the atmosphere would largely have frozen solid, but otherwise meaningless: A planet's average atmospheric pressure is unaffected by temperature. A characterization of *Watts Up With That?* by science blogger Joseph Romm could hardly be more accurate:

> In general, you can assume that if Watts has reprinted a piece, it is filled with anti-scientific disinformation. It's kind of like the laws of thermodynamics. If someone tells you they have a perpetual motion machine, you don't actually have to look at the design closely to know that, in fact, they don't.[46]

It's in the *Watts Up With That?*'s reader comments that one discovers true comedy gold—and outright malevolence toward science and scientists. A useful counterblast is the blog *Wott's Up With That* (http://wottsupwiththat.com/), run by an earth sciences-trained IT worker who for obvious reasons prefers to give his identity only as Ben. Also recommended is *The Video Climate Deniers Tried to Ban—Climate Denial Crock of the Week. Anthony Watts* by Peter Sinclair, which can be accessed on YouTube.[47] An account of Watts's attempts to suppress the video is "Climate Crock of the Week: What's Up with Anthony Watts [take 2]" by Kevin Grandia.[48]

In his February 15, 2009, column for the *Washington Post*, "Dark Green Doomsayers," conservative columnist **George F. Will** claimed: "Nine decades hence, our great-great-grandchildren will add the disappearance of California artichokes to the list of predicted planetary calamities that did not happen." He produced some science: "According to the University of Illinois' Arctic Climate Research Center, global sea ice levels now equal those of 1979" and "according to the U.N. World Meteorological Organization, there has been no recorded global warming for more than a decade."

Within hours, the Arctic Climate Research Center posted a statement on its website to the effect that sea ice area has decreased considerably since 1979: "We do not know where George Will is getting his information."[49]

In a follow-up column, "Climate Science in a Tornado,"[50] Will dug his heels in, claiming to have accurately quoted a January 1, 2009, article in the science blog *Daily Tech*: "Sea Ice Ends Year at Same Level as 1979"[51] by Michael Asher. He also indicated that, on January 12, the center had confirmed the information.

Unfortunately for Will, the online version of his column[52] included a link to the "confirmation," which—surprise!—is no such thing, but rather a clarification[53] for readers of Asher's *Daily Tech* article: Arctic sea ice levels have plummeted. So what Will hoped his readers would accept as supporting his AGW denial proved on inspection to be the opposite.

Will's second "fact"—that the World Meteorological Organization says there's been no global warming for a decade—likewise clashed with the reality. Its likely source was a BBC report about the cooling effect of La Niña written (somewhat sloppily, and since revised) by Roger Harrabin.[54]

Will's overall thesis is that, since nothing came of the consensus in the 1970s that the threat was global cooling, this proves nothing will come of the modern consensus about global warming. There are two main problems with this argument. First, it's illogical: Science was wrong about phlogiston; this doesn't imply it's wrong about Relativity. The second flaw was outlined by Chris Mooney in a March 21, 2009, rebuttal of Will that the *Washington Post* was finally shamed into publishing, "Climate Change Myths and Facts." Essentially, it is this: "*What* global cooling consensus in the 1970s?"

> Just last year, the *Bulletin of the American Meteorological Society* published a peer-reviewed study examining media coverage at the time and the contemporary scientific literature. While some media accounts did hype a cooling scare, others suggested more reasons to be concerned about warming. As for the published science? Reviewing studies between 1965 and 1979, the authors found that "emphasis on greenhouse warming dominated the scientific literature even then."

To this day, Will has refused to retract his assertions.

Another journalist to climb upon the denialist bandwagon is ultraconservative **Patrick J. Buchanan**. On March 1, 2010, Buchanan, who backs up his anti-climate-science crankery with a staunch refusal to accept evolution, combined his two denialist enthusiasms in a single column for the *WorldNetDaily*, "Hoax of the Century." His argument was that evolution could be thrown out because Piltdown Man was a hoax and Nebraska Man a mistake; both stories, particularly the former, in fact offer validations of Darwinism. What's more startling is that Buchanan seems to think this has anything to do with climate change. He dutifully recites the three minor errors in the IPCC's massive 2007 report (has the man ever wondered what errata slips are for?), claims Antarctic ice coverage is growing rapidly, says the Climategate farrago undermined the validity of the IPCC report, and mysteriously adds:

> Though America endured one of the worst winters ever, while the 2009 hurricane season was among the mildest, the warmers say this proves nothing. But when our winters were mild and the 2005 hurricane season brought four major storms to the U.S. coast, Katrina among them, the warmers said this validated their theory.
>
> You can't have it both ways.[55]

In fact, climate scientists were in general studiously careful *not* to claim any of the 2005 hurricanes as evidence of warming, while the snowy US winter of 2009–10 is entirely in keeping with predictions of extreme weather events and high rates of precipitation.

The journalist **Alexander Cockburn**, coeditor with Jeffrey St. Clair of the political newsletter *CounterPunch*, is unusual in being a leftish denier of climate science, for years using his column in the *Nation* to propagate views on the subject that were obviously bitterly resented by many of that magazine's readers. He upholds the crank theory of retired explosives expert Martin Hertzberg, who has described climate change as "the scientific hoax of the century" and claims global warming and cooling are a result of variations in the ellipticity of the earth's orbit, heightened CO_2 levels being a symptom, not a cause, of warming. Cockburn also believes greenhouse warming is impossible, as it would violate the second law of thermodynamics. What silly billies those scientists are to get these things wrong.

Christopher Booker is a UK journalist whose writings cater to a range of denialisms, notably that mad-cow disease, asbestos, and passive smoking pose no threat to health, that evolution by natural selection is an unproven theory (he likes ID), and of course that there's no such thing as global warming—in short, the ill informed antiscience prejudices of the daft elderly relative you try to avoid at weddings. Booker proclaims this crankery from the bully pulpit of a weekly column in the high-circulation UK newspaper the *Daily Telegraph*, and is author of influential antiscience books like *Scared to Death: From BSE to Global Warming, Why Scares are Costing Us the Earth* (2007; with Richard North), *The Real Global Warming Disaster* (2009), and *Climategate to Cancún: The Real Global Warming Disaster Continues* (2010; with North). He has played a major role in attempts to smear Rajendra Pachauri (see pages 252ff.).

In *The Real Global Warming Disaster* Booker cites Sir John Houghton as having said, in the latter's book *Global Warming: The Complete Briefing* (1994) that "[u]nless we announce disasters, no one will listen." Houghton has stated this neither in the book nor anywhere else: It is completely contrary to his beliefs. The "quote" appears to have been invented in November 2006 by journalist Piers Akerman in a column for the Australian *Sunday Telegraph*.

Jonathan Leake of the UK's high-circulation *Sunday Times* seems to be an example of a journalist deceived by ExxonMobil-funded scientific AGW deniers into believing their shtick. If we go back to articles like "The Climate of Fear" (with Jonathan Milne), which appeared in the newspaper on February 19, 2006, we find a perfectly competent piece on the current state of the science. His *Sunday Times* piece of January 31, 2010, "UN Climate Panel Shamed by Bogus Rainforest Claim," was something else entirely. This was the origin of the farrago soon named "Amazongate," the lie—based on an activist pamphlet that falsely cited a letter to *Nature*[56]—that the IPCC 2007 report's claim that up to 40 percent of the Amazon rainforest could disappear because of climate change was false.

In fact, the lead author of the *Nature* letter, Dan Nepstad, had e-mailed Leake two days before the article ran, saying there was nothing wrong with the IPCC's statement of the facts.[57] Leake went ahead and published anyway. Nepstad was angry enough that within four days his account was reported in *Science*. It's at this point that any responsible newspaper would retract the story. Not the *Sunday Times*, which is owned by Rupert Murdoch and syndicates Leake's stories to other Murdoch newspapers around the globe. Even as denialists were thinking "Amazongate" was the big story, many media reporters thought the *Sunday Times*'s stubbornness was the real cause célèbre. When finally a retraction appeared a full five months later (only after an official complaint[58] to the Press Complaints Commission, the UK media's policeman) the climbdown was reported in media including *Newsweek*, the *Washington Post*, and the *New York Times*.

Leake's article had run a credit, "Research by Richard North." As pointed out in the complaint to the PCC, "Richard North is a writer who steadfastly refuses to accept the mainstream scientific results relating to climate change science."[59] In other words, he writes an AGW-denialist blog while collaborating with Christopher Booker on various antiscience books (see above).

Having started down the path of AGW denialism, Leake, with the encouragement of his employer, went from a walk to a run. February 14, 2010, saw a piece from him called "World May Not Be Warming, Say Scientists" that produced a roster of sources for its skepticism that made this reader laugh out loud when he saw it: John Christy (see page 229), Ross McKitrick (see page 304), Anthony Watts (see page 297) and "Terry Mills, professor of applied statistics and econometrics at Loughborough University,"[60] whose qualifications may be fine but are not in climate science.

One of those consistently holding the *Sunday Times*'s toes to the fire over what he dubbed "Leakegate" has been Tim Lambert, whose *Deltoid* should be required reading for anyone presuming to comment on climate-change issues.[61] As he reported on March 2, 2010,[62] the *Sunday Times* response has been to try to stitch him up.

A March 14, 2010, piece by Leake, "Ed Miliband's Adverts Banned for Overstating Climate Change," reported that the UK's Advertising Standards Authority (ASA) had clamped down on four government public-service ads: "The ASA has ruled that the claims made in the newspaper adverts were not supported by solid science and has told the Department of Energy and Climate Change (DECC) that they should not be published again."[63] All told, there were ten points at issue. The ASA rejected *nine out of the ten*. Two of the four ads were exonerated; the other two required editing. "Banned"?

David Rose achieved distinction at *Time Out*, the *Guardian/Observer* and *Vanity Fair*, and his investigative true-crime books, like *A Climate of Fear* (1992), have merit. However, in the lead up to the invasion of Iraq in 2003 he was spectacularly wrong about Saddam Hussein's possession of weapons of mass destruction—he was in effect Tony Blair's journalistic useful idiot in the same way Judith Miller at the *New York Times* was George W. Bush's—and since joining the rightwing tabloids *Daily Mail/Mail on Sunday*, which have a strict AGW-denialist editorial policy, he has published numerous similarly misguided articles on the subject of climate change.

One was "What Happened to the 'Warmest Year on Record': The Truth is Global Warming has Halted," which appeared in the *Mail on Sunday* on December 5, 2010, and was fairly accurately described by George Monbiot as "his longest list of errors yet."[64] With the help of a recently formed organization of climate scientists, the Climate Science Rapid Response Team, Monbiot was able to catalog the most egregious errors—including Rose's reliance on information from the Global Warming Policy Foundation, a denialist organization whose funding is, according to SourceWatch, "mired in controversy."[65] Their claim, regurgitated by Rose, was that 2010 was not an exceptionally warm year—no warmer than, say, 1998 (the warmest year on record to that date!). Rose castigated the UK Met Office for suggesting otherwise; as we now know, the Met Office was right. Rose claimed water vapor is a much more important greenhouse gas than CO_2, a hypothesis comprehensively debunked. He claimed Michael Mann had "made an extraordinary admission" that the world might have been hotter during the Medieval Warm Period, something Mann has never said. There's more. Monbiot concluded his analysis: "The question now emerging for Rose is very simple. Just how many mistakes does he have to make before the thesis that these are innocent errors starts to collapse?"

The journalist, broadcaster, and blogger **Andrew Bolt** might be considered the Australian equivalent of Anthony Watts. His blog on the website of the Murdoch-owned *Melbourne Herald Sun* is reportedly one of the most popular in Australia; he uses it to promote rightwing ideology and anti-science, particularly AGW denial.[66]

His column of March 23, 2010,[67] is a classic example of Bolt's technique. Australia's Commonwealth Scientific and Research Organization (CSIRO) and the nation's Bureau of Meteorology had recently released a briefing, *The State of the Climate*, showing how Australia was being affected by global warming. Primed by denialist blogger Jo Nova, Bolt pounced:

> So why did the CSIRO choose to show [maps of] only the rainfall changes from 1960, when the BoM's records go back many decades earlier?
>
> Answer: perhaps because if it showed the rainfall changes from, say, 1900, you'd see that most of Australia has got wetter over the century[.]
>
> And then you might not panic.

In fact, the two maps show the exact opposite of what Bolt claims: Australia is indeed getting drier. Furthermore, if you look at the Bureau of Meteorology's entire series,[68] you can see this trend even more clearly. But how many of Bolt's readers looked?

The writer **Michael Crichton**'s main literary contribution to AGW denialism was his grossly irresponsible 2004 thriller *State of Fear*, whose plot centers on a theme of environmentalists committing mass murder in order to advance their own views and suppress contrary ones. He made use of standard conspiracy-theory tropes: Scientists alter their findings to please the organizations that fund them; the Secret Masters of the World are manufacturing crises, such as the threat of imminent climate catastrophe, in order to keep us all distracted from what they're *really* doing; the science on global warming is scanty; and so on. Sort of like a Syfy Channel original movie. Unfortunately, it seems Crichton actually believed this stuff. Even more unfortunately, so did many of the book's millions of readers. The Union of Concerned Scientists was so alarmed it created a special rebuttal page, "Crichton Thriller *State of Fear*,"[69] identifying Crichton's worst errors.

As an aside, Crichton was responsible for enunciating a classic example of muddled thinking about science: "There is no such thing as consensus science. If it's consensus, it isn't science. If it's science, it isn't consensus. Period."[70]

Two AGW contrarians frequently cited as representing the true, scientific face of anticonsensus thought are the Canadian former minerals prospector and mining consultant **Steve McIntyre**, who runs the denialist blog *Climate Audit*, and his frequent collaborator **Ross McKitrick**, a Canadian economist. McIntyre is associated with the Heartland Institute and the Marshall Institute; McKitrick is associated with those as well as with the Fraser Institute, the Competitive Enterprise Institute, the John Deutsch Institute, Friends of Science, the Interfaith Stewardship Alliance, and the Global Warming Policy Foundation. According to a long piece about them in the blog *Deep Climate*,[71]

> McIntyre and McKitrick have published exactly one . . . peer-reviewed article in a scientific journal.[72] (Besides [this], Ross McKitrick's misleading list of so-called "peer-reviewed science journal articles" also includes two pieces in the contrarian social science journal *Energy and Environment*, a comment letter to PNAS and a pair of replies to comments on [that solitary peer-reviewed] article!)

It seems McIntyre, with time on his hands, began sometime in 2003 to reanalyze the data Michael Mann et al. had used to produce the famous Hockey Stick graph, and to post about his results on a denialist website. He then hooked up with McKitrick, who was already deeply involved in fossil-fuel-funded AGW contrarianism; McKitrick helped him ready his results for publication as a scientific paper; the rest is history. Somewhere along the line they also began a liaison with PR man Tom Harris, whose APCO Worldwide has been profoundly involved in tobacco denialism (see pages 94ff. and 270ff.) as well as AGW denialism.

What seems extraordinary about M&M (as they like to call themselves) is that there's so little to say about them. McIntyre produced a paper in 2005 that generated a very brief conversation in scientific circles and a huge kerfuffle outside those circles, and that's about it. Yet M&M are cited favorably, as if qualified climate scientists, in books like *Physics for Future Presidents: The Science Behind the Headlines* (2008) by Richard A. Muller.

The Danish economist **Bjørn Lomborg**'s reputation as a leading AGW skeptic rests on his two bestsellers *The Skeptical Environmentalist* (2001) and, tightening the focus a bit, *Cool It: The Skeptical Environmentalist's Guide to Global Warming* (2007). In the former he conducted a purportedly rigorous re-examination of the claims of many of the foremost environmental advocacy groups and came to the conclusion that things really aren't that bad—to the contrary, in most areas of the environment there are actually signs of modest improvement. An adjunct that especially appealed to US Republicans was Lomborg's attack on Al Gore's *An Inconvenient Truth* (2006); alas, almost all of Lomborg's complaints have been refuted, and the remaining few are trivial.

In some specific instances Lomborg's assertions are undoubtedly correct—for example, London's Thames and New York's Hudson are both paradigms of cleanliness by comparison with their polluted filth of just a few decades ago. Yet this is to ignore that the reason the fish have returned to these rivers is that concerted campaigns have been conducted, and a lot of public money spent, to clean them up. They're just two of untold ecosystems around the world, some immeasurably larger. While many individual ecosystems have been reclaimed thanks to similar special efforts, these represent just, ahem, a drop in the ocean.

There's another sense in which Lomborg's arguments are justified. He's correct in saying some environmental groups exaggerate their claims, distort evidence, and—ironically!—treat science cavalierly. What he doesn't stress is that most don't.

Scientific American was concerned enough about the accusations in *The Skeptical Environmentalist* that it commissioned not one but four reviews of the book. All four reviewers—none of them lightweights—were little pleased by what they read. Later, one of them, Thomas Lovejoy, wrote: "I remember my frustration at inadequate citations, so much so that I characterized them in the review as a 'mirage in the desert.'" Lovejoy made this remark in his Foreword to Howard Friel's 2010 book *The Lomborg Deception* (p. vii), which performed the same exercise, although on an even larger scale, that John Abraham in the same year performed on claims by Christopher Monckton. Friel tracked down and read for himself the sources Lomborg had offered as support, and very often discovered the "sources" didn't say what Lomborg claimed they did.

In 2002 various environmental scientists complained about *The Skeptical Environmentalist* to the Danish Committees on Scientific Dishonesty (DCSD), a part of Denmark's Ministry of Science, Technology, and Innovation (MSTI) that other governments might think to emulate. In early 2003 the DCSD ruled the book scientifically dishonest but said Lomborg himself should not necessarily be considered so since, essentially, as an economist rather than a scientist, he didn't know what he was talking about. Lomborg appealed over the DCSD's head to the MSTI, which ruled in his favor, referring the case back to the DCSD for a second examination. In March 2004 the DCSD announced it wasn't going to bother with such a re-examination, because there was no reason to think it would change its opinion. As one might guess, this has been hailed in AGW-denialist circles as a triumph.

In 2010 the movie *Cool It* appeared, based on Lomborg's 2007 book and directed by Ondi Timoner. It was given a gushy, gullible review in the *New York Times* by Jeannette Catsoulis,[73] full of praise for Lomborg's supposed wisdom in taking a midway position between climate scientists and AGW deniers—rather like taking a halfway position between a cardiac surgeon and your plumber when planning a bypass operation. It's depressing she didn't bother to check the valuable Lomborg-Errors website, one of whose stated raisons d'être is "because in the handling of errors, Lomborg does not act like most persons would do. A normal person would apologize or be ashamed if concrete, factual errors or misunderstandings were pointed out—and would correct the errors at the first opportunity given. Lomborg does not do that."[74] Nor was she apparently worried about Lomborg's latest book, *Smart Solutions to Climate Change* (2010), in which he seemed abruptly to change his tune, now describing climate change (p. 2) as "undoubtedly one of the chief concerns facing the world today . . . a challenge humanity must confront." (Cynical minds have suggested it was the publication of Friel's *The Lomborg Deception* that prompted this sudden recantation.) Unfortunately, Lomborg is less interested in such boring means of dealing with global warming as reducing CO_2 emissions or developing clean energy sources—although he does put such measures forward as part of his eight-point plan—than in grandiose schemes of geoengineering, such as whitening clouds so they reflect more of the sun's radiation back into space.

Lomborg is not alone in believing geoengineering offers salvation; Bill Gates has poured many millions into related research. It seemed cruelly ironic, though, that mere moments after Lomborg had nailed his colors to the geoengineering mast, a study appeared in the *Proceedings of the National Academy of Sciences* declaring that, not only was geoengineering incapable of doing the job of restricting sea-level rises, it was also an extraordinarily dangerous approach.[75] The authors found that various measures, such as carbon storage and the spraying of the upper atmosphere with reflective aerosols, might bring short-term relief, but that none of them would make the basic problem go away: There's already too much CO_2 in the atmosphere and we're constantly adding more.

AFTERWORD

A firm candidate for Hero of the Civilized World is software developer Nigel Leck. In 2010, weary of the constant stream of antiscientific nonsense proliferating all over Twitter, he devised a chatbot whose name, for fear of denialist hackers, will not be mentioned here. Every five minutes, the 'bot combs Twitter in search of the catchphrases customarily regurgitated by AGW denialists. Analyzing the post, the 'bot then fires back a fitting comment drawn from its own hundreds-strong databank of suitable responses; if the original tweet rehearses the debunked argument about the sunspot cycle, for example, the response provides first a brief refutation and then links to where the matter has been more thoroughly ironed out. Should the denialist choose to counter, the 'bot's databank is big enough, and the 'bot itself flexible enough, to continue the argument for some while without repetition. Leck is hoping to expand its capabilities to allow it to gather anti-AGW arguments from elsewhere on Twitter and appropriate them for its own use.

Occasionally it makes mistakes. Sarcasm causes problems, for example: During the record heat wave in LA and environs in the summer of 2010, Leck had to send out multiple apologies to perfectly rational folk who'd been happily making "no such thing as global warming" jokes.

The irony here is that anyone who has ever argued on the Internet with science deniers of any stripe—climate change, vaccination/autism, evolution, you name it—will assure you that after a while there grows an uncanny sense, as the same old arguments appear again and again despite countless debunkings, that you're arguing not with a human being but with . . .

NOTES

INTRODUCTION: UNLESS WE THINK, WE AREN'T

1. Steve Striffler, "Why We Should Take Jared Loughner's Politics Seriously," *Truthout*, January 14, 2011. Accessed at http://www.truth-out.org/why-we-should-take-jared-loughners-politics-seriously66864. My italics.

2. Anthropogenic (human-generated) global warming.

3. Justin Kruger and David Dunning, "Unskilled and Unaware of It: How Difficulties in Recognizing One's Own Incompetence Lead to Inflated Self-Assessments," *Journal of Personality and Social Psychology* 77, no. 6 (December 1999): 1121–34. Accessed at http://www.wepapers.com/Papers/70939/Unskilled_and_Unaware_of_It_-_How_Difficulties_in_Recognizing_One%27s_Own_Incompetence_Lead_to_Inflated_Self-Assessments.

4. See the discussion of "chronic Lyme disease" on page 80 for one example.

5. National Science Board, *Science & Engineering Indicators—2000* (Arlington, VA: National Science Foundation, 2000), 2:549.

6. Harris Interactive Poll, "The Religious and Other Beliefs of Americans," *BusinessWire*, November 29, 2007. Accessed at http://www.businesswire.com/portal/site/home/permalink/?ndmViewId=news_view&newsId=20071129005072&newsLang=en.

7. Will Dunham, "Many Americans Turning to Unconventional Medicine," *Reuters*, December 10, 2008. Accessed at http://www.reuters.com/article/idUSTRE4B95XO20081210.

8. Jeffrey M. Jones, "In US, Many Environmental Issues at 20-Year-Low Concern," *Gallup Poll*, March 16, 2010. Accessed at http://www.gallup.com/poll/126716/Environmental-Issues-Year-Low-Concern.aspx.

9. Pamela Oldham, "Texas Children: Canaries in the Coal Mine," *Miller–McCune*, January 25, 2011. Accessed at http://www.miller-mccune.com/health/texas-children-canaries-in-the-coal-mine-27640/.

10. Ross Ramsey, "Meet the Flintstones," *Texas Tribune*, February 17, 2010. Accessed at http://www.texastribune.org/stories/2010/feb/17/meet-flintstones/#.

11. Craig Timberg, "Williams Aims to Be Mayor of a Bigger DC," *Washington Post*, January 2, 2003, cited by John Glad in *Future Human Evolution* (2006), p. 53.

12. Alfred McCoy, "How America Will Collapse (by 2025)," *Salon*, December 6, 2010. Accessed at http://www.salon.com/news/feature/2010/12/06/america_collapse_2025.

13. Accessed at http://nobelprize.org/nobel_prizes/medicine/laureates/1973/tinbergen-lecture.pdf.

14. See, for example, Chris Morgan and David R. Langford, *Facts and Fallacies* (Exeter: Webb & Bauer, 1981).

15. The term "Science Wars" is something of a misnomer, in that it implies that both warring factions base their arguments in science—that it's a matter of science vs. science. Very obviously, this isn't the case.

16. Cited in Ron Suskind, "Faith, Certainty, and the Presidency of George W. Bush," *New York Times Magazine*, October 17, 2004.

17. I leave the mathematical proof as an exercise for the reader.

18. Reprinted in Mario Bunge, *The Sociology–Philosophy Connection* (Piscataway, NJ: Transaction, 1999), p. 209.

19. Neal Koblitz, "Mathematics as Propaganda: A Tale of Three Equations; or, The Emperors Have No Clothes," *Mathematical Intelligencer* 10, no. 1 (1988): 4–10.

20. Meera Nanda, "Postmodernism, Hindu Nationalism, and 'Vedic Science,'" *Frontline*, December 21, 2003. Accessed at http://www.mukto-mona.com/Articles/vedic_science_Mira.htm.

21. Mary R. Lefkowitz, "Not Out of Africa," *New Republic*, February 10, 1992.

22. Nicholas Graham, "O'Reilly: God Causes the Tides, Not the Moon," *Huffington Post*, January 6, 2011. Accessed at http://www.huffingtonpost.com/2011/01/06/oreilly-god-causes-tides_n_805262.html.

23. Gayle A. Buck, Diandra Leslie-Pelecky, and Susan K. Kirby, "Bringing Female Scientists into the Elementary Classroom: Confronting the Strength of Elementary Students' Stereotypical Images of Scientists," *Journal of Elementary Science Education*, 14, no. 2 (September 2002): 1–10. Accessed at http://digitalcommons.unl.edu/teachlearnfacpub/15/.

24. For a fuller discussion of this aspect, see Susan Jacoby, *The Age of American Unreason*, chapter 5 (New York: Pantheon, 2008).

25. Deborah E. Lipstadt, *Denying the Holocaust: The Growing Assault on Truth and Memory* (New York: Free Press, 1993), p. 2.

26. Scientists don't always help, frequently using "theory" where they really mean "hypothesis," as in "string theory"—somehow "string hypothesis" doesn't quite cut it.

27. James B. Campbell, Jason W. Busse, and H. Stephen Injeyan, "Chiroprac-

tors and Vaccination: A Historical Perspective," *Pediatrics* 105 (2000): e43. Accessed at http://www.pediatrics.org/cgi/content/full/105/4/e43.

28. Bizarrely, Gingrich, author of these travesties, has more recently tried to reinvent himself as a champion of science and the environment in books such as *A Contract with the Earth* (Baltimore, MD: Johns Hopkins University Press, 2007; with Terry L. Maple). Such efforts sit oddly alongside other books of his, like *Drill Here, Drill Now, Pay Less: A Handbook for Slashing Gas Prices and Solving Our Energy Crisis* (Washington, DC: Regnery, 2008; with Vince Haley).

29. The other three left standing by this stage were Hillary Clinton, Mike Huckabee, and John McCain.

30. Russell Shorto, "How Christian Were the Founders?" *New York Times* magazine, February 14, 2010. Accessed at http://www.nytimes.com/2010/02/14/magazine/14texbooks-t.html?pagewanted=1&em.

31. Texas Freedom Network, "Blogging the Social Studies Debate IV," *TFN Insider*, March 11, 2010. Accessed at http://tfninsider.org/2010/03/11/blogging-the-social-studies-debate-iv/.

32. Texas Freedom Network, "Blogging the Social Studies Debate III," *TFN Insider*, March 10, 2010. Accessed at http://tfninsider.org/2010/03/10/blogging-the-social-studies-debate-iii/.

33. Abby Rapoport, "More Attention on Texas SBOE Has Mixed Results," *Texas Tribune*, March 15, 2011. Accessed at http://www.texastribune.org/texas-education/state-board-of-education/more-attention-on-texas-sboe-has-mixed results/.

34. "The Heart of the Matter," *Gin and Tacos*, September 29, 2010. Accessed at http://www.ginandtacos.com/2010/09/29/the-heart-of-the-matter/.

Chapter 1: God Told Me to Deny

1. Bertrand Russell, "Is There a God?" (1952). Accessed at http://www.cfpf.org.uk/articles/religion/br/br_god.html.

2. Cited by Andrew McLemore in "Limbaugh: Volcanic Ash Cloud Is God's Punishment for Health Care," *The Raw Story*, April 17, 2010. Accessed at http://rawstory.com/rs/2010/04/17/limbaugh-volcanic-ash-cloud-gods-punishment-health-care/. One comment: "I'm disappointed with Obama. He pumiced this wouldn't happen!"

3. Joe Kovacs, "Is 2nd Coming of Jesus Etched in Night Sky?" *WorldNetDaily*, April 18, 2010. Accessed at http://www.wnd.com/index.php?fa=PAGE.view&pageId=138997.

4. "Sangh Parivar, the Pizza-Maker," *Outlook India*, July 5, 2003. Accessed at http://www.outlookindia.com/article.aspx?220622.

5. Meera Nanda, "Postmodernism, Hindu Nationalism, and 'Vedic Science,'" *Frontline*, December 20, 2003. Accessed at http://www.mukto-mona.com/Articles/vedic_science_Mira.htm.

6. Meera Nanda, "Postmodernism, Hindu Nationalism, and 'Vedic Science,'" *Frontline*, January 3, 2004. Accessed at http://www.hinduonnet.com/fline/fl2101/stories/20040116001408700.htm.

7. Some writers maintain the heyday lasted until later than this, even into the sixteenth century, some that Muslim science went into decline after about 1100.

8. Pervez Hoodbhoy, "Science and the Islamic World: The Quest for Rapprochement," *Physics Today* (August 2007): 8.

9. *Karachi Herald*, January 1988; expanded as an appendix in Hoodbhoy's book *Islam and Science: Religious Orthodoxy and the Battle for Rationality* (London: Zed Books, 1991). The version to which I've referred is the latter.

10. And much later, in 2001, arrested for passing nuclear secrets to the Taliban.

11. This book is still, bizarrely, in print; it was most recently, as far as I can establish, reissued in 2005.

12. Indians cannot mock their Pakistani counterparts too roundly over this. In 2003 Meera Nanda reported that the Indian Ministry of Defence was "sponsoring research and development of weapons and devices with magical powers mentioned in the ancient epics" ("Postmodernism, Hindu Nationalism and Vedic Science," *Frontline*, December 20, 2003). At the same time, astrology was "flourishing" as a course subject in Indian universities.

13. Pervez Hoodbhoy, *Islam and Science*, p. 144.

14. Ziauddin Sardar, "Islam and Science," The Royal Society, December 12, 2006. Accessed at http://royalsociety.org/Islam-and-Science-12-December-2006/.

Chapter 2: "The Law Is a Ass"

1. The UK was where the most damage was done. The FDA banned Thalidomide in the US, although by then millions of samples had been distributed for purposes of clinical trials.

2. Abigail Foerstner, *James Van Allen: The First Eight Billion Miles* (Iowa City: University of Iowa Press, 2007), p. 251.

3. Some explanations can be more exotic. Joe Schwarcz, discussing MCS in his *Science, Sense, and Nonsense: 61 Nourishing, Healthy, Bunk-Free Commentaries on the Chemistry That Affects Us All* (Toronto: Doubleday Canada, 2009),

reports: "One woman, for example, claims that her body's electromagnetic polarity runs counterclockwise instead of clockwise."

4. Stephen Barrett, MD, "A Close Look at 'Multiple Chemical Sensitivity.'" Quackwatch, 1998. Accessed at http://www.quackwatch.org/01QuackeryRelatedTopics/mcs.pdf.

5. Sherine Gabriel et al., "Risk of Connective-Tissue Diseases and Other Disorders after Breast Implantation," *New England Journal Of Medicine* 330 (1994): 1697.

6. In addition, Gabriel was libeled with charges of scientific corruption, in that her report was partly financed by an educational arm of the American Society of Plastic and Reconstructive Surgeons, the Plastic Surgery Educational Foundation. Such charges she robustly repudiated.

7. This was the Gabriel et al. study.

8. Jorge Sanchez-Guerrero et al., "Silicone Breast Implants and the Risk of Connective-Tissue Diseases and Symptoms," *New England Journal of Medicine* 332 (1995): 1666–70.

9. American College of Rheumatology, "Statement on Silicone Breast Implants," October 22, 1995. Accessed at http://www.pbs.org/wgbh/pages/frontline/implants/medical/positionstate.html.

10. New Jersey Supreme Court Decision, June 23, 1994. Accessed at http://law2.umkc.edu/faculty/projects/ftrials/mcmartin/michaelsdecision.HTM.

11. Quoted in Lona Manning, "Nightmare at the Daycare: The Wee Care Case," *Crime Magazine*, last updated January 14, 2007. Accessed at http://www.crimemagazine.com/nightmare-day-care-wee-care-case.

12. Cited by Mitchell Landsberg, "McMartin Defendant Who 'Lost Everything' in Abuse Case Dies at 74," *Los Angeles Times*, December 17, 2000.

13. Hubert H. Humphrey III, *Report on Scott County Investigations*, Minnesota Attorney General's Office, February 12, 1985. Accessed at http://www.leg.state.mn.us/docs/pre2003/other/850763.pdf.

14. "Repress-Memory Doc May Lose License," Associated Press, August 13, 1998. Accessed at http://www.rickross.com/reference/false_memories/fsm7.html.

15. Esther Addley, "Interview: Esther Addley Meets 'Karen': 'I Could not Stop Crying,'" *Guardian*, October 21, 2006. Accessed at http://www.guardian.co.uk/commentisfree/2006/oct/21/comment.children.

16. Keith Fraser, "B.C. Woman Suing over Alleged False Memories," Can West News Service, October 12, 2006. Accessed at http://www.rickross.com/reference/false_memories/fsm114.html.

17. Gavin Madeley, "£20,000 Payout for Woman Who Falsely Accused Her Father of Rape after 'Recovered Memory' Therapy," *Daily Mail*, October 19, 2007. Accessed at http://www.dailymail.co.uk/news/article-488623/20-000-payout-woman-falsely-accused-father-rape-recovered-memory-therapy.html.

18. Two useful lists of the wrongfully incarcerated are held by the websites *Imaginary Crimes* (see http://members.shaw.ca/imaginarycrimes/othercases.htm) and *Religious Tolerance* (see http://www.religioustolerance.org/ra_case.htm).

19. Elizabeth Loftus and Katherine Ketcham, *The Myth of Repressed Memory: False Memories and Allegations of Sexual Abuse* (New York: St. Martin's, 1994), p. 36.

20. David Grann, "Trial by Fire: Did Texas Execute an Innocent Man?" *New Yorker*, September 7, 2009. Accessed at http://www.newyorker.com/reporting/2009/09/07/090907fa_fact_grann.

21. Craig L. Beyer, "Analysis of the Fire Investigation Methods and Procedures Used in the Criminal Arson Cases against Ernest Ray Willis and Cameron Todd Willingham," Texas Forensic Science Commission, August 17, 2009. Accessed at http://www.docstoc.com/docs/document-preview.aspx?doc_id=10401390.

22. Steve Mills, "Report Questions If Fire Was Arson," *Chicago Tribune*, August 25, 2009. Accessed at http://articles.chicagotribune.com/2009-08-25/news/chi-090825willingham_1_texas-forensic-science-commission-willingham-case-willinghams-house.

23. Matt Smith, "Judge to Review Conviction in Texas Arson-Murder Case," CNN, October 5, 2010. Accessed at http://news.blogs.cnn.com/2010/10/05/court-to-review-conviction-in-texas-arson-murder-case/.

24. Jordan Smith, "Appeals Court Rules in Willingham Court of Inquiry Recusal Issue," *The StandDown Texas Project*, December 21, 2010. Accessed at http://standdown.typepad.com/weblog/2010/12/appeals-court-rules-in-willingham-court-of-inquiry-recusal-issue.html.

25. For a number of such cases, see Maurice Possley, "Arson Myths Fuel Errors," *Chicago Tribune*, October 18, 2004. Accessed at www.chicagotribune.com/news/watchdog/chi-0410180222oct18,0,1571511.story.

26. Committee on Identifying the Needs of the Forensic Sciences Community, National Research Council, *Strengthening Forensic Science in the United States: A Path Forward* (Washington, DC: National Academies Press, 2009). Accessed at http://www.nap.edu/openbook.php?record_id=12589&page=1.

27. Radley Balko, "North Carolina's Corrupted Crime Lab," *Reason*, August 23, 2010. Accessed at http://reason.com/archives/2010/08/23/north-carolinas-corrupted-crim.

28. Innocence Project, "Injustice in Texas: The Claude Jones Case." Accessed at http://www.innocenceproject.org/Content/Injustice_in_Texas_The_Claude_Jones_Case.php.

29. Lisa Black and Steve Mills, "What Causes People to Give False Confessions?" *Chicago Tribune*, July 11, 2010. Accessed at www.chicagotribune.com/news/local/ct-met-forced-confessions-20100711,0,5509713.story.

30. Matt Kelley, "After Five Years in Jail, Set Free Without an Apology," Inno-

cence Project, August 6, 2010. Accessed at http://criminaljustice.change.org/blog/view/after_five_years_in_jail_set_free_without_an_apology.

31. "Facts and Figures," False Confessions. Accessed at http:www.falseconfessions.org/fact-a-figures.

32. Legal cases to stop the teaching of Creationism in the schools are about the separation of church and state, not about the science.

33. For an overview of the case, with plentiful links to media coverage, see Sile Lane, "Plastic Surgeon Threatened for Comment on 'Boob Job' Cream" on the excellent Sense About Science site, which has come to serve as a nexus for information about the abuse of the UK libel laws to stifle scientific debate. Accessed at http://www.senseaboutscience.org.uk/index.php/site/other/537/.

34. Ian Sample, "Government Abandons Lie Detector Tests for Catching Benefit Cheats," *Guardian*, November 9, 2010. Accessed at http://www.guardian.co.uk/science/2010/nov/09/lie-detector-tests-benefit-cheats.

35. Peter Wilmshurst, "Obstacles to Honesty in Medical Research," *HealthWatch*, August 2003. Accessed at http://www.healthwatch-uk.org/awardwinners/peterwilmshurst.html.

36. Clare Dyer, "Cardiologist Is Sued for Comments on Potential Migraine Device," *British Medical Journal*, November 3, 2008. Accessed at http://www.healthwatch-uk.org/news.htm.

37. Cited by Pallab Ghosh in "Science Writer Wins Libel Appeal," BBC, April 1, 2010. Accessed at http://news.bbc.co.uk/go/pr/fr/-/2/hi/uk_news/8598472.stm.

38. For more on the topic, see Arthur Allen, "Treating Autism as If Vaccines Caused It: The Theory May Be Dead, but the Treatments Live On," *Slate*, April 1, 2009. Accessed at http://www.slate.com/id/2215128. Also Stephen Barrett, "How the 'Urine Toxic Metals' Test Is Used to Defraud Patients," Quackwatch, latest update December 7, 2010.

39. Orac, "More Legal Thuggery against a Defender of Science-Based Medicine," *Respectful Insolence*, June 30, 2010. Accessed at http://scienceblogs.com/insolence/2010/06/more_legal_thuggery.php.

40. Ray Weymann, Barry Bickmore, John Abraham, Michael Mann, and Winslow Briggs, *Climate Scientists Respond: Response to the Written Testimony of Christopher Monckton in Connection with the May 2010 Hearing Before the Select Committee on Energy Independence and Global Warming*, September 2010. Accessed at http://bbickmore.files.wordpress.com/2010/09/response-to-monckton-web-1.pdf.

41. Christopher Monckton, "Climate Sensitivity Reconsidered," *Physics and Society*, July 2008. Accessed at http://www.aps.org/units/fps/newsletters/200807/monckton.cfm.

42. Arthur Smith, "A Detailed List of the Errors in Monckton's July 2008

Physics and Society Article," September 6, 2008. Accessed at http://altenergy action.org/Monckton.html#updates.

43. John Abraham, "A Scientist Replies to Christopher Monckton," University of St. Thomas, St. Paul, MN, 2009. Accessed at http://courseweb.stthomas .edu/jpabraham/global_warming/Monckton/Original%20Presentation/index.htm. A slightly shorter revised version was later issued; accessed at http://www.stthomas .edu/engineering/jpabraham/.

44. Christopher Monckton, "Monckton: At Last, the Climate Extremists Try to Debate Us!" *Pajamas Media*, June 4, 2010. Accessed at http://www.webcitation .org/5rEP8Y4VI.

45. Christopher Monckton, "Abraham Climbs Down," *Watts Up With That*, July 14, 2010. Accessed at http://wattsupwiththat.com/2010/07/14/abraham-climbs -down/.

Chapter 3: Thoroughly Uncomplementary

1. "Alumni Career Spotlight: Daniel Rubin, ND, FABNO," Association of Accredited Naturopathic Medical Colleges. Accessed at http://www.aanmc.org/ careers/alumni-leaders-in-the-field/daniel-rubin-profile.

2. Orac, "Holiday Weekend Reader Mailbag: In Which Orac Is Chastised for 'condemning someone who helps people,'" *Respectful Insolence,* November 26, 2010. Accessed at http://scienceblogs.com/insolence/2010/11/holiday_weekend _reader_mailbag_in_which.php.

3. D. M. Eisenberg et al., "Trends in Alternative Medicine Use in the United States, 1990–1997: Results of a Follow-Up National Survey," *Journal of the American Medical Association* (November 11, 1998). Accessed at http://www.journal-club.org/vol2/a68.html.

4. James D. Cooper and Henry M. Feder Jr., "Inaccurate Information about Lyme Disease on the Internet," *Pediatric Infectious Disease Journal* 23, no. 12 (December 2004). Accessed at http://www.cdc.gov/ncidod/dvbid/lyme/resources/ LD_Internet.pdf.

5. The authorship requires some unscrambling. The early editions claimed: "By a nuclear physicist and a medical doctor." At some fairly early stage, the name of L. Ron Hubbard was substituted for the description "a nuclear physicist" and by the late 1970s the "medical doctor" was being named on the book's cover as Richard Farley. More recent editions name not one but two "medical doctors"—Dr. Gene Denk and Dr. Farley R. Spink.

6. It's an important 15 percent, though.

7. Luana Colloca et al., "Overt Versus Covert Treatment for Pain, Anxiety, and

Parkinson's Disease," *Lancet Neurology* 3, no. 11 (November 2004). My thanks to Michael Brooks, author of *13 Things that Don't Make Sense* (2008), for sending me this paper.

8. Ted J. Kaptchuk et al., "Placebos without Deception," *PLoS ONE*, December 22, 2010. Accessed at http://www.plosone.org/article/info:doi/10.1371/journal.pone.0015591.

9. For a far more complete analysis, see the medical blogger Orac's "More Dubious Statements about Placebo Effects," *Respectful Insolence*, December 23, 2010. Accessed at http://scienceblogs.com/insolence/2010/12/more_dubious_statements_about_placebo_ef.php.

10. Ann Helm, "Truth-Telling, Placebos, and Deception," *Aviation, Space, and Environmental Medicine* (January 1985).

11. Gary Null et al., "Death by Medicine," *Life Extension*, March 2004. Accessed at http://www.lef.org/magazine/mag2004/mar2004_awsi_death_01.htm.

12. Centers for Disease Control, "Ten Great Public Health Achievements—United States, 1900–1999," *Morbidity and Mortality Weekly Report*, April 1999. Accessed at http://cdc.gov/mmwr/preview/mmwrhtml/00056796.htm.

13. Cited by Dan Agin in his *Junk Science* (New York, Thomas Dunne Books, 2006), p. 115.

Chapter 4: Puffing the Product

1. Fake grassroots, astroturf—geddit?

2. "Smoking Conspiracy: Second Hand Smoke," *Smoking Aloud*. Accessed at http://www.smokingaloud.com/ets.html.

3. Karen Wilson et al., "Tobacco-Smoke Exposure in Children Who Live in Multiunit Housing," *Pediatrics* 127, no. 1 (January 2011): 85–92. Accessed at http://pediatrics.aappublications.org/cgi/content/abstract/127/1/85.

4. Cited in "Flat-Dwelling Children Exposed to Neighbours' Smoke Too," BBC, December 13, 2010. Accessed at http://www.bbc.co.uk/news/health-11969074.

5. Legacy Tobacco Documents Library (San Francisco, CA: University of California, San Francisco). Accessed at http://legacy.library.ucsf.edu/.

6. Legacy Tobacco Documents Library. Accessed at http://legacy.library.ucsf.edu/tid/vde14f00/pdf.

7. Brown & Williamson, "Smoking and Health Proposal" (1969), *Tobacco Documents Online*. Accessed at http://tobaccodocuments.org/landman/332506.html.

8. Horace R. Kornegay, "New Directions" (1981), Legacy Tobacco Documents Library. Accessed at http://legacy.library.ucsf.edu/tid/akc20e00/pdf.

9. Monique E. Muggli, Richard D. Hurt, and Lee B. Becker, "Turning Free Speech into Corporate Speech: Philip Morris' Efforts to Influence US and European Journalists Regarding the US EPA Report on Secondhand Smoke," *Preventive Medicine* 39, no. 3 (September 2004). Abstract accessed at http://www.ncbi.nlm.nih.gov/pubmed/15313097.

10. US Environmental Protection Agency, *Respiratory Health Effects of Passive Smoking: Lung Cancer and Other Disorders* (Washington, DC: US Environmental Protection Agency, December 1992). Accessed via http://cfpub2.epa.gov/ncea/cfm/recordisplay.cfm?deid=2835.

11. Cited by Naomi Oreskes and Erik M. Conway in *Merchants of Doubt: How a Handful of Scientists Obscured the Truth on Issues from Tobacco Smoke to Global Warming* (New York: Bloomsbury, 2010), pp. 154–55.

12. Oreskes and Conway, *Merchants of Doubt*, p. 141.

13. Gallup News Service, "Many Americans Still Downplay Risk of Passive Smoking," July 21, 2006. Summary accessed at http://www.gallup.com/poll/23851/Many-Americans-Still-Downplay-Risk-Passive-Smoking.aspx.

14. Accessed at http://www.justice.gov/civil/cases/tobacco2/amended%20opinion.pdf.

15. For more on this, see Duff Wilson, "Cigarette Giants in Global Fight on Tighter Rules," *New York Times*, November 13, 2010.

16. Mattias Öberg et al., "Worldwide Burden of Disease from Exposure to Second-Hand Smoke: A Retrospective Analysis of Data from 192 Countries," *Lancet*, early online publication November 26, 2010. Accessed at http://www.who.int/quantifying_ehimpacts/publications/smoking.pdf.

Chapter 5: Paying with Their Lives

1. Cited in Mensah M. Dean, "First, Do No Harm: Prayer or Medicine?" *Philadelphia Daily News*, December 7, 2010. Accessed at http://www.philly.com/philly/news/20101207_First__do_no_harm__Prayer_or_medicine_.html?viewAll=y&c=y.

2. This and following quotes are compiled in CHILD, Inc., "Victims of Religion-Based Medical Neglect." Accessed at http://childrenshealthcare.org/?page_id=132.

3. Quoted in Associated Press, "Girl's Death Probed after Parents Rely on Prayer," 2008. Accessed at http://www.msnbc.msn.com/id/23832053/

4. Alissa Lim et al., "Adverse Effects Associated with the Use of Complementary and Alternative Medicine in Children," *Archives of Disease in Childhood*,

advance online publication December 22, 2010. Abstract accessed at http://adc.bmj.com/content/early/2010/11/24/adc.2010.183152.short.

5. Both Hunt and Olmsted are cited by Amanda Lee Myers in "Ariz. Hospital Loses Catholic Status over Surgery," *USA Today*, December 22, 2010. Accessed at http://www.usatoday.com/news/religion/2010-12-21-phoenix-catholic-hospital_N.htm.

6. Center for Reproductive Rights, "L.C. v. Peru (UN Committee on the Elimination of Discrimination against Women)," June 18, 2009. Accessed at http://reproductiverights.org/en/case/lc-v-peru-un-committee-on-the-elimination-of-discrimination-against-women.

Chapter 6: The Antivaxers

1. Anne Dachel, "Dr. Mayer Eisenstein's New Book Merits Close Attention," *Age of Autism*, June 15, 2010. Accessed at http://www.ageofautism.com/2010/06/dr-mayer-eisensteins-new-book-merits-close-attention.html.

2. From the various equally legitimate spellings in use, I've chosen "antivax" and "antivaxer."

3. N. P. Thompson, R. E. Pounder, A. J. Wakefield, and S. M. Montgomery, "Is Measles Vaccination a Risk Factor for Inflammatory Bowel Disease?" *Lancet* 345, no. 8957 (April 29, 1995): 1071–74. Abstract accessed at http://www.thelancet.com/journals/lancet/article/PIIS0140-6736%2895%2990816-1/abstract.

4. N. Chadwick, I. J. Bruce, S. Schepelmann, R. E. Pounder, and A. J. Wakefield, "Measles Virus RNA Is Not Detected in Inflammatory Bowel Disease Using Hybrid Capture and Reverse Transcription Followed by the Polymerase Chain Reaction," *Journal of Medical Virology* 55, no. 4 (August 1998): 305–311. Abstract accessed at http://www.ncbi.nlm.nih.gov/pubmed/9661840.

5. A. J. Wakefield et al., "Ileal-Lymphoid-Nodular Hyperplasia, Non-Specific Colitis, and Pervasive Developmental Disorder in Children," *Lancet* 351, no. 9103 (February 28, 1998): 637–41. Abstract accessed at http://www.ncbi.nlm.nih.gov/pubmed/9500320.

6. Quoted by Paul A. Offit in *Autism's False Prophets: Bad Science, Risky Medicine, and the Search for a Cure* (New York: Columbia University Press, 2008), p. 22.

7. See Ben Goldacre, *Bad Science* (London, Fourth Estate, 2008), pp. 299–312, for a blistering account of the UK press's irresponsible MMR–autism sensationalism.

8. Kristina Chew, "Wakefield's Study Linking Vaccines to Autism was 'Deliberate Fraud,'" *Care2*, January 5, 2011. Accessed at http://www.care2.com/causes/

health-policy/blog/wakefields-study-linking-vaccines-to-autism-was-deliberate-fraud/. *Care2* has published almost as much baloney on the autism/vaccination farrago as *Huffington Post* (which is saying something), so it was pleasing to find Chew's objective piece.

9. Dick Taverne, *The March of Unreason: Science, Democracy, and the New Fundamentalism* (Oxford: Oxford University Press, 2005), p. 53.

10. This is not to imply dishonesty. We all, very frequently, and often quite unconsciously, persuade ourselves that our preferred version of the past is the true one.

11. N. Andrews et al., "Recall Bias, MMR, and Autism," *Archives of Disease in Childhood* 87 (August 2002): 493–94. Abstract accessed at http://adc.bmj.com/content/87/6/493.abstract.

12. Brian Deer, "How the Case against the MMR Vaccine Was Fixed," *British Medical Journal*, January 5, 2011. Accessed at http://www.bmj.com/content/342/bmj.c5347.full. The second part of Deer's series, "How the Vaccine Crisis Was Meant to Make Money," published by the *British Medical Journal* on January 11, 2011, was accessed at http://www.bmj.com/content/342/bmj.c5258.full. The third part, "The *Lancet*'s Two Days to Bury Bad News," published by the *British Medical Journal* on January 18, 2011, was accessed at http://www.bmj.com/content/342/bmj.c7001.full.

13. Russell L. Blaylock, "Blaylock: Big Pharma Vilified Researcher for Threatening Vaccine Program," *Newsmax Health*, January, 13, 2011. Accessed at http://www.newsmaxhealth.com/health_stories/big_pharma_vaccine_autism/2011/01/13/371064.html.

14. Jenny McCarthy, "In the Vaccine-Autism Debate, What Can Parents Believe?" *Huffington Post*, January 10, 2011. Accessed at http://www.huffington-post.com/jenny-mccarthy/vaccine-autism-debate_b_806857.html.

15. Her other qualification to pronounce on medical matters is that she's the mother of an autistic child. This was her second medical diagnosis of his condition. Her first was that he was an Indigo Child—see her essay "Insights of an Indigo Mom" at http://childrenofthenewearth.com/free.php?page=articles_free/mccarthy_jenny/article1.

16. Accessed at http://web.archive.org/web/20010124042500/www.house.gov/burton/pr102600.htm. Thimerosal is a trade name for a substance more correctly called thiomersal. I've chosen to use the more familiar form.

17. See "About Dental Amalgam Fillings," US Food and Drug Administration. Accessed at http://www.fda.gov/medicaldevices/productsandmedicalprocedures/dentalproducts/dentalamalgam/ucm171094.htm.

18. "Mercury in Health Care," WHO Policy Paper, August 2005. Accessed at http://www.who.int/water_sanitation_health/medicalwaste/mercurypolpaper.pdf.

19. Quoted in Offit, *Autism's False Prophets*, p. 31.

20. Ibid., p. 34.

21. Miriam Falco, "10 Infants Dead in California Whooping Cough Outbreak," CNN, October 20, 2010. Accessed at http://www.cnn.com/2010/HEALTH/10/20/california.whooping.cough/index.html.

22. Center for Mind–Body Medicine, "Complimentary and Alternative Medicine (CAM) in Cancer Care: A Progress Report," June 11, 1989. Accessed at http://www.cmbm.org/mind_body_medicine_RESEARCH/1999-Transcripts/burton.html.

23. *Rolling Stone*'s copublisher of the article, the online magazine *Salon*, chose to leave the article in situ, complete with numerous errata and corrigenda that quite undermined Kennedy's argument. Finally, in January 2011, the magazine's editors decided the public might be better served if the article were deleted entirely.

24. Quoted in Offit, *Autism's False Prophets*, pp. 95–96.

25. Chelation therapy is used for treating heavy metal intoxication, i.e., poisoning by arsenic, lead, or mercury. Essentially, a compound (chelating agent) is introduced to the body that will eagerly bond to the metal molecule to form a harmless compound that the body can then flush away.

26. Complaint accessed at http://scienceblogs.com/insolence/upload/2010/01/suppression_of_speech_anti-vaccine_editi/FishervOffit1.pdf.

27. Mark R. Geier and David A. Geier, "Thimerosal in Childhood Vaccines, Neurodevelopment Disorders, and Heart Disease in the United States," *Journal of American Physicians and Surgeons* 8, no. 1 (Spring 2003): 6–11. Accessed at http://www.jpands.org/vol8no1/geier.pdf.

28. Stephanie Mencimer, "The Tea Party's Favorite Doctors," *Mother Jones*, November 18, 2009. Accessed at http://motherjones.com/politics/2009/11/tea-party-doctors-american-association-physicians-surgeons#comment-243833.

29. Arthur Krigsman et al., "Clinical Presentation and Histologic Findings at Ileocolonoscopy in Children with Autistic Spectrum Disorder and Chronic Gastrointestinal Symptoms," *Autism Insights*, January 27, 2010. Accessed at http://www.la-press.com/clinical-presentation-and-histologic-findings-at-ileo colonoscopy-in-ch-article-a1816.

30. "*Autism Insights*, Another Journal for Questionable Autism Research?" *Left Brain Right Brain*, January 30, 2010. Accessed at http://leftbrainrightbrain.co.uk/2010/01/autism-insights-another-journal-for-questionable-autism-research/. Wakefield and Jepson have since vanished from the list, while Krigsman, like Wakefield, has vanished from Thoughtful House.

31. Samantha Poling, "Doctors Warn over Homeopathic 'Vaccines,'" BBC, September 13, 2010. Accessed at http://www.bbc.co.uk/news/uk-scotland-11277990.

32. Simon Singh and Edzard Ernst, *Trick or Treatment: Alternative Medicine on Trial* (London: Bantam, 2008), pp. 184–85.

33. Kreesten Meldgaard Madsen et al., *New England Journal of Medicine* 347 (November 7, 2002): 1477–82. Accessed at http://www.nejm.org/doi/full/10.1056/NEJMoa021134#t=article.

34. "Nigeria: Kano State Resumes Polio Vaccinations after 10-Month Ban," *Irin*, August 2, 2004. Accessed at http://www.irinnews.org/report.aspx?reportid=50902.

35. Singh and Ernst, *Trick or Treatment*, p. 186.

36. "Vapositori in Climb-Down over Immunisation," *Zimbabwe Standard*, May 22, 2010. Accessed at http://www.thestandard.co.zw/local/24763-vapositori-in-climb-down-over-immunisation.html.

37. Christopher Wanjek, *Bad Medicine: Misconceptions and Misuses Revealed, from Distance Healing to Vitamin O* (Hoboken, NJ: Wiley, 2003), p. 198.

Chapter 7: The AIDS "Controversy"

1. Nathan D. Wolfe and Tony Goldberg, "HIV-1 Origins: What We Don't Know," *Cell Science Reviews* 3 (April 2007): 9–14.

2. "Discussing Swine Flu, Limbaugh Asserts, 'AIDS Was Going to Get Really Bad But It Didn't,'" *Media Matters*, April 28, 2009. Accessed at http://mediamatters.org/mmtv/200904280018.

3. Dean Powers, "'Polio Is Not a Virus' Rush Limbaugh Said on Radio Today," *OpEd News*, April 24, 2008. Accessed at http://www.opednews.com/articles/opedne_dean_pow_080424__22polio_is_not_a_viru.htm.

4. Anthony S. Fauci, Margaret I. Johnston, and Gary J. Nabel, "Statement on National HIV Vaccine Awareness Day," National Institute of Allergy and Infectious Diseases, National Institutes of Health, May 18, 2010. Accessed at http://www.niaid.nih.gov/news/newsreleases/2010/Pages/HVAD2010.aspx.

5. Azidothymidine, also called zidovudine, is a primary member of the spectrum of antiretroviral drugs used in AIDS treatment.

6. See, for example, "Christine Maggiore Died of AIDS," *Aids Truth*, March 9, 2009. Accessed at http://www.aidstruth.org/news/2009/christine-maggiore-died-aids.

7. Casper Schmidt, "The Group-Fantasy Origins of AIDS," *Journal of Psychohistory* (Summer 1984). Accessed at http://www.virusmyth.com/aids/hiv/csfantasy.htm.

8. Michael Specter, "The Denialists," *New Yorker*, March 12, 2007. Accessed at http://www.business.highbeam.com/410951/article-1G1-161266659/denialists.

9. See *TAC Electronic Newsletter*, May 9, 2006. Accessed at http://www.tac.org.za/community/node/2214.

10. As reported, for example, by the Global Health Council, "The Impact of HIV/AIDS." Accessed at http://www.globalhealth.org/hiv_aids/.

11. Damian Thompson, *Counterknowledge: How We Surrendered to Conspiracy Theories, Quack Medicine, Bogus Science, and Fake History* (New York: W. W. Norton & Company, 2008), pp. 129–30.

12. Nancy Padian et al., "Heterosexual Transmission of Human Immunodeficiency Virus (HIV) in Northern California: Results from a Ten-Year Study," *American Journal of Epidemiology* 146 (1997). Accessed at http://www.aidstruth.org/denialism/misuse/padian.

13. Hank Barnes, "The Padian Waffle!" *You Bet Your Life*, August 10, 2006. Accessed at http://barnesworld.blogs.com/barnes_world/2006/08/more_on_african.html.

14. Barnes's site carries the note: "Comments are regarded as letters to the editor. They are subject to the same policies as the *NY Times* and *Nature*, and are not published until after editorial review." Unsurprisingly, voices dissenting from Barnes's views seem noticeably absent.

15. Michael Worobey et al., "Contaminated Polio Vaccine Theory Refuted," *Nature* 428 (April 22, 2004): 820. Accessed at http://www.uow.edu.au/~/bmartin/dissent/documents/AIDS/Worobey04.pdf.

16. Orac, "The Strange Science and Ethics of the Anti-Vaccine Movement," *Respectful Insolence*, November 2, 2010. Accessed at http://scienceblogs.com/insolence/2010/11/the_strange_science_and_ethics_of_the_an.php.

17. Jake Crosby, "How Vaccine Damage Deniers Threaten Us All," *Age of Autism*, October 28, 2010. Accessed at http://www.ageofautism.com/2010/10/how-vaccine-damage-deniers-threaten-us-all.html.

18. Meera Nanda, "Postmodernism, Hindu Nationalism, and 'Vedic Science,'" *Frontline*, December 20, 2003. Accessed at http://www.mukto-mona.com/Articles/vedic_science_Mira.htm.

19. "No Charges against Ubhejane Seller—Court," *IOL News*, March 21, 2006. Accessed at http://www.iol.co.za/news/south-africa/no-charges-against-ubhejane-seller-court-1.270171.

20. See http://www.robertogiraldo.com/.

21. "President Jammeh Gives Ultimatum for Homosexuals to Leave," *Gambia News*, May 15, 2008. Accessed at http://www.gambianow.com/news/News/Gambia-News-President-Jammeh-Gives-Ultimatum-for-Homosexuals-to-.html.

22. "The Gambia Country Profile," BBC, last updated October 7, 2010. Accessed at http://news.bbc.co.uk/2/hi/africa/country_profiles/1032156.stm#leaders.

23. "President's 'HIV Cure' Condemned," BBC, February 2, 2007. Accessed at http://news.bbc.co.uk/2/hi/africa/6323449.stm.

24. Accessed at http://www.ion.ac.uk/about_ION.htm.

25. Ben Goldacre, "Vitamin Deficiency," *Guardian*, January 6, 2005. Accessed at http://www.guardian.co.uk/science/2005/jan/06/badscience.science.

26. Accessed at http://www.hootervillegazette.com/mofb.html.

27. For a detailed analysis of the AIDS-denialism situation in South Africa, see Nicoli Nattrass's *Mortal Combat: AIDS Denialism and the Struggle for Antiretrovirals in South Africa* (Scottsville, South Africa: University of KwaZulu-Natal Press, 2007).

28. "Remarks by President Thabo Mbeki at the First Meeting of the Presidential Advisory Panel on AIDS," *South African Government Information*, May 6, 2000. Transcript accessed at http://www.info.gov.za/speeches/2000/0005311255p1003.htm.

29. Paranoid AIDS conspiracy theories along these lines are far from uncommon elsewhere in the world.

30. "The Durban Declaration," *Nature*, July 6, 2000. Accessed at http://www.nature.com/nature/journal/v406/n6791/full/406015a0.html.

31. Mark Schoofs, "Tanzanian Military Helped Company Skirt Drug Regulations to Test Virodene," *Wall Street Journal*, July 19, 2001. Accessed at http://www.aegis.com/news/wsj/2001/WJ010704.html.

Chapter 8: Selfish Help

1. "Secret Life of Fugitive Karadzič," BBC, July 25, 2008. Accessed at http://news.bbc.co.uk/2/hi/7520661.stm.

2. It's also a movie. When I tried to watch this, however, the universe read my innermost desires. My DVD player refused to function.

3. "ASIC Bans Victorian Man from Providing Financial Services for Life," Australian Securities and Investments Commission, June 23, 2010. Accessed at http://tinyurl.com/4q7tcyp.

Chapter 9: Dissent about Descent

1. Colleen Thomas, "Pleadian Shipmates," *Wild Colleen*, September 18, 2010. Accessed at http://wildcolleen.com/blog/09-18-2010/pleadian-shipmates.

2. Colleen Thomas, "The Next Paradigm Shift," *Intellectually Honest Science*. Accessed at http://www.intellectuallyhonestscience.com/.

3. Opinions vary as to whether Kammerer himself was the faker.

4. For all its anachronistic notions, the book remains surprisingly interesting.

It can be found free online at Project Gutenberg or at Electronic Scholarly Publishing. Its sequel, *Explanations: A Sequel to "Vestiges of the Natural History of Creation" By the Author of That Work* (1845) can be found at http://www.darwin-online.org.

5. This is a point much ignored by religious Creationists in their campaigns to demonize Darwin.

6. Whenever I hear this, I wonder about human beings evolving into idiots.

7. According to pangenesis, which Darwin first put forward in *The Variation of Animals and Plants under Domestication* (1868), body cells are constantly shedding entities called gemmules. These collect in the sex cells. Thus, every cell of the bodies of both parents has an influence on the offspring of the union of the parents' sex cells.

8. Quoted in Peter Stiles, "Darwin and Philip Henry Gosse," Darwin 200 in Devon. Accessed at http://www.devonhumanists.org.uk/d200dev/?page_id=703.

9. The term "genetics" was coined by the Cambridge biologist William Bateson, whose team did much to establish the new science around the turn of the century.

10. Thomas Bell, Presidential Address given May 24, 1859, *Proceedings of the Linnean Society* (1859), p. viii.

11. Pope John Paul II, "Truth Cannot Contradict Truth," *Address of Pope John Paul II to the Pontifical Academy of Sciences*, October 22, 1996. Accessed at http://www.newadvent.org/library/docs_jp02tc.htm.

12. Christoph Schönborn, "Finding Design in Nature," *New York Times*, July 7, 2005. Accessed at http://www.nytimes.com/2005/07/07/opinion/07schonborn.html.

13. Fr. George V. Coyne, SJ, "Science Does Not Need God, or Does It? A Catholic Scientist Looks at Evolution," Palm Beach Atlantic University, West Palm Beach, Florida, January 31, 2006. Accessed at http://www.catholic.org/national/national_story.php?id=18504.

14. Conservapedia, "Evolutionists Who Have Had Problems with Being Overweight and/or Obese." Accessed at http://www.conservapedia.com/Evolutionists_who_have_had_problems_with_being_overweight_and/or_obese.

15. Conservapedia, "Atheism and Obesity." Accessed at http://www.conservapedia.com/Atheism_and_obesity.

16. Conservapedia, "Essay: Does Richard Dawkins Have Machismo?" Accessed at http://www.conservapedia.com/Essay:_Does_Richard_Dawkins_have_machismo%3F.

17. "Facts FTW," *Failbook: Too Funny to Unfriend*. Accessed at http://failbook.failblog.org/2010/12/22/funny-facebook-fails-facts-ftw/. The various typos, etc., are *sic*.

18. Henry Morris, "Evolution, Thermodynamics, and Entropy," Institute for Creation Research, 1973. Accessed at http://www.icr.org/article/evolution-thermodynamics-entropy/. Among other papers generously placed online by the Institute

for Creation Research, a favorite is 1985's "Oceans of Piffle in Evolutionary Indoctrination" by Thomas G. Barnes (http://www.icr.org/article/oceans-piffle-evolutionary-indoctrination/).

19. Carl E. Baugh, "Creation Model Session 8," Creation Evidence Museum. Accessed at http://www.creationevidence.org/index.php?option=com_content&task=view&id=66&Itemid=10 on October 26, 2010.

20. Jim Gardner, "Kent Hovind's Dissertation" (blog post), *How Good is That?* December 12, 2009. Accessed at http://howgoodisthat.wordpress.com/2009/12/12/kent-hovinds-dissertation/.

21. "Saturday hate mail-a-palooza," *Daily Kos*, October 30, 2010. Accessed at http://www.dailykos.com/story/2010/10/30/915134/-Saturday-hate-mail-a-palooza.

22. David W. Cloud, "Pope Supports Evolution," Fundamental Baptist Information Service, October 24, 1996. Accessed at http://theonemediator.com/Catholicism/Evolution/pope_supports_evolution.htm.

23. Quoted by Dan Amira in "GOP's Delaware Senate Nominee Christine O'Donnell Not a Big Fan of Evolution," *New York Magazine*, October 15, 2010, Accessed at http://nymag.com/daily/intel/2010/09/the_gops_delaware_senate_nomin.html.

24. Harun Yahya, "Darwinism Is the Main Source Of Racism," *Harun Yahya: An Invitation to the Truth*, February 1, 2009. Accessed at http://us1.harunyahya.com/Detail/T/EDCRFV/productId/12529. Capitalization and punctuation *sic*.

25. Yes, I know, it seems odd a Muslim should be using this as an insult.

26. Oktar got his revenge in 2008 by persuading the Turkish authorities to ban Dawkins's website from Turkish access. This was part of a campaign by Oktar to have websites that criticized him banned by the courts. It is to the profound shame of those courts that they granted his requests. Among those blocked has been, since 2007, the entirety of WordPress.

27. Quoted in Dorian Jones, "Evolution Under Pressure," *Qantara.de*, August 1, 2006. Accessed at http://www.qantara.de/webcom/show_article.php/_c-478/_nr-478/i.html.

28. Quoted in "Turkey Will Be Entirely Cleansed of Belief in Evolution!" *Harun Yahya*, May 23, 2007. Accessed at http://www.harunyahyaimpact.com/haberDetay.php?haberId=229.

29. Quoted in Sevim Songün, "Turkey Evolves as Creationist Center" *Hurriyet Daily News*, March 28, 2009.

30. Mark Perakh, "Not a Very Big Bang about *Genesis*," *Talk Reason*, 1999 (revised 2001). Accessed at http://www.talkreason.org/articles/schroeder.cfm. I've relied on Perakh's essay for this discussion of Schroeder's hypotheses.

31. Meera Nanda, "Postmodernism, Hindu Nationalism, and 'Vedic Science,'" *Frontline*, December 20, 2003. Accessed at http://www.mukto-mona.com/Articles/vedic_science_Mira.htm.

Chapter 10: We're (Badly) Designed

1. Michael Powell, "Doubting Rationalist: 'Intelligent Design' Proponent Phillip Johnson, and How He Came to Be," *Washington Post*, May 15, 2005. Accessed at http://www.washingtonpost.com/wp-dyn/content/article/2005/05/14/AR2005051401222.html.

2. William Paley, *Natural Theology, or Evidences of the Existence and Attributes of the Deity* (London: Richardson & Co., 1821), pp. 9–10.

3. Sir Isaac Newton, "A Short Schem[e] of the True Religion," Keynes MS 7, King's College, Cambridge, UK. Accessed at http://www.newtonproject.sussex.ac.uk/texts/viewtext.php?id=THEM00007&mode=normalized.

4. Michael Behe, *Darwin's Black Box* (New York: Free Press, 1996), p. 97.

5. Quoted in Adel Ziadat, *Western Science in the Arab World: The Impact of Darwinism, 1860–1930* (New York: St. Martin's, 1986), p. 100.

6. Phillip E. Johnson, "The Intelligent Design Movement: Challenging the Modernist Monopoly on Science," in *Signs of Intelligence: Understanding Intelligent Design*, ed. William A. Dembski and James M. Kushiner (Ada, MI: Brazos Press, 2001).

7. Neil Ormerod, "How Design Supporters Insult God's Intelligence," *Sydney Morning Herald*, November 15, 2005. Accessed at http://www.smh.com.au/news/opinion/how-design-supporters-insult-gods-intelligence/2005/11/14/1131951095200.html.

8. Quoted in David L. Allen, "A Reply to Tom Nettles' Review of William A. Dembski's *The End of Christianity: Finding a Good God in an Evil World,*" Center for Theological Research, February 2010. Accessed at http://www.baptisttheology.org/documents/AReplytoTomNettlesReviewofDembskisTheEndofChristianity.pdf.

9. Michelangelo D'Agostino, "In the Matter of Berkeley v. Berkeley," *Berkeley Science Review* 10 (Spring 2006): 35. Accessed at http://sciencereview.berkeley.edu/articles/issue10/evolution.pdf.

Chapter 11: No Safe Classroom?

1. State Senator Josh Brecheen, "Brecheen Discusses Evolution and Darwinian Theory," *Durant Daily Democrat*, December 19, 2010. Accessed at http://www.durantdemocrat.com/view/full_story/10717736/article-Brecheen-discusses-evolution-and-Darwinian-Theory?instance=secondary_opinion_left_column.

2. Not surprisingly, "annettejohnson" removed her comment in the face of

ridicule, or whyever; but at least as late as July 2011, when I checked, its ghost still existed on Google's cached version of the page.

3. "Public Acts of the State of Tennessee Passed by the Sixty-Fourth General Assembly, 1925," chap. 27, House Bill 185, 1925. Accessed at http://www.law.umkc.edu/faculty/projects/ftrials/scopes/tennstat.htm.

4. The results here were beyond their dreams. The attention was international, not just national; and Dayton is *still* milking the trial as a tourist attraction.

5. The play was inspired not so much by the actual trial as by the simplistic account of it given in Frederick Lewis Allen's *Only Yesterday: An Informal History of the Nineteen-Twenties* (1931). According to Allen (quoted by Edward J. Larson in *Summer for the Gods* [New York: Basic Books, 1997], p. 226), "In the eyes of the public, the trial was a battle between Fundamentalists on the one hand and twentieth-century skepticism (assisted by Modernism) on the other." As Larson notes, "The defense's fight for individual liberty and the prosecution's appeal to majoritarianism disappeared from Allen's version of events. . . ."

6. H. L. Mencken, "The Scopes Trial—Aftermath," *Baltimore Evening Sun*, September 14, 1925. Accessed at http://www.positiveatheism.org/hist/menck 05.htm.

7. The trial documents, including a full transcript, are available online at http://www.talkorigins.org/faqs/dover/kitzmiller_v_dover.html.

8. "The Clergy Letter—from American Christian Clergy—An Open Letter Concerning Religion and Science," *Evolution Weekend: The Clergy Letter Project*. Accessed July 3, 2010, at http://blue.butler.edu/~mzimmerm/Christian_Clergy/ChrClergyLtr.htm.

9. *Institute for Creation Research Graduate School v. Texas Higher Education Coordinating Board*. The judgment is available online at http://courtweb.pamd.uscourts.gov/courtwebsearch/txwd/06454397.pdf.

Chapter 12: Evilution

1. Kenneth M. Pierce, J. Madeleine Nash, and D. L. Coutu, "Education: Putting Darwin Back in the Dock," *Time*, May 16, 1981. Accessed at http://www.time.com/time/magazine/article/0,9171,922471-5,00.html#ixzz0mt5lNQ7h.

2. Brock Lee, "Pushing Teens into Being Sexually Active," *Owatanna People's Press*, July 10, 2010. Accessed at http://www.owatonna.com/news.php?viewStory=118653.

3. Max Nordau, "The Philosophy and Morals of War," *North American Review* 169 (1889): 794.

4. I was amused to discover, when checking in May 2010, that amazon.co.uk classifies this book as fiction. And quite right, too.

5. Although not as long before as we might think. While apologists for color prejudice assure us it's hardwired into the human psyche, etc., it appears to be a relatively recent phenomenon. The ancient Greeks and Romans, while being highly aware of the *culture* from which a person came—and they could certainly be racist in that respect, particularly the Greeks—regarded skin color as no more important than eye color or hair color.

Chapter 13: Eugenically Speaking

1. He did so by 1883; before then he had used the word "viriculture," by analogy with "agriculture."

2. Francis Galton, "Hereditary Character and Talent," *Macmillan's Magazine* 12 (November 1864 and April 1965): 157–66 and 318–27; this extract is from pp. 165–66.

3. Quoted in Jonathan Spiro, *Defending the Master Race: Conservation, Eugenics, and the Legacy of Madison Grant* (Lebanon, NH: University of Vermont Press, 2008), p. 158.

4. Walter Lippmann, "Tests of Hereditary Intelligence," *New Republic*, November 22, 1922, p. 330.

5. *Official Proceedings of the Second National Conference on Race Betterment,* Volume II (1915). Cited by Steven Selden in *Inheriting Shame: The Story of Eugenics and Racism in America* (New York: Teachers College Press, 1999), p. 11.

6. Quoted in Harry Hamilton Laughlin, *Eugenical Sterilization in the United States* (Chicago: Psychopathic Laboratory of the Municipal Court of Chicago, 1922), p. 22.

7. *Buck v. Bell*, 274 US 200 (1927).

Chapter 14: Social Darwinism

1. Cited widely; the earliest version I could find was (incomplete) in William James Ghent's *Our Benevolent Feudalism* (1903; reprint, Whitefish, MT: Kessinger Publishing, 2010), p. 29.

2. When Darwin first used the expression, in the 5th (1869) edition of *Origin of Species*, it was at the urging of Wallace, who'd explained he was unhappy with the term "natural selection."

3. It's not coincidental that the phrase "reciprocal altruism" has become a popular term in academia to describe what you or I would call cooperation.

4. Quoted in M. J. Savage, "Professor Swing and Herbert Spencer," *Weekly Magazine of Chicago*; reprinted in *Unity* 13, no. 4 (March 1884).

5. William James, *Memories and Studies* (1912; reprint, Charleston, SC: BiblioBazaar, 2007), p. 112.

6. William Graham Sumner, *The Challenge of Facts and Other Essays* (1914; reprint, Charleston, SC: BiblioBazaar, 2009), p. 57.

CHAPTER 15: IT'S THE ECOLOGY, STUPID

1. Ann Coulter, "Global Warming: The French Connection," *WorldNetDaily*, May 28, 2003. Accessed at http://www.wnd.com/index.php?pageId=19019.

2. "Don Young: Gulf Spill 'Not an Environmental Disaster,'" *Anchorage Daily News*, June 2, 2010. Accessed at http://www.adn.com/2010/06/02/1304209/don-young-gulf-oil-spill-not-an.html.

3. Brook R. Corwin, "Independents Take Center Stage in Obama Era: Section 9: The Environment and the Economy," Pew Research Center for the People and the Press. Accessed at http://brookcorwin.com/environment/researchpaper.docx.

4. Reported, like some of the other information here, by Julia Whitty, in "The Fate of the Ocean," *Mother Jones*, March/April 2006. Accessed at http://motherjones.com/politics/2006/03/fate-ocean.

5. Although the relationship is not direct, CO_2 being less soluble in warmer water.

6. See, for example, Elizabeth Kolbert, "The Acid Sea," *National Geographic*, April 2011. Accessed at http://ngm.nationalgeographic.com/2011/04/ocean-acidification/kolbert-text.

7. Although not all have yet signed up to all of its various revisions, in 1990, 1991, 1992, 1993, 1995, 1997, and 1999.

8. Quoted in "DuPont: A Case Study in the 3D Corporate Strategy," Greenpeace Position Paper, September 1997. Accessed at http://archive.greenpeace.org/ozone/greenfreeze/moral97/6dupont.html.

9. Quoted in "Saving the Ozone Layer: The Montreal Protocol and Moral Priorities—A Greenpeace Position Paper," Greenpeace, September 1997. Accessed at http://archive.greenpeace.org/ozone/greenfreeze/moral97/index.html.

10. "Join the Discussion," *American Thinker*, December 19, 2010. Both accessed at http://comments.americanthinker.com/read/42323/731481.html.

CHAPTER 16: SO, WHAT WAS THE WEATHER LIKE IN 2010?

1. I'm indebted for much of the information here to the NOAA National Climatic Data Center of the US Department of Commerce. See http://lwf.ncdc.noaa.gov/oa/reports/weather-events.html#2010. Where figures differ from the NOAA ones they're from more detailed reports and/or represent updates.

2. Quoted in Matt Corley, "Utah State Representative Claims Climate Change Is a 'Conspiracy' Aimed at Population Control," *Alternet*, February 7, 2010. Accessed at http://blogs.alternet.org/speakeasy/2010/02/07/utah-state-representative-claims-climate-change-is-a-conspiracy-aimed-at-population-control/.

3. Mark Kinver, "Food Chains Disrupted by Earlier Arrival of Spring," BBC News, February 9, 2010. Accessed at http://news.bbc.co.uk/2/hi/science/nature/8506363.stm.

4. National Science Foundation press release, "Methane Releases from Arctic Shelf May Be Much Larger and Faster Than Anticipated," National Science Foundation, March 4, 2010. Accessed at http://www.nsf.gov/news/news_summ.jsp?cntn_id=116532&org=NSF&from=news.

5. Christopher Skinner, Arthur DeGaetano, and Brian Chabot, "Implications of Twenty-First Century Climate Change on Northeastern United States Maple Syrup Production: Impacts and Adaptations," *Climatic Change* 100, no. 3 (June 1, 2010).

6. Nurfika Osman, "'Super-Extreme' Weather Is the Worst on Record," *Jakarta Globe*, August 19, 2010. Accessed at http://www.thejakartaglobe.com/news/super-extreme-weather-is-the-worst-on-record/391736.

7. Curtis Brainard, "Temperate Coverage of Extreme Weather," *Columbia Journalism Review*, August 12, 2010. Accessed at http://www.cjr.org/the_observatory/temperate_coverage_of_extreme.php.

8. Quoted in Justin Gillis, "In Weather Chaos, a Case for Global Warming," *New York Times*, August 14, 2010. Accessed at http://www.nytimes.com/2010/08/15/science/earth/15climate.html.

9. Quoted in Suzanne Bohan, "State Nursery Falls Under Budget Ax," *Contra Costa Times*, October 17, 2010. Accessed at http://www.contracostatimes.com/top-stories/ci_16350405?nclick_check=1.

10. National Oceanic and Atmospheric Administration, "State of the Climate: Global Hazards, September 2010." National Climate Data Center, September 2010. Accessed at http://www.ncdc.noaa.gov/sotc/?report=hazards&year=2010&month=9&submitted=Get+Report.

11. Stephen Leahy, "Arctic Ice in Death Spiral," *IPS News*, September 20, 2010. Accessed at http://ipsnews.net/news.asp?idnews=52896.

12. "Severe Drought Afflicts Brazilian Amazon," BBC, October 22, 2010. Accessed at http://www.bbc.co.uk/news/world-latin-america-11610382.

13. John M. Broder, "Climate Change Doubt Is Tea Party Article of Faith," *New York Times*, October 20, 2010.

14. Roger Harrabin, "Rich Nations 'Failing to Deliver Climate Cash,'" BBC, October 8, 2010. Accessed at http://www.bbc.co.uk/news/science-environment-11502019.

15. Roger Harrabin, "Met Office says 2010 'Among Hottest on Record,'" BBC, November 26, 2010. Accessed at http://www.bbc.co.uk/news/uk-11841368. A complaint to the BBC drew no response.

16. Associated Press, "2 People Die in Severe Floods in Southwestern Belgium," November 14, 2010. Accessed at http://www.google.com/hostednews/canadianpress/article/Aleqm5gplb82e8efacd5w1fr2s5cD-Tmoq?docId=5131138.

17. Oxfam International, "Now More Than Ever: Climate Talks That Work for Those Who Need Them Most." Oxfam International Media Briefing, November 2010. Accessed at http://www.oxfam.org/sites/www.oxfam.org/files/oxfam-cancun-media-briefing-2010.pdf.

18. "Israel Fire Near Haifa Kills Dozens of Prison Guards," BBC, December 2, 2010. Accessed at http://www.bbc.co.uk/news/world-middle-east-11901750.

19. Jane Qiu, "Global Warming May Worsen Locust Swarms," *Nature News*, October 7, 2009. Accessed at http://www.nature.com/news/2009/091007/full/news.2009.978.html.

20. Michael Heath and Angus Whitley, "Australian Floods Prompt New South Wales Disaster Declaration, Crop Alert," Bloomberg, December 3, 2010. Accessed at http://www.bloomberg.com/news/2010-12-04/australian-floods-prompt-new-south-wales-disaster-declaration-crop-alert.html.

21. "Australia Swaps Summer for Christmas Snow," AFP, December 20, 2010. Accessed at http://www.google.com/hostednews/afp/article/Aleqm5jv1dn_Camrxqckewrmczpkryjpp A?docId=CNG.fa1b2905c40572e9934b2e3a6b52d6f4.611.

22. Quoted in "Pastor Says Kevin Rudd to Blame for Floods," *News.Com.Au*, January 11, 2011. Accessed at http://www.news.com.au/breaking-news/pastor-says-kevin-rudd-to-blame-for-floods/story-e6frfku0-1225985730895#ixzz1ApUe8RVi .

23. "Pressing the Silence: At the U.N. Climate Change Conference, the Media Center Is Oddly Quiet," *Democracy Now!*, December 6, 2010. Accessed at http://www.democracynow.org/2010/12/6/pressing_the_silence_at_the_un.

24. Daniel G. Boyce et al., "Global Phytoplankton Decline over the Past Century," *Nature*, July 29, 2010. A good summary can be found in Steve Connor, "The Dead Sea: Global Warming Blamed for 40 Per Cent Decline in the Ocean's Phytoplankton," *Independent*, July 29, 2010. Accessed at http://ww.w.independent.co.uk/environment/climate-change/the-dead-sea-global-warming-blamed-for-40-per-cent-decline-in-the-oceans-phytoplankton-2038074.html.

25. A few months earlier, the election of a Coalition government in the UK

brought to Parliament many new Conservative Party MPs who denied climate change. Luckily, their leadership had more sense.

26. Neela Banerjee, "GOP Plans Attacks on the EPA and Climate Scientists," *Los Angeles Times*, October 30, 2010. Accessed at http://articles.latimes.com/2010/oct/30/nation/la-na-epa-battle-ahead-20101030.

27. Ronald Brownstein, "GOP Gives Climate Science a Cold Shoulder," *National Journal*, October 9, 2010. Accessed at http://nationaljournal.com/columns/political-connections/gop-gives-climate-science-a-cold-shoulder-20101009.

28. Mireya Navarro, "States Diverting Money from Climate Initiative," *New York Times*, November 28, 2010. Accessed at http://www.nytimes.com/2010/11/29/nyregion/29greenhouse.html.

29. Bruce Usher, "On Global Warming, Start Small," *New York Times*, November 27, 2010. Accessed at http://www.nytimes.com/2010/11/28/opinion/28usher.html.

30. Cathal Kelly, "God Will Save Us from Climate Change: U.S. Representative," *Toronto Star*, November 14, 2010. Accessed at http://www.thestar.com/news/world/article/888472–god-will-save-us-from-climate-change-u-s-representative.

31. Andrew Restuccia, "Shimkus' Greatest Hits: Climate Change Edition," *Washington Independent*, November 9, 2010. Accessed at http://washingtonindependent.com/103079/shimkus-greatest-hits-climate-change-edition.

32. Douglas Fischer, "2010 in Review: The Year Climate Coverage 'Fell Off the Map,'" *Daily Climate*, January 3, 2011. Accessed at http://wwwp.dailyclimate.org/tdc-newsroom/2011/01/climate-coverage.

Chapter 17: Global Weirding

1. Peter Hitchens, "God Help Us All if We Get these Lawless EU Robocops," *Mail on Sunday*, July 29, 2001. Accessed at http://www.dailymail.co.uk/debate/columnists/article-117626/God-help-lawless-EU-robocops.html#ixzz1Ce1e1vRs.

2. Except that it's not quite that simple—nothing ever is in climate science. The increased levels of atmospheric water vapor are likely to result in greater cloud cover. The cloudier the world is, the less infrared gets through to the surface to take part in the greenhouse effect. On the other hand, the less likely it becomes that any heat can escape from *beneath* the cloud layer. There's a balance point here somewhere, but no one as yet is sure how to calculate it.

3. Environment News Service, "Climate Change Could Bankrupt World by 2065," *Albion Monitor* 82, December 2000. Accessed at http://www.albionmonitor.com/0012a/climatebankrupt.html.

4. Energy Information Administration, "Per Capita Total Carbon Dioxide Emis-

sions from the Consumption of Energy, Most Countries, 1980–2006," *International Energy Annual 2006*. Accessed at http://www.eia.doe.gov/emeu/international/carbondioxide.html.

5. Quoted in Fred Krupp and Miriam Horn, *Earth: The Sequel* (New York: Norton, 2008), p. 11.

6. Quoted in part 1 of Elizabeth Kolbert, "The Climate of Man," *New Yorker*, April 25, 2005. Accessed at http://www.newyorker.com/archive/2005/04/25/050425fa_fact3.

7. National Research Council, *Advancing the Science of Climate Change* (Washington, DC: National Academies Press, 2010). Executive summary accessed at http://dels.nas.edu/Report/Advancing-Science-Climate-Change/12782.

8. Pew Research Center, "Little Change in Opinions about Global Warming: Increasing Partisan Divide on Energy Policies," Pew Research Center for the People and the Press, 2010. Accessed at http://people-press.org/report/669/.

9. Naomi Oreskes, "The Scientific Consensus on Climate Change," *Science*, December 3, 2004 Accessed at http://www.sciencemag.org/content/306/5702/1686 .full.

10. William R. L. Anderegg, James W. Prall, Jacob Harold, and Stephen H. Schneider, "Expert Credibility in Climate Change," *Proceedings of the National Academy of Sciences*, April 9, 2010. Abstract accessed at http://www.pnas.org/content/early/2010/06/22/1003187107.abstract.

11. Joanne Nova, "PNAS: Witchdoctors of Science," *JoNova*, June 23, 2010. Accessed at http://joannenova.com.au/2010/06/pnas-witchdoctors-of-science/. With no apparent sense of irony, Nova adds that "the article is tagged with '*Climate Denier*.' In doing so, the NAS officially steps across that ugly line into outright name-calling."

12. John H. Mercer, "West Antarctic Ice Sheet and CO_2 Greenhouse Effect: A Threat of Disaster," *Nature*, January 26, 1978. Abstract accessed at http://www.nature.com/nature/journal/v271/n5643/abs/271321a0.html.

13. Fen Montaigne, "The Warming of Antarctica: A Citadel of Ice Begins to Melt," *Yale Environment 360*, November 22, 2010. Accessed at http://e360.yale.edu/feature/the_warming_of_antarctica_a_citadel_of_ice_begins_to_melt_/2342/.

14. Mark Hawthorne, "Rains Place State at Risk of Mosquito-Borne Diseases," *The Age*, January 17, 2011. Accessed at http://www.theage.com.au/environment/weather/rains-place-state-at-risk-of-mosquitoborne-diseases-20110116-19sji.html.

15. Paul Jay, "The Beetle and the Damage Done," CBC, April 23, 2008. Accessed at http://www.cbc.ca/news/background/science/beetle.html.

16. These figures from the 2009 Quadrennial Fire Review are cited in Frances Beinecke, *Clean Energy Common Sense*, with Bob Deans (Lanham, MD: Rowman & Littlefield, 2010), pp. 24–26.

17. World Health Organization, "Climate and Health," World Health Organiza-

tion, fact sheet, July 2005. Accessed at http://www.who.int/globalchange/news/fsclimandhealth/en/index.html.

18. Quoted in Beinecke, *Clean Energy Common Sense*, p. 34.

19. Hans Joachim Schellnhuber, Wolfgang Cramer, Nebojsa Nakicenovic, Tom Wigley, and Gary Yohe, eds., *Avoiding Dangerous Climate Change* (Cambridge: Cambridge University Press, 2006), pp. 155–61.

20. For more, see the Royal Society's press release, "Impact of Climate Change on Crops Worse than Previously Thought," Royal Society, 2005. Accessed at http://royalsociety.org/General_WF.aspx?pageid=7317&terms=.

21. "Rising CO_2 Levels Threaten Crops and Food Quality," *UC Davis News and Information*, May 13, 2010. Accessed at http://www.news.ucdavis.edu/search/news_detail.lasso?id=9479.

22. Steve Connor, "Climate Emails Hacked by Spies," *Independent*, February 1, 2010.

23. Richard Girling, "The Leak Was Bad. Then Came the Death Threats," *Sunday Times*, February 7, 2010.

24. K. R. Briffa et al., "Reduced Sensitivity of Recent Tree-Growth to Temperature at High Northern Latitudes," *Nature*, February 12, 1998. Abstract accessed at http://www.nature.com/nature/journal/v391/n6668/abs/391678a0.html.

25. Perhaps because of increased CO_2 reducing the trees' ability to process nitrates?

26. James Delingpole, "Climategate: The Final Nail in the Coffin of 'Anthropogenic Global Warming'?" *Daily Telegraph*, November 20, 2009. Accessed at http://blogs.telegraph.co.uk/news/jamesdelingpole/100017393/climategate-the-final-nail-in-the-coffin-of-anthropogenic-global-warming/.

27. Kate Ravilious, "Hacked Email Climate Scientists Receive Death Threats," *Guardian*, December 8, 2009. A selection of the threats is available online at http://www.guardian.co.uk/environment/2010/jul/06/hacked-climate-science-emails-sceptics-abuse.

28. *Government Response to the House of Commons Science and Technology Committee 8th Report of Session 2009–10: The Disclosure of Climate Data from the Climatic Research Unit at the University of East Anglia, Presented to Parliament by the Secretary of State for Energy and Climate Change by Command of Her Majesty* (London: Stationery Office, September 2010). Accessed at http://www.official-documents.gov.uk/document/cm79/7934/7934.pdf.

29. Leo Hickman, "US Climate Scientists Receive Hate Mail Barrage in Wake of UEA Scandal," *Guardian*, July 5, 2010. Accessed at http://www.guardian.co.uk/environment/2010/jul/05/hate-mail-climategate.

30. Henry C. Foley, Alan W. Scaroni, and Candice A. Yekel, "Concerning the Allegations of Research Misconduct against Dr. Michael E. Mann, Department of Meteorology, College of Earth and Mineral Sciences, The Pennsylvania State Univer-

sity," *RA-10 Inquiry Report*, preamble, Pennsylvania State University, February 3, 2010. Accesssed at http://www.research.psu.edu/orp/Findings_Mann_Inquiry.pdf .

31. Newsvine, "Are You Satisfied with the British Panel's Conclusion That While 'Climategate' Scientists Were Not Always Forthcoming, Their Science Was Sound?" July 7, 2010. Accessed December 2, 2010 at http://tinyurl.com/4rr6nd4.

32. Andrew Bolt, "All Clear, Expect [*sic*] for One or Two Big Exaggerations," *Herald Sun*, April 16, 2010. Accessed at http://blogs.news.com.au/heraldsun/andrewbolt/index.php/heraldsun/comments/all_clear_expect_for_one_or_two_big_exaggerations/.

33. Scott Mandia, "Climategate Coverage: Unfair & Unbalanced," *Prof-Mandia*, April 18, 2010. Accessed at http://profmandia.wordpress.com/2010/04/18/climategate-coverage-unfair-unbalanced/.

34. The difference in the periods covered reflects restrictions in Google's advanced search facilities.

35. Editorial, "A Climate Change Corrective," *New York Times*, July 9, 2010. Accessed at http://www.nytimes.com/2010/07/11/opinion/11sun2.html.

36. Quoted in Brad Johnson, "Stumped by Science: Michele Bachmann Calls CO_2 'Harmless,' 'Negligible,' 'Necessary,' 'Natural,'" *Think Progress*, Wonkroom, April 24, 2009. Accessed at http://wonkroom.thinkprogress.org/2009/04/24/bachmann-harmless-co2/.

37. Michael E. Mann, Raymond S. Bradley, and Malcolm K. Hughes, "Global-Scale Temperature Patterns and Climate Forcing over the Past Six Centuries," *Nature*, April 23, 1998.

38. Michael E. Mann, Raymond S. Bradley, and Malcolm K. Hughes, "Northern Hemisphere Temperature During the Past Millennium: Inferences, Uncertainties, and Limitations," *Geophysical Research Letters* 26, no. 6 (1999).

39. A further paper by Michael E. Mann et al., "Proxy-Based Reconstructions of Hemispheric and Global Surface Temperature Variations over the Past Two Millennia," *Proceedings of the National Academy of Sciences*, September 9, 2008, extended the coverage to 2000 years.

40. National Research Council, *Surface Temperature Reconstructions for the Last 2,000 Years* (Washington, DC: National Academies Press, 2006).

41. "Steve McIntyre and Ross McKitrick, Part 2: The Story Behind the Barton–Whitfield Investigation and the Wegman Panel," *Deep Climate*, February 8, 2010. Accessed at http://deepclimate.org/2010/02/08/steve-mcintyre-and-ross-mckitrick-part-2-barton-wegman/.

42. Dan Vergano, "Experts Claim 2006 Climate Report Plagiarized," *USA Today*, November 22, 2010. Accessed at http://www.usatoday.com/weather/climate/globalwarming/2010-11-21-climate-report-questioned_N.htm.

43. John R. Mashey, "Strange Inquiries at George Mason University," *DeSmog*

Blog, December 13, 2010. Accessed at http://www.desmogblog.com/sites/beta.desmogblog.com/files/strange%20inquiries%20v1%200.pdf.

44. John R. Mashey, "Strange Scholarship in the Wegman Report: A Facade for the Climate Anti-Science PR Campaign," *Deep Climate*, September 26, 2010. Accessed at http://deepclimate.files.wordpress.com/2010/09/strange-scholarship-v1-02.pdf .

45. A Delhi-based charity of which Pachauri became a director in 1981 and director general in 2001.

46. Christopher Booker and Richard North, "Questions over Business Deals of UN Climate Change Guru Dr Rajendra Pachauri," *Sunday Telegraph*, December 20, 2009. Accessed at http://www.prisonplanet.com/questions-over-business-deals-of-un-climate-change-guru-dr-rajendra-pachauri.html.

47. George Monbiot, "Rajendra Pachauri Innocent of Financial Misdealings but Smears Will Continue," *Guardian Online*, August 26, 2010. Accessed at http://www.guardian.co.uk/environment/georgemonbiot/2010/aug/26/rajendra-pachauri-financial-relationships.

48. The first reference that came up when I googled for it was at the Orwellianly named *Climate Realists* site (http://climaterealists.com/index.php?id=4713). Sure enough, not a hint of a retraction.

49. Christopher Booker, "The 'Anomalies' of Dr. Rajendra Pachauri's Charity Accounts," *Sunday Telegraph*, October 2, 2010. Accessed at http://www.telegraph.co.uk/comment/columnists/christopherbooker/8039035/The-anomalies-of-Dr-Rajendra-Pachauris-charity-accounts.html#disqus_thread.

50. Robert Mendick and Amrit Dhillon, "Revealed: The Racy Novel Written by the World's Most Powerful Climate Scientist," *Sunday Telegraph*, January 30, 2010. Accessed at http://www.telegraph.co.uk/earth/environment/climatechange/7111068/Revealed-the-racy-novel-written-by-the-worlds-most-powerful-climate -scientist.html.

51. Richard Black, "Climate Panel Agrees 'Milestone' Reforms, Defers Others," BBC, October 14, 2010. Accessed at http://www.bbc.co.uk/news/science-environment-11541056.

52. Ann Coulter, "Global Warming: The French Connection," *WorldNetDaily*, May 28, 2003. Accessed at http://www.wnd.com/news/article.asp?ARTICLE_ID=32802.

53. Luddhunter, "Oreskes on 'The American Denial of Global Warming,'" *Deltoid*, February 8, 2008. Accessed at http://scienceblogs.com/deltoid/2008/02/oreskes_on_the_american_denial.php.

54. John Cook, "Skeptic Arguments and What the Science Says," *Skeptical Science*. Accessed at http://www.skepticalscience.com/argument.php.

55. Coby Beck, "How to Talk to a Climate Skeptic," *A Few Things Ill Consid-*

ered, July 6, 2008. Accessed at http://scienceblogs.com/illconsidered/2008/07/how_to_talk_to_a_sceptic.php.

56. Rebecca Townsend, "Vicky Hartzler, Republican," *Missouri News Horizon*, October 28, 2010. Accessed at http://monewshorizonblog.org/2010/10/vicky-hartzler/.

57. Cited in Steve Connor, "Carbon Emissions Set to be Highest in History," *Independent*, November 22, 2010. Accessed at http://www.ind.ependent.co.uk/environment/climate-change/carbon-emissions-set-to-be-highest-in-history-2140291.html.

58. Frank Luntz, "The Environment: A Cleaner, Safer, Healthier America," November 2002. Accessed at http://watchingthedeniers.files.wordpress.com/2010/04/cleaneramerica.pdf .

59. Ross Gelbspan, *Boiling Point* (New York: Basic Books, 2004), p. 103.

60. George Monbiot, *Heat* (Cambridge, MA: South End Press, 2007), p. v.

61. Kofi Annan, "Introduction," *Human Impact Report: Climate Change—The Anatomy of a Silent Crisis* (Geneva: Global Humanitarian Forum, 2009), p. iii. Accessed at http://www.bb.undp.org/uploads/file/pdfs/energy_environment/CC%20human%20impact%20report.pdf.

62. Ibid., p. 21.

63. Jonathan Overpeck, Director of the Institute for the Study of Planet Earth, University of Arizona in Tucson. This comment is quoted in "Warming to Cause Catastrophic Rise in Sea Level?" *National Geographic News*, April 26, 2004. Accessed at http://news.nationalgeographic.com/news/2004/04/0420_040420_earthday_2.html.

64. Stefan Lovgren, "Warming to Cause Catastrophic Rise in Sea Level?" *National Geographic News*, April 26, 2004. Accessed at http://news.nationalgeographic.com/news/2004/04/0420_040420_earthday_2.html.

65. Quoted in Andrew Restuccia, "Osama Bin Laden Lectures Obama on Climate Change," *Washington Independent*, October 1, 2010. Accessed at http://washingtonindependent.com/99352/osama-bin-laden-lectures-obama-on-climate-change.

66. Steve Maley, "The Kinder, Gentler Osama bin Laden," *Redstate*, January 29, 2010. Accessed at http://www.redstate.com/vladimir/2010/01/29/the-kinder-gentler-osama-bin-laden/.

67. Ed Brayton, "Osama bin Laden: Environmentalist," *Dispatches from the Culture Wars*, October 4, 2010. Accessed at http://scienceblogs.com/dispatches/2010/10/osama_bin_laden_environmentali.php.

68. Suzanne Goldenberg, "Osama bin Laden Lends Unwelcome Support in Fight against Climate Change," *Guardian*, January 29, 2010. Accessed at http://www.guardian.co.uk/environment/2010/jan/29/osama-bin-laden-climate-change.

69. John Deutch et al., "The Future of Coal: Options for a Carbon-Constrained World," Massachusetts Institute of Technology, 2007. Accessed at http://web.mit.edu/coal/The_Future_of_Coal_Summary_Report.pdf..

70. Fred Pearce, "Can Coal Live up to its Clean Promise?" *New Scientist*, March 27, 2008. Accessed at climatechange.flinders.edu.au/NSClean%20coal.doc.

71. Much of the information here is from SourceWatch, "Coal Mining Disasters." Accessed at http://www.sourcewatch.org/index.php?title=Coal_mining_disasters.

72. Figures from Frances Beinecke, *Clean Energy Common Sense*, with Bob Deans (Lanham, MD: Rowman & Littlefield, 2010), pp. 51–52.

73. Ibid., p. 58.

74. Fred Krupp and Miriam Horn, *Earth: The Sequel* (New York: Norton, 2008), p. 30.

75. Quoted in David Roberts, "Bidding a Fond Farewell to ExxonMobil CEO Lee Raymond," *Grist*, August 6, 2005. Accessed at http://www.grist.org/article/roberts-raymond/.

76. Richard A. Muller, *Physics for Future Presidents* (New York: Norton, 2008), p. 329.

77. Elisabeth Rosenthal, "Europe Finds Clean Energy in Trash, but U.S. Lags," *New York Times*, April 12, 2010. Accessed at http://www.nytimes.com/2010/04/13/science/earth/13trash.html?nl=todaysheadlines&emc=a3.

78. David Fermin, quoted in "Making Car Fuel from Thin Air," *Science Daily*, March 29, 2010. Accessed at http://www.sciencedaily.com/releases/2010/03/100324184556.htm.

79. Jonalex, "Ice Age Solution to Global Warming," *All Voices*, November 28, 2010. Accessed at http://www.allvoices.com/contributed-news/7461490-ice-age-solution-to-global-warming.

80. Gidon Eshel and Pamela A. Martin, "Diet, Energy, and Global Warming," *Earth Interactions* 10 (2006). Accessed at http://journals.ametsoc.org/doi/pdf/10.1175/EI167.1.

81. Hansen, *Storms of My Grandchildren* (2009), pp. 209–22.

Chapter 18: Marketing Climate Denialism

1. Miranda Devine, "Beware the Church of Climate Alarm," *Sydney Morning Herald*, November 27, 2008. Accessed at http://www.smh.com.au/articles/2008/11/26/1227491635989.html.

2. Go to http://www.exxonsecrets.org/index.php?mapid=612. Click on "LOAD MAP" in the list at top left, and then choose, for example, "Biggest Exxon $$ Winners" for an immediate graphic representation that names names.

3. A more complete, regularly updated list can be found at http://www.exxonsecrets.org.

4. For copies of the e-mails, see the OfcomSwindleComplaint website

(Ofcom is the UK's Office of Communications). Accessed at http://www.ofcomswindlecomplaint.net/emails/foia-ebell-to-cooney_obtained-by-greenpeace-under-freedom-of-inforrmation-act.pdf.

5. Antonio Regalado, Dionne Searcey, and Jeffrey Ball, "Where Did That Video Spoofing Gore's Film Come From?" *Wall Street Journal*, August 3, 2006. Accessed at http://www.mail-archive.com/medianews@twiar.org/msg12590.html.

6. "Fox News Repeatedly Advanced CEI Fellow's Accusations NASA Is 'Manipulating Data on Climate Change' without Noting CEI Received Millions from Oil Industry," *Media Matters*, December 04, 2009. Accessed at http://mediamatters.org/research/200912040022.

7. James Hoggan and Richard Littlemore, *Climate Cover-Up* (Vancouver: Greystone, 2009), p. 53.

8. "ICSC Mission Statement," International Climate Science Coalition. Accessed at http://www.climatescienceinternational.org/index.php?option=com_content&view=article&id=5&Itemid=6.

9. Greenpeace, "Koch Industries: Secretly Funding the Climate Denial Machine," Greenpeace, 2010, and "Koch Industries: Still Fueling Climate Denial (2011 Update)," Greenpeace, 2011. Accessed at http://www.greenpeace.org/usa/en/campaigns/global-warming-and-energy/polluterwatch/koch-industries/.

10. Jane Mayer, "Covert Operations," *New Yorker*, August 30, 2010. Accessed at http://www.newyorker.com/reporting/2010/08/30/100830fa_fact_mayer.

11. The Claude R. Lambe Foundation, the Charles G. Koch Foundation, and the David H. Koch Foundation. The list of organizations funded indirectly by the Koch brothers can be found at http://www.greenpeace.org/usa/campaigns/global-warming-and-energy/polluterwatch/koch-industries-secretly-fund/.

12. Greenpeace, "Koch Industries: Secretly Funding the Climate Denial Machine," pp. 32–33.

13. Ibid., p. 33.

14. "For the Health of the Nation: An Evangelical Call to Civic Responsibility," National Association of Evangelicals, 2004. Accessed at http://www.nae.net/images/content/For_The_Health_Of_The_Nation.pdf.

15. "Factsheet: Interfaith Stewardship Alliance," *ExxonSecrets*. Accessed at http://www.exxonsecrets.org/html/orgfactsheet.php?id=142. A partial list of signatories to the associated "Evangelical Declaration on Global Warming" can be found at http://www.cornwallalliance.org/blog/item/prominent-signers-of-an-evangelical-declaration-on-global-warming/.

16. "Cornwall Declaration on Environmental Stewardship," *Veritas—A Quarterly Journal of Public Policy in Texas* (Summer 2000). Accessed at http://www.texaspolicy.com/pdf/2000-veritas-1-2-cornwall.pdf.

17. "An Evangelical Declaration on Global Warming," Cornwall Alliance,

2009. Accessed at http://www.cornwallalliance.org/articles/read/an-evangelical-declaration-on-global-warming/.

18. Quoted in John Collins Rudolf, "An Evangelical Backlash against Environmentalism," *New York Times*, December 30, 2010.

19. Clay Ramsay, Steven Kull, Evan Lewis, and Stefan Subias, "Misinformation and the 2010 Election: A Study of the US Electorate," *World Public Opinion*, December 10, 2010. Accessed at http://www.worldpublicopinion.org/pipa/pdf/dec10/Misinformation_Dec10_rpt.pdf .

20. Quoted in Ben Dimiero, "FOXLEAKS: Fox Boss Ordered Staff to Cast Doubt on Climate Science," *Media Matters for America*, December 15, 2010. Accessed at http://mediamatters.org/blog/201012150004.

21. Maxwell T. Boykoff and Jules M. Boykoff, "Balance as Bias: Global Warming and the US Prestige Press," *Global Environmental Change*, no.14 (2004): 125–36. Accessed at http://www.eci.ox.ac.uk/publications/downloads/boykoff04 -gec.pdf.

22. Quoted in George Monbiot, *Heat* (Cambridge, MA: South End Press, 2007), p. 23.

23. Peter J. Jacques, Riley E. Dunlap, and Mark Freeman, "The Organisation of Denial: Conservative Think Tanks and Environmental Scepticism," *Environmental Politics* 17, no. 3 (June 2008). Accessed at http://www.informaworld.com/smpp/section?content=a793291693&fulltext=713240928.

Chapter 19: Climate Denialism: Dramatis Personae

1. Freeman J. Dyson, *Selected Papers of Freeman Dyson with Commentary* (Providence, RI: American Mathematical Society, 1996), p. 44.

2. James Hansen, *Storms of My Grandchildren: The Truth About the Coming Climate Catastrophe and Our Last Chance to Save Humanity* (New York: Bloomsbury, 2009), p. 55.

3. Richard Lindzen, "Climate Science in Denial: Global Warming Alarmists Have Been Discredited, but You Wouldn't Know It from the Rhetoric This Earth Day," *Wall Street Journal*, April 22, 2010.

4. This particular wording by Ian Plimer from "Legislative Time Bomb," *The Drum*, Australian Broadcasting Corporation, August 13, 2009. Accessed at http://www.abc.net.au/unleashed/29320.html.

5. Quoted in Bill Nicholas, "Don't Hold Your Breath on CO_2," *Independent Weekly*, March 22, 2008.

6. Ian Plimer, letter to the editor, *Spectator*, July 18, 2009. Accessed at http://www.spectator.co.uk/politics/all/5186003/letters.thtml.

7. "Plimer, Monbiot Cross Swords in Climate Debate," Australian Broadcasting Corporation, December 15, 2009. Accessed at http://www.abc.net.au/lateline/content/2009/s2772906.htm.

8. James Hoggan, *Climate Cover-Up: The Crusade to Deny Global Warming*, with Richard Littlemore (Vancouver: Greystone, 2009), p. 41.

9. Quoted at ExxonSecrets. Accessed at http://www.exxonsecrets.org/html/orgfactsheet.php?id=65#src3.

10. S. Fred Singer, "Secondhand Smoke, Lung Cancer, and the Global Warming Debate," *American Thinker*, December 19, 2010. Accessed at http://www.americanthinker.com/2010/12/second_hand_smoke_lung_cancer.html.

11. Frederick Seitz, "Presentation to Operating Committee, R. J. Reynolds Industries, Inc.," August 8, 1979. Accessed at http://tobaccodocuments.org/rjr/504779244-9250.html.

12. Oreskes and Conway, *Merchants of Doubt*, p. 28.

13. S. Fred Singer, Roger Revelle, and Chauncey Starr, "What To Do About Greenhouse Warming: Look Before You Leap," *Cosmos: A Journal of Emerging Issues* 5, no. 2 (summer 1992). Accessed at http://ruby.fgcu.edu/courses/ twimberley/envirophilo/lookbeforeyouleap.pdf.

14. Sherwood B. Idso, "The Climatological Significance of a Doubling of Earth's Atmospheric Carbon Dioxide Concentration," *Science*, March 28, 1980. Abstract accessed at http://www.sciencemag.org/content/207/4438/1462.abstract.

15. Josh Harkinson, "The Dirty Dozen of Climate Change Denial, #8: Center for the Study of Carbon Dioxide and Global Change (A.K.A. The Idso Family)," *Mother Jones*, December 4, 2009. Accessed at http://motherjones.com/environment/2009/12/dirty-dozen-climate-change-denial-11-idso-family.

16. Sherwood B. Idso, "Real-World Constraints on Global Warming," Fraser Institute, 1999. Accessed at http://oldfraser.lexi.net/publications/books/g_warming/real_world.html.

17. "Center for the Study of Carbon Dioxide and Global Change—Koch Industries Climate Denial Front Group," Greenpeace. Accessed at http://www.greenpeace.org/usa/en/campaigns/global-warming-and-energy/polluterwatch/koch-industries/center-for-the-study-of-carbon/.

18. "Dr. Sallie Baliunas," George C. Marshall Institute. Accessed at http://www.marshall.org/experts.php?id=38.

19. "Factsheet: Willie Soon," *ExxonSecrets*. Accessed at http://www.exxonsecrets.org/html/personfactsheet.php?id=860#src10.

20. M. G. Dyck et al., "Polar Bears of Western Hudson Bay and Climate Change: Are Warming Spring Air Temperatures the 'Ultimate' Survival Control Factor?" *Ecological Complexity* 4, no. 3. Accessed at http://ruby.fgcu.edu/courses/twimberley/EnviroPhilo/HudsonBay.pdf .

21. Willie Soon and Sallie Baliunas, "Lessons and Limits of Climate History: Was the 20th Century Climate Unusual?" George C. Marshall Institute, 2003. Accessed at http://www.marshall.org/pdf/materials/136.pdf.

22. One Climategate accusation is that the resignations were the result of pressure from Michael Mann and members of the CRU. All concerned have denied this.

23. Floor speech delivered July 28, 2003. Accessed at http://inhofe.senate.gov/pressreleases/climate.htm .

24. "Factsheet: Sen. James Inhofe," *ExxonSecrets*. Accessed at http://www.exxonsecrets.org/html/personfactsheet.php?id=945.

25. Chris Mooney, "Earth Last," *American Prospect*, April 13, 2004.

26. Calculation by Tim Lambert in "Inhofe: Less Honest than the Discovery Institute," *Deltoid*, December 13, 2008. Accessed at http://scienceblogs.com/deltoid/2008/12/inhofe_less_honest_than_the_di.php.

27. Quoted in Andrew Dessler, "Today: George Waldenberger," *Grist*, January 15, 2008. Accessed at http://www.grist.org/article/the-inhofe-400-skeptic-of-the-day3/.

28. Accessed via Inhofe's own website at http://epw.senate.gov/public/index.cfm?FuseAction=Files.View&FileStore_id=83947f5d-d84a-4a84-ad5d-6e2d71db52d9&CFID=31000783&CFTOKEN=64876380.

29. Jonathan Karl and Z. Byron Wolf, "Amid Heat Wave, Senator Talks 'Global Cooling,'" ABC News, July 23, 2010. Accessed at http://abcnews.go.com/Politics/amid-heat-wave-senator-talks-global-coolilng/story?id=11237381.

30. Quoted in Dave Michaels, "Barton a Steadfast Skeptic on Climate Change," *Dallas Morning News*, June 1, 2008.

31. Aron Pilhofer and Bob Williams, "Big Oil Protects its Interests," Center for Public Integrity, revised March 31, 2006. Accessed at http://projects.publicintegrity.org/oil//report.aspx?aid=345&sid=100.

32. "Joe Barton," *Open Secrets*. Accessed at http://www.opensecrets.org/politicians/summary.php?cid=N00005656&cycle=Career.

33. Quoted in Kate Sheppard, "House Republicans Bring Strange Theories and Wacky Witnesses to Climate Hearings," *Grist*, April 20, 2009. Accessed at http://www.grist.org/article/2009-04-20-house-republicans-bring.

34. Judy Fahys, "Debate on Climate Heats up Online," *Salt Lake Tribune*, April 9, 2010. Accessed at http://www.sltrib.com/sltrib/home/49374964-73/monckton-bickmore-university-climate.html.csp.

35. Or, strictly speaking, by "someone" who just happened to be using Monckton's e-mail address. Monbiot has posted the very amusing correspondence between the two on the subject at http://www.monbiot.com/archives/2007/10/03/did-lord-monckton-fabricate-a-claim-on-his-wikipedia-page/.

36. Christopher Monckton, "More in Sorrow Than in Anger: An Open Letter from The Viscount Monckton of Brenchley to Senator John McCain about Climate

Science and Policy," Science and Public Policy Institute. Accessed at http://science andpublicpolicy.org/images/stories/papers/reprint/Letter_to_McCain.pdf .

37. Barry Bickmore, "Lord Monckton: 3rd Viscount of Brenchley, King of Fantasyland," Barry Bickmore's website, Climate, 2010. Accessed at http://home .comcast.net/~bbickmore/Climate/KingOfFantasyland.htm.

38. "Committee for a Constructive Tomorrow," Sourcewatch. Accessed at http://www.sourcewatch.org/index.php?title=Committee_For_A_Constructive _Tomorrow.

39. Kevin Grandia, "Copenhagen Climate Talks: Monckton 'Hitler Youth' Video," *DeSmog Blog*, December 10, 2009. Accessed at http://www.desmogblog .com/christopher-monckton-copenhagen-i-will-not-shake-hand-hitler-youth.

40. Not to be confused with the Science and Public Policy Institute founded in 1994 by George Carlo, whose concern was public health related to tobacco use and mobile phone use.

41. There is discussion of the Heartland Institute's funding at http://www .sourcewatch.org/index.php?title=Heartland_Institute and a listing of its known donors at http://mediamattersaction.org/transparency/organization/Heartland _Institute/funders.

42. The summary is available online: Martin Robbins, "Ukip: Science Test Results," *Guardian*, April 27, 2010. Accessed at http://www.guardian.co.uk/ science/2010/apr/27/ukip-science-policy-election. The full answers are available here: "Ukip Answers Questions about Its Science Policy," *Guardian*, April 27, 2010. Accessed at http://www.guardian.co.uk/science/2010/apr/27/ukip-science-policy -general-election.

43. John Abraham, "Monckton Myths," *Skeptical Science*. Accessed at http://www.skepticalscience.com/Monckton_Myths.htm.

44. Steve Goddard, "Hyperventilating on Venus," *Watts Up with That?* May 6, 2010. Accessed at http://wattsupwiththat.com/2010/05/06/hyperventilating-on-venus/.

45. Steve Goddard, "Venus Envy," *Watts Up with That?* May 8, 2010. Accessed at http://wattsupwiththat.com/2010/05/08/venus-envy/. A good analysis of Goddard's two postings can be found in Chris Colose, "Goddard's World" *Climate Change*, May 12, 2010. Accessed at http://chriscolose.wordpress.com/2010/05/ 12/goddards-world/.

46. Joseph Romm, "Revkin's DotEarth Hypes Disinformation Posted on an Anti-Science Website," *Climate Progress*, February 10, 2010. Accessed at http:// climateprogress.org/2010/02/10/revkin-dotearth-science-wattsupwiththat-climate -sensitivity-jerome-ravetz/.

47. Peter Sinclair, *The Video Climate Deniers Tried to Ban—Climate Denial Crock of the Week: Anthony Watts*, YouTube. Accessed at http://www.youtube .com/watch?v=P_0-gX7aUKk.

48. Kevin Grandia, "Climate Crock of the Week: What's Up with Anthony Watts [take 2]," *DeSmog Blog*, July 28, 2009. Accessed at http://www.desmog blog.com/climate-crock-week-whats-anthony-watts-take-2.

49. The Center seems not to archive such material. This statement has, however, been widely cited—as a quick Google search shows.

50. George Will, "Climate Science in a Tornado," *Washington Post*, February 27, 2009.

51. Michael Asher, "Sea Ice Ends Year at Same Level as 1979," *DailyTech*, January 1, 2009. Accessed at http://www.dailytech.com/Article.aspx?newsid=13834.

52. Accessed at http://www.washingtonpost.com/wp-dyn/content/article/2009/02/26/AR2009022602906.html.

53. Accessed at http://arctic.atmos.uiuc.edu/cryosphere/global.sea.ice.area.pdf.

54. Roger Harrabin, "Global Temperatures 'To Decrease,'" BBC, April 4, 2008.

55. Patrick J. Buchanan, "The Hoax of the Century," *WorldNetDaily*, March 1, 2010. Accessed at http://www.wnd.com/index.php?pageId=126661.

56. Daniel C. Nepstad et al., "Large-Scale Impoverishment of Amazonian Forests by Logging and Fire," *Nature*, April 8, 1999. Abstract accessed at http://www.nature.com/nature/journal/v398/n6727/full/398505a0.htmla.

57. Cited in Eli Kintisch, "Scientist Disputes Claim of 'Bogus' IPCC Reference on Threatened Rainforests," *Science*, February 3, 2010.

58. Accessed at http://climateprogress.org/wp-content/uploads/2010/03/Lewis_S_Times_PCC_Complaint_As_Sent1.pdf.

59. Ibid.

60. Accessed at http://www.timesonline.co.uk/tol/news/environment/article7026317.ece. The rest of the data are here in the text.

61. It's at http://scienceblogs.com/deltoid/.

62. Tim Lambert, "Leakegate: Jonathan Leake Strikes Back," *Deltoid*, March 2, 2010. Accessed at http://scienceblogs.com/deltoid/2010/03/leakegate_jonathan_leake _strik.php.

63. Jonathan Leake, "Ed Miliband's Adverts Banned for Overstating Climate Change," *Sunday Times*, March 14, 2010. Accessed at http://www.timesonline.co.uk/tol/news/environment/article7061162.ece.

64. George Monbiot, "David Rose's Climate Science Writing Shows He Has Not Learned from Previous Mistakes," *Guardian Online*, December 8, 2010. Accessed at http://www.guardian.co.uk/environment/georgemonbiot/2010/dec/08/david-rose-climate-science.

65. "Global Warming Policy Foundation," *Sourcewatch*. Accessed at http://www.sourcewatch.org/index.php?title=Global_Warming_Policy_Foundation.

66. The newspaper as a whole seems to be in the business of denying science.

See "*Herald Sun* War on Science" at *Watching the Deniers* (http://watchingthedeniers.wordpress.com/herald-sun-war-on-science/) for a sample.

67. Andrew Bolt, "CSIRO 'Forgets' We Were Once Drier," *Herald Sun* blog, March 23, 2010. Accessed at http://blogs.news.com.au/heraldsun/andrewbolt/index.php/heraldsun/comments/csiro_forgets_we_were_once_drier/.

68. Australian Government, Bureau of Meteorology, "Australian Climate Variability and Change—Trend Maps," Bureau of Meteorology. Accessed at http://www.bom.gov.au/cgi-bin/climate/change/trendmaps.cgi?map=rain&area=aus&season=0112&period=1970. A good analysis of this column is "Lies, Damned Lies and BoM Maps! How Australia's Denial Movement Can't Read a Map," *Watching the Deniers*, March 23, 2010. Accessed at http://watchingthedeniers.wordpress.com/2010/03/23/lies-damned-lies-and-statistics-how-australias-denial-movement-cant-read-a-map/.

69. Union of Concerned Scientists, "Crichton Thriller *State of Fear*." Accessed at http://www.ucsusa.org/global_warming/science_and_impacts/global_warming_contrarians/crichton-thriller-state-of.html.

70. Michael Crichton, "'Aliens Cause Global Warming,'" *Wall Street Journal*, November 7, 2008, based on a lecture delivered at the California Institute of Technology on January 17, 2003.

71. John R. Mashey, "Steve McIntyre and Ross McKitrick, part 1: In the Beginning," February 4, 2010. Accessed at http://deepclimate.org/2010/02/04/steve mcintyre-and-ross-mckitrick-part-1-in-the-beginning/#comments.

72. Stephen McIntyre and Ross McKitrick, "Hockey Sticks, Principal Components and Spurious Significance," *Geophysical Research Letters* 32, no. 3 (February 12, 2005).

73. Jeannette Catsoulis, "Global Warming and Common Sense," *New York Times*, November 11, 2010. Accessed at http://movies.nytimes.com/2010/11/12/movies/12cool.html.

74. Accessed at http://lomborg-errors.dk/.

75. J. C. Moore, S. Jevrejeva and A. Grinsted, "Efficacy of Geoengineering to Limit 21st-Century Sea-Level Rise," *Proceedings of the National Academy of Sciences*, August 23, 2010. Accessed at http://www.pnas.org/content/early/2010/08/20/1008153107.

BIBLIOGRAPHY

Agin, Dan. *Junk Science: An Overdue Indictment of Government, Industry, and Faith Groups That Twist Science for Their Own Gain*. New York: Thomas Dunne Books, 2006.

Angell, Marcia. *Science on Trial: The Clash of Medical Evidence and the Law in the Breast Implant Case*. New York: Norton, 1996.

Armstrong, Karen. *Islam: A Short History*. New York: Modern Library, 2000.

Ashby, Eric. *Technology and the Academics: An Essay on Universities and the Scientific Revolution*. London: Macmillan, 1959.

Ayala, Francisco J. *Darwin's Gift to Science and Religion*. Washington, DC: Joseph Henry Press, 2007.

Bannister, Robert C. *Social Darwinism: Science and Myth in Anglo-American Social Thought*. Philadelphia: Temple University Press, 1979.

Bartholomew, Robert E., and Benjamin Radford. *Hoaxes, Myths, and Manias: Why We Need Critical Thinking*. Amherst, NY: Prometheus Books, 2003.

Beck, Glenn. *An Inconvenient Book: Real Solutions to the World's Biggest Problems*. New York: Threshold/Simon & Schuster, 2007.

Beinecke, Frances. *Clean Energy Common Sense: An American Call to Action on Global Climate Change*. With Bob Deans. Lanham MD: Rowman & Littlefield, 2010.

Black, Edwin. *War Against the Weak: Eugenics and America's Campaign to Create a Master Race*. New York: Four Walls Eight Windows, 2003.

Brookes, Martin. *Extreme Measures: The Dark Visions and Bright Ideas of Francis Galton*. London: Bloomsbury, 2004.

Brooks, Michael. *13 Things that Don't Make Sense: The Most Baffling Scientific Mysteries of Our Time*. Toronto: Doubleday Canada, 2008.

Brown, James Robert. *Who Rules in Science?: An Opinionated Guide to the Wars*. Cambridge, MA: Harvard University Press, 2001.

Browne, Janet. *Darwin's* Origin of Species*: A Biography*. New York: Atlantic Monthly Press, 2006.

Bruinius, Harry. *Better for All the World: The Secret History of Forced Sterilization and America's Quest for Racial Purity*. New York: Knopf, 2006.

Bunge, Mario. "In Praise of Intolerance to Charlatanism in Academia." *Annals of the New York Academy of Sciences* 775 (June 24, 1996): 96–115.

Byrne, Rhonda. *The Secret*. New York: Atria, 2006.

Carlson, Elof Axel. *The Unfit: A History of a Bad Idea*. Cold Spring Harbor, NY: Cold Spring Harbor Laboratory Press, 2001.

Carroll, Sean B. *The Making of the Fittest: DNA and the Ultimate Forensic Record of Evolution*. New York: Norton, 2006.

Charpak, Georges, and Henri Broch. *Devenez Sorciers, Devenez Savants*. Paris: Odile Jacob, 2002. Translated by Bart K. Holland as *Debunked!: ESP, Telekinesis, and Other Pseudoscience*. Baltimore, MD: Johns Hopkins University Press, 2004.

Cohen, I. Bernard. *Revolution in Science*. Cambridge, MA: Belknap Press/Harvard University Press, 1985.

Cohen, Stewart J., and Melissa W. Waddell. *Climate Change in the 21st Century*. Montreal & Kingston, ON: McGill–Queen's University Press, 2009.

Cornwell, John. *Hitler's Scientists: Science, War, and the Devil's Pact*. New York: Viking, 2003.

Coulter, Ann. *Godless: The Church of Liberalism*. New York: Crown, 2006.

———. *If Democrats Had Any Brains, They'd Be Republicans*. New York: Crown Forum, 2007.

Cremo, Michael A., and Richard L. Thompson. *The Hidden History of the Human Race*. Badger, CA: Govardhan Hill, 1994.

Darwin, Charles. *The Origin of Species by Means of Natural Selection; or The Preservation of Favoured Races in the Struggle for Life*. 6th ed. London: John Murray, 1872.

Dawkins, Richard. *The Blind Watchmaker: Why the Evidence of Evolution Reveals a Universe Without Design*. New York: Norton, 1986.

Dean, Jodi. *Aliens in America: Conspiracy Cultures from Outerspace to Cyberspace*. Ithaca, NY: Cornell University Press, 1998.

de Camp, L. Sprague. *The Ragged Edge of Science*. Philadelphia: Owlswick, 1980.

Dennett, Daniel. *Darwin's Dangerous Idea: Evolution and the Meanings of Life*. New York: Simon & Schuster, 1995.

Desmond, Adrian, and James Moore. *Darwin's Sacred Cause: How a Hatred of Slavery Shaped Darwin's Views on Human Evolution*. Boston: Houghton Mifflin Harcourt, 2009.

Diamond, Jared. *Collapse: How Societies Choose to Fail or Succeed*. New York: Viking Penguin, 2005.

Diamond, John. *Snake Oil, and Other Preoccupations*. New York: Vintage, 2001.

Dowie, Mark. *Losing Ground: American Environmentalism at the Close of the Twentieth Century*. Cambridge, MA: MIT Press, 1995.

Dyer, Gwynne. *Climate Wars: The Fight for Survival as the World Overheats*. Oxford: Oneworld, 2010.

Dyson, Freeman J. *Selected Papers of Freeman Dyson with Commentary*. Providence, RI: American Mathematical Society, 1996.

Ecklund, Elaine Howard. *Science vs. Religion: What Scientists Really Think*. New York: Oxford University Press, 2010.

Edis, Taner. *An Illusion of Harmony: Science and Religion in Islam*. Amherst, NY: Prometheus Books, 2007.

Evans, Christopher. *Cults of Unreason*. London: Harrap, 1973.

Eve, Raymond A., and Francis B. Harrold. *The Creationist Movement in Modern America*. Boston: Twayne, 1991.

Feder, Kenneth L. *Frauds, Myths, and Mysteries: Science and Pseudoscience in Archaeology*. 3rd ed. Mountain View, CA: Mayfield, 1999.

Fitzgerald, A. Ernest. *The High Priests of Waste*. New York: Norton, 1972.

Flank, Lenny. *Deception by Design: The Intelligent Design Movement in America*. St. Petersburg, FL: Red and Black, 2007.

Forrest, Barbara, and Paul R. Gross. *Creationism's Trojan Horse: The Wedge of Intelligent Design*. Oxford: Oxford University Press, 2004.

Frankfurt, Harry G. *On Bullshit*. Princeton, NJ: Princeton University Press, 2005.

Friedlander, Michael W. *At the Fringes of Science*. Boulder, CO: Westview, 1995.

Friel, Howard. *The Lomborg Deception: Setting the Record Straight about Global Warming*. New Haven, CT: Yale University Press, 2010.

Fritz, Ben, Bryan Keefer, and Brendan Nyhan. *All the President's Spin: George W. Bush, the Media, and the Truth*. New York: Touchstone, 2004.

Futuyma, Douglas J. *Science on Trial: The Case for Evolution*. New York: Pantheon, 1982.

Galton, Francis. "Hereditary Character and Talent." Pts. 1 and 2. *Macmillan's Magazine*. (November 1864): 157–66; (April 1965): 319–27.

———. *Hereditary Genius: An Inquiry into Its Laws and Consequences*. London: Macmillan, 1869.

Gardner, Martin. *Did Adam and Eve Have Navels? Debunking Pseudoscience*. New York: Norton, 2000.

———. *Fads and Fallacies in the Name of Science*. New York: Dover, 1957.

———. *The New Age: Notes of a Fringe Watcher*. Buffalo, NY: Prometheus, 1988.

———. *Science: Good, Bad and Bogus*. Amherst, NY: Prometheus, 1989.

Gelbspan, Ross. *Boiling Point: How Politicians, Big Oil and Coal, Journalists, and Activists Have Fueled the Climate Crisis—and What We Can Do to Avert Disaster*. New York: Basic Books, 2004.

Glad, John. *Future Human Evolution: Eugenics in the Twenty-First Century*. Schuylikill Haven, PA: Hermitage, 2006.

Glick, Thomas F., ed. *The Comparative Reception of Darwinism*. Austin: University of Texas Press, 1975.

Goldacre, Ben. *Bad Science*. Rev. ed. London: Fourth Estate, 2009.

Goldin, Claudia, and Lawrence F. Katz. *The Race Between Education and Technology*. Cambridge, MA: Belknap Press, 2008.

Gore, Al. *An Inconvenient Truth: The Planetary Emergency of Global Warming and What We Can Do About It*. New York: Rodale, 2006.

Gosling, David L. *Science and the Indian Tradition: When Einstein Met Tagore*. London: Routledge, 2007.

Gosse, Philip Henry. *Omphalos: An Attempt to Untie the Geological Knot*. London: John Van Voorst, 1857.

Gould, Stephen Jay. *The Mismeasure of Man*. New York: Norton, 1981.

Grant, Edward. *Science & Religion, 400 B.C.–A.D. 1550: From Aristotle to Copernicus*. Westport, CT: Greenwood, 2004.

Grant, John. *Bogus Science: Some People Really Believe These Things*. Wisley, UK: AAPPL, 2009.

———. *Corrupted Science: Fraud, Ideology, and Politics in Science*. London: AAPPL, 2007.

———. *Discarded Science: Ideas That Seemed Good at the Time*. London: AAPPL, 2006.

Gratzer, Walter. *The Undergrowth of Science: Delusion, Self-Deception, and Human Frailty*. Oxford: Oxford University Press, 2000.

Gross, Paul R., and Norman Levitt. *Higher Superstition: The Academic Left and Its Quarrels with Science*. Baltimore, MD: Johns Hopkins University Press, 1994.

Gruber, Howard E. *Darwin on Man: A Psychological Study of Human Creativity*. London: Wildwood, 1974.

Hansen, James. *Storms of My Grandchildren: The Truth about the Coming Climate Catastrophe and Our Last Chance to Save Humanity*. New York: Bloomsbury, 2009.

Haught, John F. *God after Darwin: A Theology of Evolution*. 2nd ed. Boulder, CO: Westview, 2008.

Heard, Alex. *Apocalypse Pretty Soon: Travels in End-Time America*. New York: Norton, 1999.

Hendricks, Stephenie. *Divine Destruction: Wise Use, Dominion Theology, and the Making of American Environmental Policy*. Brooklyn, NY: Melville House, 2005.

Hofstadter, Richard. *Social Darwinism in American Thought*. Rev. ed. New York: Braziller, 1959.

Hoggan, James. *Climate Cover-Up: The Crusade to Deny Global Warming*. With Richard Littlemore. Vancouver: Greystone, 2009.

Holmes, Richard. *The Age of Wonder: How the Romantic Generation Discovered the Beauty and Terror of Science*. London: Harper Press, 2008.

Holton, Gerald. *Science and Anti-Science*. Cambridge, MA: Harvard University Press, 1993.

Hoodbhoy, Pervez. *Islam and Science: Religious Orthodoxy and the Battle for Rationality*. London: Zed Books, 1991.

———. "Science and the Islamic World: The Quest for Rapprochement." *Physics Today*, August 2007.

Huber, Peter W. *Galileo's Revenge: Junk Science in the Courtroom*. New York: Basic Books, 1993.

Hurley, Dan. *Natural Causes: Death, Lies, and Politics in America's Vitamin and Herbal Supplement Industry*. New York: Broadway, 2006.

Iqbal, Muzaffar. *Science and Islam*. Westport, CT: Greenwood, 2007.

Irvine, William. *Apes, Angels, and Victorians*. New York: McGraw-Hill, 1955.

Jacoby, Susan. *The Age of American Unreason*. New York: Pantheon, 2008.

Kalichman, Seth. *Denying AIDS: Conspiracy Theories, Pseudoscience, and Human Tragedy*. New York: Copernicus/Springer, 2009.

Kaminer, Wendy. *Sleeping with Extra-Terrestrials: The Rise of Irrationalism and Perils of Piety*. New York: Pantheon, 1999.

Kaplan, Michael, and Ellen Kaplan. *Bozo Sapiens: Why to Err Is Human*. New York: Bloomsbury, 2009.

Keen, Andrew. *The Cult of the Amateur: How Today's Internet Is Killing Our Culture*. New York: Doubleday/Currency, 2007.

Kitcher, Philip. *Abusing Science: The Case Against Creationism*. Cambridge, MA: MIT Press, 1982.

Knight, David. *The Nature of Science: The History of Science in Western Culture Since 1600*. London: Deutsch, 1976.

Koertge, Noretta, ed. *A House Built on Sand: Exposing Postmodernist Myths about Science*. Oxford: Oxford University Press, 1998.

Koestler, Arthur. *The Case of the Midwife Toad*. London: Hutchinson, 1971.

Kolbert, Elizabeth. "The Climate of Man." Parts 1, 2, and 3. *New Yorker*, April 25, May 2, May 9, 2005.

Krupp, Fred, and Miriam Horn. *Earth: The Sequel—The Race to Reinvent Energy and Stop Global Warming*. New York: Norton, 2008.

Lanning, Kenneth V. *Investigator's Guide to Allegations of "Ritual" Child Abuse*. Rev. ed. Quantico VA: Federal Bureau of Investigation, 1997. Available online at http://www.religioustolerance.org/ra_rep03.htm#fbi.

Larson, Edward J. *Summer for the Gods: The Scopes Trial and America's Continuing Debate over Science and Religion*. New York: Basic Books, 1997.

Lefkowitz, Mary. *Not Out of Africa: How Afrocentrism Became an Excuse to Teach Myth as History*. New York: New Republic Books/Basic Books, 1996.

Lipstadt, Deborah E. *Denying the Holocaust: The Growing Assault on Truth and Memory*. New York: Free Press, 1993.

Loftus, Elizabeth, and Katherine Ketcham. *The Myth of Repressed Memory: False Memories and Allegations of Sexual Abuse*. New York: St Martin's, 1994.

Martin, Brian. *Information Liberation: Challenging the Corruptions of Information Power*. London: Freedom Press, 1998.

Miller, Kenneth R. *Finding Darwin's God: A Scientist's Search for Common Ground between God and Evolution*. New York: Cliff Street, 1999.

Mitroff, Ian I., and Warren Bennis. *The Unreality Industry: The Deliberate Manufacturing of Falsehood and What It Is Doing to Our Lives*. New York: Oxford University Press, 1989.

Monbiot, George. *Heat: How to Stop the Planet Burning*. Cambridge, MA: South End Press, 2007.

Mooney, Chris. *The Republican War on Science*. New York: Basic Books, 2005.

Mooney, Chris, and Sheril Kirshenbaum. *Unscientific America: How Scientific Illiteracy Threatens Our Future*. New York: Basic Books, 2009.

Moore, Kathleen Dean, and Michael P. Nelson, eds. *Moral Ground: Ethical Action for a Planet in Peril*. San Antonio, TX: Trinity University Press, 2010.

Moskovits, Martin, ed. *Science and Society*. Concord, ON: Anansi, 1995.

Muller, Richard A. *Physics for Future Presidents: The Science Behind the Headlines*. New York: Norton, 2008.

Nanda, Meera. "Postmodernism, Hindu Nationalism, and 'Vedic Science,'" Parts 1 and 2. *Frontline*, December 20, 2003, and January 3, 2004.

Nathan, Debbie, and Michael Snedker. *Satan's Silence: Ritual Abuse and the Making of a Modern American Witch Hunt*. New York: Basic Books, 1995.

Nattrass, Nicoli. *Mortal Combat: AIDS Denialism and the Struggle for Antiretrovirals in South Africa*. Scottsville, ZA: University of KwaZulu-Natal Press, 2007.

Numbers, Ronald L. *The Creationists: The Evolution of Scientific Creationism*. New York: Knopf, 1992.

Ofshe, Richard, and Ethan Watters. *Making Monsters: False Memories, Psychotherapy, and Sexual Hysteria*. New York: Scribner, 1994.

Oreskes, Naomi, and Erik M. Conway. *Merchants of Doubt: How a Handful of Scientists Obscured the Truth on Issues from Tobacco Smoke to Global Warming*. New York: Bloomsbury, 2010.

Orzel, Chad. *How to Teach Physics to Your Dog*. New York: Scribner, 2009.

Park, Robert L. *Superstition: Belief in the Age of Science*. Princeton NJ: Princeton University Press, 2008.

———. *Voodoo Science: The Road from Foolishness to Fraud*. Oxford: Oxford University Press, 2000.

Pennock, Robert T., ed. *Intelligent Design Creationism and Its Critics: Philosophical, Theological, and Scientific Perspectives*. Cambridge, MA: MIT Press, 2001.

Peters, Shawn Francis. *When Prayer Fails: Faith Healing, Children, and the Law.* Oxford: Oxford University Press, 2008.

Plotz, David. *The Genius Factory: The Curious History of the Nobel Prize Sperm Bank.* New York: Random House, 2005.

Press, Bill. *Spin This! All the Ways We Don't Tell the Truth.* New York: Pocket Books, 2001.

Pringle, Heather. *The Master Plan: Himmler's Scholars and the Holocaust.* New York: Hyperion, 2006.

Radner, Daisie, and Michael Radner. *Science and Unreason.* Belmont, CA: Wadsworth, 1982.

Rees, Martin. *Just Six Numbers: The Deep Forces that Shape the Universe.* London: Weidenfeld & Nicolson, 1999.

Regal, Brian. *Human Evolution: A Guide to the Debates.* Santa Barbara, CA: ABC-CLIO, 2004.

Sagan, Carl. *The Demon-Haunted World: Science as a Candle in the Dark.* London: Headline, 1996.

Salerno, Steve. *SHAM: How the Self-Help Movement Made America Helpless.* New York: Crown 2005.

Schwarcz, Joe. *Science, Sense, and Nonsense: 61 Nourishing, Healthy, Bunk-Free Commentaries on the Chemistry That Affects Us All.* Toronto: Doubleday Canada, 2009.

Seethaler, Sherry. *Lies, Damned Lies, and Science: How to Sort Through the Noise around Global Warming, the Latest Health Claims, and Other Scientific Controversies.* Upper Saddle River, NJ: FT Press Science, 2009.

Selden, Steven. *Inheriting Shame: The Story of Eugenics and Racism in America.* New York: Teachers College Press, 1999.

Shanks, Niall. *God, the Devil, and Darwin: A Critique of Intelligent Design Theory.* Oxford: Oxford University Press, 2006.

Shermer, Michael. *The Borderlands of Science: Where Science Meets Nonsense.* New York: Oxford University Press, 2001.

———. *Why People Believe Weird Things: Pseudo-Science, Superstition, and Bogus Notions of Our Time.* New York: Freeman, 1997.

Shulman, Seth. *Undermining Science: Suppression and Distortion in the Bush Administration.* Berkeley, CA: University of California Press, 2006.

Singer, Peter. *A Darwinian Left: Politics, Evolution, and Cooperation.* New Haven, CT: Yale University Press, 1999.

Singh, Simon, and Edzard Ernst. *Trick or Treatment: Alternative Medicine on Trial.* London: Bantam, 2008.

Singham, Mano. *God vs. Darwinism: The War between Evolution and Creationism in the Classroom.* Lanham, MD: Rowman & Littlefield, 2009.

Sladek, John. *The New Apocrypha: A Guide to Strange Sciences and Occult Beliefs*. St. Albans, UK: Hart-Davis, MacGibbon, 1974.

Snow, C. P. *The Two Cultures, and A Second Look*. Cambridge: Cambridge University Press, 1964.

Snowden, Frank M. Jr. *Before Color Prejudice*. Cambridge, MA: Harvard University Press, 1983.

Sokal, Alan, and Jean Bricmont. *Fashionable Nonsense: Postmodern Intellectuals' Abuse of Science*. New York: Picador, 1998. Cut translation of *Impostures Intellectuelles*. Paris: Odile Jacob, 1997.

Specter, Michael. *Denialism: How Irrational Thinking Hinders Scientific Progress, Harms the Planet, and Threatens Our Lives*. New York: Penguin, 2009.

———. "The Denialists: The Dangerous Attacks on the Consensus about H.I.V. and AIDS." *New Yorker*, March 12, 2007.

Stiebing, William H. Jr. *Ancient Astronauts, Cosmic Collisions, and Other Popular Theories about Man's Past*. Amherst, NY: Prometheus Books, 1984.

Swanwick, Michael. *Bones of the Earth*. New York: Eos, 2002.

Taibbi, Matt. *The Great Derangement: A Terrifying True Story of War, Politics, and Religion at the Twilight of the American Empire*. New York: Spiegel & Grau, 2008.

Taverne, Dick. *The March of Unreason: Science, Democracy, and the New Fundamentalism*. Oxford: Oxford University Press, 2005.

Thompson, Damian. *Counterknowledge: How We Surrendered to Conspiracy Theories, Quack Medicine, Bogus Science, and Fake History*. New York: Norton, 2008.

Toumey, Christopher P. *God's Own Scientists: Creationists in a Secular World*. New Brunswick, NJ: Rutgers University Press, 1994.

US Global Change Research Program. *Global Climate Change Impacts in the United States*. Edited by Thomas R. Karl, Jerry M. Melillo, and Thomas C. Peterson. Cambridge: Cambridge University Press, 2009.

Vorzimmer, Peter J. *Charles Darwin: The Years of Controversy—The* Origin of Species *and Its Critics 1859–82*. London: University of London Press, 1972.

Wanjek, Christopher. *Bad Medicine: Misconceptions and Misuses Revealed, from Distance Healing to Vitamin O*. Hoboken, NJ: Wiley, 2003.

Webb, George E. *The Evolution Controversy in America*. Lexington, KY: University Press of Kentucky, 1994.

Weston, Anthony. *How to Re-Imagine the World: A Pocket Guide for Practical Visionaries*. Gabriola Island, BC: New Society, 2007.

Wheen, Francis. *How Mumbo-Jumbo Conquered the World: A Short History of Modern Delusions*. London: Fourth Estate, 2004.

Wicker, Christine. *Not in Kansas Anymore: A Curious Tale of how Magic Is Transforming America*. San Francisco: HarperSanFrancisco, 2005.

Wright, Lawrence. *Remembering Satan: A Case of Recovered Memory and the Shattering of an American Family*. New York: Knopf, 1994.
Ziadat, Adel A. *Western Science in the Arab World: The Impact of Darwinism, 1860–1930*. New York: St Martin's, 1986.

INDEX